# SOCIAL COGNITION IN PSYCHOSIS

# SOCIAL COGNITION IN PSYCHOSIS

Edited by

KATHRYN EVE LEWANDOWSKI
AHMED A. MOUSTAFA

ELSEVIER

ACADEMIC PRESS
An imprint of Elsevier

Academic Press is an imprint of Elsevier
125 London Wall, London EC2Y 5AS, United Kingdom
525 B Street, Suite 1650, San Diego, CA 92101, United States
50 Hampshire Street, 5th Floor, Cambridge, MA 02139, United States
The Boulevard, Langford Lane, Kidlington, Oxford OX5 1GB, United Kingdom



**Library of Congress Cataloging-in-Publication Data**
A catalog record for this book is available from the Library of Congress

**British Library Cataloguing-in-Publication Data**
A catalogue record for this book is available from the British Library

ISBN 978-0-12-815315-4



*Publisher:* Nikki Levy
*Acquisition Editor:* Joslyn T. Chaiprasert-Paguio
*Editorial Project Manager:* Tracy I. Tufaga
*Production Project Manager:* Nirmala Arumugam
*Cover Designer:* Christian J. Bilbow

Typeset by SPi Global, India

# Dedication

Kathryn Eve Lewandowski dedicates this book with all her gratitude to Christina, Reece, and Simone, and to Jean and John Lewandowski.

Ahmed A. Moustafa dedicates this book to Marwa, Rasha, Angelina, Hasan, Haneen, and Kristina.

# Contents

# Contributors

**Robin Achterhof**
KU Leuven, Department of Neurosciences, Research Group Psychiatry, Center for Contextual Psychiatry, Leuven, Belgium

**Cali F. Bartholomeusz**
Melbourne Neuropsychiatry Centre, Department of Psychiatry, Carlton; Orygen, The National Centre of Excellence in Youth Mental Health; The Centre for Youth Mental Health, The University of Melbourne, Parkville, VIC, Australia

**Isabelle E. Bauer**
The University of Texas Health Science Center at Houston, McGovern Medical School, Department of Psychiatry and Behavioral Sciences, Houston, TX, United States

**Margherita Bechi**
Department of Clinical Neurosciences, IRCCS San Raffaele Scientific Institute, Milan, Italy

**Caridad Benavides**
Department of Psychiatry, Icahn School of Medicine at Mount Sinai, New York, NY, United States

**Andrea Bulbena-Cabre**
Department of Psychiatry, Icahn School of Medicine at Mount Sinai, New York, NY, United States

**Katherine Elizabeth Burdick**
Brigham and Women's Hospital, Department of Psychiatry, Harvard Medical School, Boston, MA, United States

**Alex S. Cohen**
Psychology Department, Louisiana State University, Baton Rouge, LA, United States

**Tovah Cowan**
Psychology Department, Louisiana State University, Baton Rouge, LA, United States

**Jacob J. Crouse**
Youth Mental Health Team, Brain & Mind Centre, The University of Sydney, Camperdown, NSW, Australia

**Charles A. Davidson**
Department of Clinical Psychology, Mercer University College of Health Professions, Atlanta, GA, United States

**Jarrett Fastman**
Department of Psychiatry, Icahn School of Medicine at Mount Sinai, New York, NY, United States

**Joanna M. Fiszdon**
VA Connecticut Healthcare System, Department of Psychology, West Haven; Department of Psychiatry, Yale University School of Medicine, New Haven, CT, United States

**Sharely Fred-Torres**
Department of Psychiatry, Icahn School of Medicine at Mount Sinai, New York, NY, United States

**Eleni P. Ganella**
Melbourne Neuropsychiatry Centre, Department of Psychiatry, The University of Melbourne, Carlton, VIC, Australia

**Philip D. Harvey**
University of Miami Miller School of Medicine; Bruce W. Carter VA Medical Center, Miami, FL, United States

**Karlijn Hermans**
KU Leuven, Department of Neurosciences, Research Group Psychiatry, Center for Contextual Psychiatry, Leuven, Belgium

**Daniel F. Hermens**
Youth Mental Health Team, Brain & Mind Centre, The University of Sydney, Camperdown, NSW; Sunshine Coast Mind and Neuroscience—Thompson Institute, University of the Sunshine Coast, Birtinya, QLD, Australia

**Daphne J. Holt**
Massachusetts General Hospital, Department of Psychiatry; Harvard Medical School, Boston; The Athinoula A. Martinos Center for Biomedical Imaging, Charlestown, MA, United States

**Zuzana Kasanova**
KU Leuven, Department of Neurosciences, Research Group Psychiatry, Center for Contextual Psychiatry, Leuven, Belgium

**Szabolcs Kéri**
Department of Cognitive Science, Budapest University of Technology and Economics; Nyírő Gyula Hospital, National Institute of Psychiatry and Addictions, Budapest, Hungary

**Matcheri S. Keshavan**
Harvard Medical School, Department of Psychiatry; Massachusetts Mental Health Center Public Psychiatry Division, Beth Israel Deaconess Medical Center, Boston, MA, United States

**Olivia Kirtley**
KU Leuven, Department of Neurosciences, Research Group Psychiatry, Center for Contextual Psychiatry, Leuven, Belgium

**Emmett Larsen**
Department of Psychiatry, Icahn School of Medicine at Mount Sinai, New York, NY, United States

**Thanh P. Le**
Psychology Department, Louisiana State University, Baton Rouge, LA, United States

**Lauren Lepow**
Department of Psychiatry, Icahn School of Medicine at Mount Sinai, New York, NY, United States

**Kathryn E. Lewandowski**
Harvard Medical School, Department of Psychiatry, Boston; McLean Hospital, Psychotic Disorders Division, Belmont, MA, United States

**Katie Mahon**
Department of Psychiatry, Icahn School of Medicine at Mount Sinai, New York, NY, United States

**Raquelle I. Mesholam-Gately**
Harvard Medical School, Department of Psychiatry; Massachusetts Mental Health Center Public Psychiatry Division, Beth Israel Deaconess Medical Center, Boston, MA, United States

**Ahmed A. Moustafa**
School of Social Sciences and Psychology, Marcs Institute for Brain and Behaviour, Western Sydney University, Penrith, NSW, Australia

**Inez Myin-Germeys**
KU Leuven, Department of Neurosciences, Research Group Psychiatry, Center for Contextual Psychiatry, Leuven, Belgium

**Andrea Pelletier-Baldelli**
University of North Carolina at Chapel Hill, Department of Psychiatry, Chapel Hill, NC, United States

**Anna Pengue**
School of Social Sciences and Psychology, Marcs Institute for Brain and Behaviour, Western Sydney University, Penrith, NSW, Australia

**Maria Mercedes Perez-Rodriguez**
Department of Psychiatry, Icahn School of Medicine at Mount Sinai, New York, NY, United States

**Bertalan Polner**
Department of Cognitive Science, Budapest University of Technology and Economics, Budapest, Hungary

**Daniel S. Quintana**
NORMENT, KG Jebsen Centre for Psychosis Research, Division of Mental Health and Addiction, University of Oslo and Oslo University Hospital, Oslo, Norway

**Manuela Russo**
Department of Psychiatry, Icahn School of Medicine at Mount Sinai, New York, NY, United States

**Sarah Barbara Rutter**
Department of Psychiatry, Icahn School of Medicine at Mount Sinai, New York, NY, United States

**Maude Schneider**
KU Leuven, Department of Neurosciences, Research Group Psychiatry, Center for Contextual Psychiatry, Leuven, Belgium; University of Geneva, Developmental Imaging and Psychopathology Lab, Department of Psychiatry, Faculty of Medicine, Geneva, Switzerland

**Juliete M. Silberstein**
University of Miami Miller School of Medicine, Miami, FL, United States

**Marco Spangaro**
Department of Clinical Neurosciences, IRCCS San Raffaele Scientific Institute, Milan, Italy

**Allison K. Ungar**
Department of Psychiatry, Icahn School of Medicine at Mount Sinai, New York, NY, United States

**Tamsyn E. Van Rheenen**
Melbourne Neuropsychiatry Centre, Department of Psychiatry, The University of Melbourne, Carlton; Centre for Mental Health, School of Health Sciences, Faculty of Health, Arts and Design, Swinburne University of Technology, Hawthorn, VIC, Australia

**Alexis E. Whitton**
The University of Sydney, Faculty of Medicine and Health, Sydney, NSW, Australia

# Preface

Social interest and engagement are core aspects of the human experience. Further, social support is associated with greater feelings of well-being, and even with better health outcomes. Healthy social functioning depends upon effective processing of, and response to, various types of social information, known as social cognition.

Deficits in social cognition in patients with schizophrenia and related disorders have been documented since the early formulations of the illness by Emil Kraepelin and Eugen Bleuler, and have been considered a core aspect of the syndrome. Recent evidence suggests that social cognitive deficits are among the strongest predictors of community functioning and quality of life in patients with psychosis. The renewed interest in the role of social cognition in psychosis has generated exciting new avenues of research and major developments in this area.

In this book, we present the most current thinking with regard to social cognition in psychosis, including the multidimensional nature of social cognition and areas of both intact functioning and greatest need, neurobiological underpinnings of social cognitive function and dysfunction, the development of social cognitive deficits across phases of illness, the state of the art in social cognitive assessment, and exciting developments in the treatment of social cognitive deficits toward improving patient outcomes. Additionally, while the vast majority of social cognition studies have been carried out in patients with schizophrenia, this book also presents findings from the social cognition literature in related illnesses. The past several decades have demonstrated that there is considerable overlap in symptoms, biology, and heritability between schizophrenia and related psychotic disorders such as bipolar disorder; here, we offer evidence of both shared and distinct aspects of social cognitive functioning among related illnesses.

The chapters of this book were written by leading experts in the field, and we are thrilled to have a broad perspective on social cognition in psychosis by preeminent researchers from across the globe. We would like to thank the authors for their thoughtful syntheses of a vast literature, and for their pioneering work in the study of social cognition in psychosis. Their hard work and dedication have moved the field forward in both our understanding of this key aspect of psychotic illnesses, and in the development of targeted and effective interventions to help patients function and feel better in daily life. We would also like to thank the countless participants who have generously volunteered their time and efforts to the many research studies presented here, without whom we would not have these important insights.

# Acknowledgement

**Kathryn Eve Lewandowski**
I am grateful to the dedicated researchers in this field whose tireless efforts bring understanding of these illnesses and improve the lives of our patients, and to the generosity of the research volunteers without whom our work would not be possible.
I would also like to thank this book's co-editor, Ahmed, for the lively discussions and learning I've greatly enjoyed on this journey.

**Ahmed A. Moustafa**
I am grateful to all schizophrenia and bipolar patients I worked with as well as their family for the time and efforts.
I also would like to thank this book's co-editor, Eve, for being very supportive, understanding, as well as for great discussion we had on this topic.

# Editors biography

**Dr. Ahmed A. Moustafa** is an associate professor in Cognitive and Behavioural Neuroscience at Marcs Institute for Brain, Behaviour, and Development and School of Social Sciences and Psychology, Western Sydney University. Ahmed is trained in computer science, psychology, neuroscience, and cognitive science. His early training took place at Cairo University in mathematics and computer science. Before joining Western Sydney University as a lab director, Ahmed spent 11 years in America studying psychology and neuroscience. Ahmed conducts research on computational and neuropsychological studies of addiction, schizophrenia, Parkinson's disease, PTSD, and depression. He has published more than 150 papers in high-ranking journals, including Science, PNAS, Journal of Neuroscience, Brain, Neuroscience and Biobehavioral Reviews, Nature (Parkinson's disease), and Neuron, among others. Ahmed has recently published two books: (a) Computational Models of Brain and Behavior, which provides a comprehensive overview of recent advances in the field of computational neuroscience, and (b) Computational Neuroscience Models of the Basal Ganglia, which provides several models of the basal ganglia.

**Dr. Kathryn Eve Lewandowski** is an assistant professor in Psychiatry at Harvard Medical School and Director of Clinical Programming for McLean OnTrack first episode clinic. Eve received her PhD in clinical psychology from the University of North Carolina, Greensboro, where she studied cognitive, behavioral, and genetic markers of schizotypy. She completed her postdoctoral fellowship at McLean Hospital/Harvard Medical School prior to joining the faculty. Her program of research centers on cognition and cognitive interventions for people with psychotic disorders. Her federally-funded research has included efficacy and neurobiological outcomes of cognitive remediation in bipolar disorder, computational approaches to characterizing heterogeneity in cognition across the psychoses, and examination of the latent structure of motivation and reward. She is currently conducting implementation work to translate cognitive remediation into clinical practice, for which she was awarded the 2018 Connie Lieber Science to Practice Award. Eve serves as the Director of the Psychology Doctoral Internship Program Psychotic Disorders training, and provides supervision, training, and mentorship to junior faculty, fellows, residents, and interns.

CHAPTER ONE

# Characterization of social cognitive deficits on the schizophrenia-bipolar disorder spectrum: An overview of current evidence

**Tamsyn E. Van Rheenen*,†, Eleni P. Ganella*, Isabelle E. Bauer‡, Cali F. Bartholomeusz*,§,¶**

*Melbourne Neuropsychiatry Centre, Department of Psychiatry, The University of Melbourne, Carlton, VIC, Australia

†Centre for Mental Health, School of Health Sciences, Faculty of Health, Arts and Design, Swinburne University of Technology, Hawthorn, VIC, Australia

‡The University of Texas Health Science Center at Houston, McGovern Medical School, Department of Psychiatry and Behavioral Sciences, Houston, TX, United States

§Orygen, The National Centre of Excellence in Youth Mental Health, Parkville, VIC, Australia

¶The Centre for Youth Mental Health, The University of Melbourne, Parkville, VIC, Australia

## Introduction

Bipolar disorder (BD) and schizophrenia (Sz) spectrum disorders, including Sz and schizoaffective disorder, are debilitating psychiatric illnesses with shared genetic etiology and overlapping phenotypic expressions (International Schizophrenia Consortium et al., 2009; Laursen, Agerbo, & Pedersen, 2009; Van Rheenen et al., 2016). A significant amount of literature has highlighted that patients with Sz spectrum disorders, and to a lesser extent, BD, are characterized by deficits in social cognition (Mercer & Becerra, 2013; Penn, Sanna, & Roberts, 2008). Social cognition is an umbrella term describing the innate human ability to perceive, understand, and regulate information about others and ourselves. Research under this construct broadly focuses on how people process social information, with emotion processing, theory of mind, attributional style, and social perception among the domains that have been of most interest in the behavioral assessment of these disorders (Penn et al., 2008; Pinkham et al., 2014; Pinkham, Penn, Green, & Harvey, 2016).

Emotion processing broadly describes the ability to perceive and use emotions, and is typically measured in behavioral studies using perceptual

*Social Cognition in Psychosis*
https://doi.org/10.1016/B978-0-12-815315-4.00001-X

tasks assessing the speed and accuracy with which several universally identifiable emotional expressions are recognized or discriminated from facial or auditory-prosodic stimuli (Pinkham et al., 2014; Van Rheenen & Rossell, 2013a). Assessments of emotional intelligence, which measure the ability to understand and manage emotions, are considered higher-level processes under this domain. Theory of mind, or mentalizing, on the other hand, describes the ability to decode and reason about the mental states of others and ourselves. This not only allows for the generation of inferences about what others are thinking or feeling, but enables the understanding that these thoughts and feelings may be different from one's own (Premack & Woodruff, 1978). Given the multifaceted nature of theory of mind, the tasks that have been used to behaviorally quantify it in Sz-spectrum disorders and BD vary widely; some measure the extent to which one understands that another person's beliefs may be different from reality (first order false-beliefs), while others measure the extent to which one can infer what one person believes another may be thinking or feeling (second-order false beliefs). Others measure the capacity for complex emotion recognition, or the ability to infer the intentions, beliefs, or feelings of others by understanding hints, metaphors, deception, irony, or sarcasm (Samamé, Martino, & Strejilevich, 2015).

While both emotion processing and theory of mind have been thoroughly covered in the behavioral social cognitive literature of both Sz-spectrum disorders and BD, the domains of attributional style and social perception have really only been addressed in the former. Attributional style refers to the causal attribution of social events to internal or external sources, and is typically measured through questionnaires that assess responses to hypothetical scenarios (Penn et al., 2008; Pinkham et al., 2014; Pinkham, Penn, et al., 2016). In contrast, social perception refers to context processing and social knowledge that allows for the decoding and interpretation of social cues that influence ones perception of another (Penn et al., 2008; Pinkham et al., 2014; Pinkham, Penn, et al., 2016). Social perception differs from emotion perception in that it requires the use of social cues to make judgments about interpersonal interactions, such as intimacy and status (Sergi, Rassovsky, Nuechterlein, & Green, 2006), rather than the underlying mood state. It is typically measured using video-based assessments that require decoding of important interactions depicted in brief movie scenes.

In Sz-spectrum disorders, several studies have shown that poor social cognitive ability directly contributes to poor functional status (Brekke, Kay, Lee, & Green, 2005; Horan et al., 2012; Kee, Green, Mintz, &

Brekke, 2003); even acting as a mediator of the detrimental influence of neurocognitive impairment on functional outcomes (Addington, Girard, Christensen, & Addington, 2010; Addington, Saeedi, & Addington, 2006a). In BD, the relationship between social cognition and functioning has been far less studied, but recent research suggests that it may act as a moderator, rather than a mediator, for the negative influence of neurocognitive impairment on functional outcomes in this disorder (Ospina et al., 2018). Irrespective of the direct or indirect nature of the link, the relevance that social cognition and its component domains have to adapt to everyday functioning is intuitive. Facial and prosodic emotional cues, for example, are among the most readily available in the social environment, and the accurate and timely perception of them is critical for decoding underlying emotional states (Van Rheenen & Rossell, 2013a). Similarly, an ability to decode social cues to infer the thoughts and emotional states of others and situate them within a social context is critical to predicting other's actions, and supporting prosocial behaviors and interactions, which are key elements of occupational and social success.

At present, it is unclear whether the behaviorally measured social cognitive deficits evident in Sz-spectrum disorders are the same as those evident in BD. In this chapter, we provide a brief overview of the current state of the social cognitive literature on these disorders, with an aim to discuss overlaps and differences in behavioral deficit characterization, their clinical correlates, and their possible contributing factors. We begin with a summary and synthesis of key findings within each illness class, arranged by the social cognitive domains of interest to either BD or Sz-spectrum disorders, and broadly covering both the early (i.e., pediatric/ultra-high risk (UHR)/first-episode samples) and later (adult/established illness samples) periods of the illness course. We then discuss points of convergence and divergence in the social cognitive profile of both Sz-spectrum disorders and BD, before presenting direction for further study in the field.

## Early course Sz-spectrum disorders

Over the past two decades, research attention in Sz-spectrum disorders has turned toward investigating the extent of social cognitive impairments in its early illness phases, in an attempt to further characterize symptoms of the disorder and highlight areas that may be responsive to early intervention strategies. The early illness phase is generally comprised of individuals who are at UHR for developing psychosis (also termed "clinical high risk," and including those that are prodromal [~27%–35%] for a Sz-spectrum disorder) (Miller et al., 2002; Nelson et al., 2013; Simon et al., 2013; Yung et al., 1998), as well as individuals experiencing their first episode of psychosis (FEP). Although there is some variation as to the inclusion criteria of FEP across studies, general criteria consist of (a) the first presentation to psychiatric services and (b) an illness duration of less than 2 years (Healey, Bartholomeusz, & Penn, 2016). Social cognition has not been as extensively studied in UHR populations relative to FEP and established Sz-spectrum disorder groups; however, growing evidence suggests that social cognitive impairments are evident in the UHR phase of psychosis, with emotion processing and theory of mind being the most widely studied social cognitive domains.

## Ultra high risk for psychosis

### *Emotion processing*

Studies that have investigated emotion processing in UHR groups are largely consistent in their findings of impairments in individuals at UHR relative to healthy controls. The majority of work has addressed basic emotion recognition as opposed to higher-level emotional intelligence, and recent metaanalytic findings have shown significant moderate effect-sized deficits in emotion recognition of faces in individuals at UHR relative to healthy controls (van Donkersgoed, Wunderink, Nieboer, Aleman, & Pijnenborg, 2015). Most studies to date have investigated the ability to *recognize* facial affect, and results largely demonstrate impairments in UHR groups relative to healthy controls. However, it remains unclear whether UHR individuals also show an impaired ability to *discriminate* between different emotions, with some studies reporting no significant differences (Addington et al., 2006a; Corcoran et al., 2015). This may be because discriminating emotions is thought to be less difficult than identifying emotions (Addington et al., 2006a). A handful of studies have also reported deficits in the ability to recognize prosodic affects (Amminger, Schafer, Klier, et al., 2012; Thompson et al., 2012) in UHR individuals relative to controls, indicating that deficits in basic emotion processing are detectable across a variety of tasks administered.

The MSCEIT is an emotional-intelligence task that assesses the perception, understanding, and management of emotions, and taps into a more complex aspect of emotion processing. To date, two known studies have evaluated this task in UHR populations, and both reported the UHR groups to be intermediate in performance between that of healthy controls and FEP individuals (Green et al., 2012; Thompson et al., 2012). However, only one study (Green et al., 2012) reported significant deficits in the UHR group relative to controls. Taken together, deficits in emotion processing in UHR individuals may vary relative to task difficulty; however, results generally indicate emotion processing impairments in the UHR phase of psychosis. This highlights the importance of evaluating different aspects of emotion processing to further understand the extent to which this domain of social cognition is impaired in UHR (Lincoln, Norkett, Frost, Gonzalez-Heydrich, & D'Angelo, 2017).

### *Theory of mind*

Studies that have investigated theory of mind in UHR groups are fairly congruent in their findings of impaired theory of mind in individuals at UHR relative to healthy controls. An initial metaanalysis (Bora & Pantelis, 2013) showed a moderate effect size of theory of mind deficits in UHR groups in comparison with controls, and this finding has since been replicated in two later metaanalyses (Lee, Hong, Shin, & Kwon, 2015; van Donkersgoed et al., 2015). Interestingly, deficits in UHR groups appear to be most consistently reported on verbal, as opposed to nonverbal theory of mind tasks (van Donkersgoed et al., 2015), with multiple studies reporting no significant difference in visual theory of mind performance between UHR and control groups (Brune et al., 2011; Chung, Kang, Shin, Yoo, & Kwon, 2008; Hur et al., 2013). This may be related to the increased cognitive effort that is required for verbal tasks relative to nonverbal tasks (van Donkersgoed et al., 2015).

Impaired social cognition is known to contribute to poor social functioning, which is a well-established risk factor for transition to full-blown psychosis (Gee & Cannon, 2011). Longitudinal studies have shown deficits in theory of mind are more severe in UHR individuals who later transition to psychosis compared with those who do not transition (Bora & Pantelis, 2013; Healey, Penn, Perkins, Woods, & Addington, 2013; Kim et al., 2011), suggesting that theory of mind impairments may represent predictive markers for psychosis onset in UHR groups. However, at present, the longitudinal literature is insufficient to deduce whether theory of mind deficits significantly differ between UHR individuals who transition to psychosis compared with those who do not.

### *Social perception and attributional style*

Studies that have investigated social perception in UHR populations are sparse and inconsistent. Recent metaanalyses have shown social perception and attributional style to be impaired in UHR (Lee et al., 2015; van Donkersgoed et al., 2015), however, findings are not as congruent or abundant in comparison with results generated from theory of mind and emotion processing tasks. A handful of studies indicate impairments in social perception in UHR relative to controls, particularly for more complex tasks (Couture, Penn, Addington, Woods, & Perkins, 2008; Gill et al., 2016; Green et al., 2012; Healey et al., 2013). However, other studies have reported no significant differences (Adolphs, Tranel, & Damasio, 1998; Thompson et al., 2012). Interestingly, despite the few studies that have explored attributional style in UHR populations (<5), some metaanalytic findings (Lee et al., 2015) show it to be the social cognitive domain with the largest effect size. It has since been suggested that attributional bias could be the main driver of the *overall* decrease in social cognitive ability seen in UHR individuals (Kohler et al., 2003). However, further work is necessary to comprehensively understand whether these domains are truly disturbed in the UHR phase of psychosis.

## First-episode psychosis

### *Emotion processing*

Relative to healthy controls, FEP individuals also show fairly consistent impairments in basic emotion processing tasks, with the majority of studies reporting deficits in facial emotion recognition *overall* (Healey et al., 2016). With this said, a handful of studies have reported no significant differences between groups (Achim, Ouellet, Roy, & Jackson, 2012; Dean et al., 2013; Reske et al., 2009), however, the psychometric properties of the tests used in the aforementioned studies are not well-validated. Studies that investigated the identification of *specific* facial emotions found that negative emotions (i.e., fear and sadness) are most consistently misinterpreted by FEP individuals relative to controls (Amminger, Schafer, Papageorgiou, et al., 2012; Comparelli et al., 2013; Healey et al., 2016; Kucharska-Pietura, David, Masiak, & Phillips, 2005; Seiferth et al., 2009). Deficits in the ability to detect emotion in prosody are also frequently reported in FEP groups relative to controls (Amminger, Schafer, Papageorgiou, et al., 2012; Edwards, Pattison, Jackson, & Wales, 2001; Kucharska-Pietura et al., 2005), with impairments in the perception of prosodic fear and sadness again most often described. This finding of specific impairments in the ability to recognize negative emotions as opposed to positive emotions in FEP is consistent with findings in established Sz-spectrum disorder cohorts, and may reflect that positive emotions are generally easier to perceive (Kohler et al., 2003). More challenging facial and prosodic emotion recognition tasks are needed to account for ceiling effects, and thus more comprehensively evaluate emotion recognition.

Substantially fewer studies have explored higher-level emotion processing abilities in FEP. Some studies have shown significant deficits in FEP groups relative to controls, by assessing emotional responses to a series of hypothetical vignettes (Green et al., 2012; Mazza et al., 2013; Thompson et al., 2012). However, as one study (Green et al., 2012) combined samples of established Sz and FEP patients, its results may not be truly representative of deficits in FEP. Higher-level emotion processing assessed by the MSCEIT has also been shown to be impaired in FEP individuals relative to controls (Green et al., 2012; Thompson et al., 2012). Importantly, studies that have assessed the stability of lower level (Addington et al., 2006a; Hill et al., 2008) and higher level emotion processing (Horan et al., 2012) all reported no significant change in task performance over three months (Hill et al., 2008), 1 year (Addington et al., 2006a; Horan et al., 2012), and 5 years (McCleery et al., 2016) in FEP. This suggests stability of emotion processing impairments during the early illness phase of Sz-spectrum disorders, however, further longitudinal work is required.

### *Theory of mind*

The literature is largely consistent in reporting deficits in theory of mind in FEP individuals relative to healthy controls, with the most frequently reported and robust findings generated from studies that have assessed theory of mind using verbal tasks (Healey et al., 2016). Studies that used nonverbal theory of mind tasks also largely reported impaired theory of mind in FEP individuals relative to controls; however, methodologically, many studies did not control for the potential effect of neurocognition. The extent to which deficits in social cognition can be explained by poor neurocognition is a contentious topic in the field of cognitive research. Given that intact social cognition is rare in the presence of poor neurocognition in Sz-spectrum disorders, it has been suggested that general neurocognitive functions are necessary, but not entirely sufficient, for social cognition (Minor & Lysaker, 2014). Theory of mind performance has been found to correlate with other cognitive abilities, such as intelligence quotient (IQ), executive functioning, and memory (Bora & Pantelis, 2013). It can therefore be argued that theory of mind deficits may not be independently related to the disorder, but instead may be reflective of neurocognitive impairment and the general cognitive decline associated with psychosis. A recent review, however, found that deficits observed in theory of mind in FEP groups are not entirely explained by general cognitive decline (Healey et al., 2016), indicating that social and general cognition partly overlap in FEP, however, impairments in single domains are also possible (Green et al., 2008).

Theory of mind deficits appear to remain stable over time in FEP (between 6-month and 1-year follow-up periods) (Horan et al., 2012; Sullivan et al., 2014; Ventura et al., 2015). With this said, the trajectory of theory of mind ability in FEP has not been extensively compared with the trajectory of theory of mind performance in healthy controls, thus limiting the inferences that can be made as to the stability of this impairment in early psychosis (Healey et al., 2016).

### *Social perception*

Substantially less work has explored social perception in FEP, however, findings once again are largely consistent in reporting deficits in FEP groups relative to healthy controls (Healey et al., 2016). However, many studies did not control for the effect of IQ, and given that social perception has been shown to significantly correlate with neurocognition (Addington, Saeedi, & Addington, 2006b), further work is necessary to determine whether deficits in social perception exist independent of general cognitive impairment in FEP.

The few studies that have investigated the stability of social perception deficits in FEP found performance to be stably impaired over a 1-year (Addington et al., 2006b; Horan et al., 2012) and 5-year period (McCleery et al., 2016). In a study by Addington et al. (2006b), the healthy control group showed significant improvement in social perception performance over time, while the FEP group did not. This suggests that the FEP individuals are not exhibiting the same developmental improvements as seen in healthy individuals. Further work mapping the trajectory of social cognition in healthy cohorts is necessary (Healey et al., 2016).

### *Attributional style*

Findings from studies investigating attributional style in FEP have generated mixed results. Findings from one study indicated that FEP individuals perceived significantly more hostility in the intention of others compared with controls; however, groups did not differ in overall intentionality, blame, anger, or aggression (An et al., 2010). Conversely, findings from another study that focused on externalizing biases (So, Tang, & Leung, 2015) showed FEP individuals were more likely to attribute positive (not negative) events to oneself relative to healthy controls; however, this finding was not replicated in other studies (Achim, Sutliff, Samson, Montreuil, & Lecomte, 2016; Fornells-Ambrojo & Garety, 2009). There is some evidence to suggest that FEP individuals engage more in personalizing bias, that is, the tendency to attribute negative events to others rather than the environment (Fornells-Ambrojo & Garety, 2009), and this is particularly evident in FEP participants that are experiencing greater delusional beliefs (Achim et al., 2016). It is difficult to draw conclusions about attributional style in FEP groups given studies are inconsistent in the features examined. However, overall, the limited literature suggests that FEP individuals show an increased tendency to blame and perceive hostility in others when responding to ambiguous situations (Healey et al., 2016). To date, no known study has investigated attributional style in FEP longitudinally.

## Established Sz-spectrum disorders

An abundance of social cognitive research has been conducted in established Sz-spectrum disorders, with the most extensive research related to emotion processing and theory of mind. There is good evidence demonstrating that social cognition is severely and stably impaired across three of the four subdomains in established Sz-spectrum disorders. In the following section we provide an overview of the available evidence in the four key areas, drawing heavily on metaanalytic studies.

### Emotion processing

Emotion processing is the social cognitive domain for which there is the largest amount of empirical research in Sz-spectrum disorders. The first comprehensive metaanalysis in this area was focused on facial emotion processing, was comprised of 86 studies, and spanned literature published between 1970 and 2007 (Kohler, Walker, Martin, Healey, & Moberg, 2010). Sz-spectrum patients had a large deficit in emotion processing ability relative to healthy controls, with no significant difference in the magnitude of the deficit between emotion processing tasks that assessed facial affect identification (i.e., labeling an emotional expression) or discrimination (i.e., differentiating emotional expressions). Further, effect size differences were *not* evident between studies that were comprised of only Sz patients, and studies that were comprised of patients with Sz and other Sz-spectrum diagnoses. Having a later age at illness onset, experiencing a higher level of positive or negative symptoms, being an inpatient, and being treated with a first- as opposed to second-generation antipsychotic were associated with greater impairments in emotion processing. A subsequent later metaanalysis also found a small positive effect of treatment; however, concluded that it was too small to be of any clinical significance (Gabay, Kempton, & Mehta, 2015). While the original metaanalysis (Kohler et al., 2010) found that being of older age was associated with poorer performance, this was not unique to the patient group, as the same relationship was observed in healthy controls. In contrast, sex, race, duration of illness, education, and antipsychotic *dose* did not influence emotion processing performance in Sz-spectrum patients. A later metaanalysis showing large emotion processing deficits in patients also found that age, education, sex, inpatient status, and English-speaking status had no effects (Savla, Vella, Armstrong, Penn, & Twamley, 2013), although patients with a longer illness duration did demonstrate greater impairment. A more recent review of the literature confirms that sex is not a major moderator of the emotion processing deficit in Sz (Mote & Kring, 2016). These findings, along with evidence that emotion processing is impaired in first-degree relatives of patients with a Sz-spectrum disorder (Ay et al., 2016; Lavoie et al., 2013), point toward emotion processing deficits being more of a trait, rather than state characteristic of psychotic illness.

A long-standing historical debate regarding the nature of the facial emotion processing deficit in Sz has centered around the relative contribution of impaired face processing itself. Chan, Li, Cheung, and Gong (2010) conducted a metaanalysis of 28 emotion processing studies, involving 1007 Sz-spectrum patients, and found not only a large emotion processing deficit, but also a significant moderate-large performance deficit on the control task relative to healthy participants. Given that Sz-spectrum patients are shown to have basic perceptual face processing deficits (i.e., a deficit in recognizing faces previously observed, posited to be due to an abnormal pattern of scanning and encoding facial features) (Joshua, 2010; Joshua, Van Rheenen, Castle, & Rossell, 2016; Van Rheenen, Joshua, Castle, & Rossell, 2017), the generalized versus specific deficit hypothesis of emotion processing became a topic of hot debate. Similarly, the role of basic auditory perceptual processing, in reference to recognition of emotions in speech, was similarly questioned. Ventura, Wood, Jimenez, and Hellemann (2013) used metaanalytic techniques to specifically explore the degree of the relationship between face recognition and emotion processing in 102 Sz-spectrum studies; they found that face recognition was highly correlated with not only emotion processing of facial affect, but also of prosody. They proposed that face recognition, a lower-level bottom-up process, is fundamentally linked to higher order emotion processing, and as such should be considered closely linked to social cognition. Ventura et al. (2013) further demonstrated that facial recognition was moderately correlated with the majority of MATRICS Consensus Cognitive Battery (MCCB) neurocognitive domains, leading them to conclude that both face recognition and emotion processing are part of a generalized social cognitive deficit. Indeed, given that all four key social cognitive domains are impaired in Sz-spectrum disorders, there does appear to be consensus for a global social cognitive deficit (Browne et al., 2016), although this does not speak to the likely disparate underlying neurobiological mechanisms driving impairment in individual social cognitive abilities. The question remains as to what degree basic perceptual face processing/recognition and emotion processing overlap (currently suggested to be 25%) (Ventura et al., 2013), and whether these separate processes explain a unique variance in important clinical outcomes, such as interpersonal functioning, which will have important implications for targeted social cognitive treatment in the future.

In relation to emotion processing of prosody, there is relatively less literature, compared with facial emotion processing in Sz (Edwards, Jackson, & Pattison, 2002; Hoekert, Kahn, Pijnenborg, & Aleman, 2007). Of the empirical evidence that does exist, the finding of a significant deficit is consistent and robust. Hoekert et al. (2007) metaanalyzed prosodic emotion processing data from 17 studies comprised of 623 patients, and found that Sz-spectrum patients displayed large receptive emotional prosody deficits compared with healthy controls. Based on data from 11 of the 17 studies, they also found that expressive emotional prosody was impaired, and may be associated with more severe negative symptoms, although there was insufficient symptomatology data to assess this relationship. Unlike with facial emotion processing, Hoekert et al. (2007) found no moderating effects of any variables examined, including age, sex, education, illness duration, inpatient status, or medication status. They did, however, find significant heterogeneity among studies, which may be related to the varying types of emotion processing prosody tasks employed, which differed in the number of emotions used, stimuli and sex of speaker, and level of overall neurocognitive demands (i.e., attentional, semantic, working memory, etc.)

(Hoekert et al., 2007). While the majority of extant research suggests that poor recognition of negative emotions is common, deficits in the processing of happy/positive affects have been found in a study in which it has been linked to clinical symptomatology (Tseng et al., 2013). Thus, further research, including that which employs novel tasks assessing a range of positive emotions, is needed in this area. Last, akin to perceptual face processing, Sz-spectrum patients are impaired in basic pitch perception (assessed, for example, with a tone-matching task), and efforts should be made to develop better tasks that assess these disparate, yet overlapping, abilities (Petkova et al., 2014).

In the early 2000s an increasing number of Sz researchers began assessing the higher-level emotion processing construct "emotional intelligence," providing evidence for impairment across the broad domain of emotional intelligence in established Sz-spectrum disorders (Dawson, Kettler, Burton, & Galletly, 2012; Frajo-Apor, Pardeller, Kemmler, Welte, & Hofer, 2016; McCleery et al., 2016) that has been found to be stable over a 5-year follow-up period (McCleery et al., 2016). As with other measures of emotion processing, there are questions regarding the relative contribution of basic neurocognition to emotional intelligence, with Frajo-Apor and colleagues (Frajo-Apor, Pardeller, et al., 2016) finding that after controlling for general cognitive skills, only the managing emotions branch of the Mayer-Salovey Emotional Intelligence Test (MSCIET) remained significantly impaired in patients, compared with healthy controls.

## Theory of mind

Impaired theory of mind was first implicated in Sz in 1992 (Frith, 1992). Since then, there has been a wealth of evidence to demonstrate that a theory of mind deficit is a severe and inherent feature of established Sz-spectrum illnesses. The first metaanalysis covering 29 studies comprised of 831 patients found a highly significant large overall theory of mind deficit in patients relative to healthy controls (Sprong, Schothorst, Vos, Hox, & van Engeland, 2007). When patients were divided into subgroups based on symptom profile (i.e., paranoid, disorganized, nondisorganized, and remitted), disorganized patients appeared to be more impaired than the other subgroups; however, remitted Sz-spectrum patients were still significantly impaired relative to the control group. A subsequent metaanalysis by Bora, Yucel, and Pantelis (2009) comprised of 1181 patients and 36 studies similarly showed large effect size differences for both remitted and nonremitted patients, compared with controls. These metaanalyses additionally took into consideration the diverse nature of different types of theory of mind tasks, and despite the variability in the degree of difficulty, the construct targeted (e.g., first-order compared with second-order tasks), and the mode of stimulus delivery (i.e., verbal, visual), mean effect sizes were generally found to be similar, indicating that the full breadth of theory of mind subconstructs examined are impaired in the established illness stage. Nonetheless, there was some heterogeneity for effect size distributions in studies using first and/or second-order tasks (Sprong et al., 2007) and false-belief stories/sequencing (Bora et al., 2009), while indirect speech and intention-inferencing tasks demonstrated homogeneity. Bora and colleagues (Bora et al., 2009) found that only the individual tasks "Reading the Mind in the Eyes" and the "Hinting Task," produced consistent effects. There is some trend level evidence for greater impairment in second- as opposed to first-order theory of mind tasks (Bora et al., 2009), and in verbal as opposed to visual theory of mind tasks (Chung, Barch, & Strube, 2014). Overall, heterogeneity found within different types of theory of mind measures is not explained by age, sex, or education.

The finding of impaired theory of mind in remitted Sz-spectrum patients, along with evidence of impaired theory of mind in first-degree relative of patients (Ay et al., 2016; Bora & Pantelis, 2013; Lavoie et al., 2013), suggests that theory of mind deficits may be more trait- rather than state-dependent features of the illness. However, the magnitude of the deficit was somewhat dampened in the remitted patients, implicating additional state-dependent influences of psychopathology on theory of mind performance. There is some inconsistency in metaanalytic results regarding the influence of other variables, such as IQ, age at illness onset, and duration of illness on theory of mind performance; Bora et al. (2009) found that lower IQ had a significant negative effect on theory of mind performance for patients in remission, but in general, IQ has been shown to explain minimal/nonsignificant variance in the theory of mind deficit of nonremitted Sz-spectrum patients (Bora et al., 2009; Sprong et al., 2007). Patients with longer illness durations have also been found to perform worse on theory of mind tasks; however, Bora et al. (2009) found this to be only a trend level effect, and other metaanalyses similarly failed to find statistically significant effects of illness duration (Savla et al., 2013), providing greater support for theory of mind deficits being trait-driven characteristics of the disorder. Antipsychotic medication dose (Bora et al., 2009), English-speaking status, and inpatient status (Savla et al., 2013) have also been found to have no significant effects on theory of mind performance.

It should be acknowledged that the psychometric properties of many theory of mind tasks have been called into question, and establishment of a consensus battery is underway (Pinkham et al., 2014b). In addition, results from a more recent unpublished metaanalysis were presented at the 16th International Congress on Schizophrenia Research in 2017 (Gilleen, Xie, & Strelchuk, 2017), and suggest some inconsistencies with the reported metaanalytic results described herein. The preliminary findings from this metaanalysis, which were comprised of 74 theory of mind studies and 3555 Sz-spectrum patients, showed consistent large deficits across the different theory of mind task domains (i.e., verbal intentions, indirect speech, faux-pas first- and second-order false beliefs, and emotion theory of mind), with the largest deficit observed for verbal intention tasks. However, surprisingly, there was no significant impairment in performance of visual intention tasks, which is inconsistent with previous metaanalyses. It is difficult to speculate as to the reason for this incongruent finding, given the minimal information provided; however, it is possible that the method for grouping theory of mind tasks differs from previous metaanalytic studies. Given that the full manuscript has yet to be published, the preliminary findings from this metaanalysis should be viewed with caution.

## Social perception

Savla et al.'s (2013) metaanalysis of 13 social perception studies comprised of 503 patients found this social cognitive domain to be significantly impaired in Sz-spectrum patients relative to controls. Consistent with findings in emotion processing and theory of mind domains, the magnitude of this effect was large. Of the potential moderating variables examined, only inpatient status had a significant effect, where being an inpatient corresponded to poorer performance. There were no effects of age, education, illness duration, English-speaking status, or first-episode status. There was, however, significant heterogeneity between studies, and this is likely related to the differing social perception tasks employed. In 2012, a panel of experts reached a consensus on the best available social cognitive measures for use in Sz research; only the relationships across domains (RAD) was selected for the measure of social perception (Pinkham et al., 2014b). McCleery et al. (2016) recently found that social perception as measured with the RAD was stably impaired over a 5-year period in Sz-spectrum patients. Since this study was conducted, the RAD, among the other social cognitive measures put forward by the panel of experts, was further psychometrically evaluated in a sample of 179 Sz-spectrum patients. While social perception performance was significantly impaired in patients relative to controls, the RAD demonstrated weak characteristics (e.g., floor effects with 43% of patients scoring at or below chance) and was not recommended for use in clinical trials (Pinkham, Penn, et al., 2016). Two other measures that have previously been used to assess social perception include the Trustworthiness Task and the Profile of Nonverbal Sensitivity Test; these measures appeared in a metaanalysis of two studies examining social perception ability of first-degree relatives ($N=45$) of Sz-spectrum patients (Lavoie et al., 2013). First-degree relatives were moderately impaired in social perception in comparison with healthy controls. While the psychometric properties of these measures have been called into question, there is nonetheless some evidence to suggest that social perception deficits are trait characteristics of Sz-spectrum illnesses.

## Attributional style

Of the four key social cognitive abilities identified as being implicated in Sz-spectrum disorders, attributional style is the area with the least amount of empirical evidence. Savla et al.'s (2013) metaanalysis of 5 studies (comprising 255 patients) analyzed personalizing bias and externalizing bias separately, given that all included studies used the Internal, Personal, and Situational Attributions Questionnaire (IPSAQ). There were no significant differences in the degree of bias on either the personalizing bias or externalizing bias scale between Sz-spectrum patients and healthy controls. Even when the analysis was restricted to include only patients who presented with persecutory delusions, there were still no significant effects of group. Conversely, a later study consisting of 258 Sz-spectrum patients, and employing a revised version of the IPSAQ that permitted participants to rate the degree of the contribution of personal, internal, and situational factors in important life events, found that patients had a significantly reduced externalizing bias, and more often blamed themselves for negative events, compared with healthy controls (Mehl et al., 2014). In addition, having a pronounced personalizing bias (in this measure, personalizing bias specifically referred to blaming others for negative events) independently predicted persecutory delusions and symptoms of depression. Within the patient population with persecutory delusions, two subgroups were identified; one that had a pronounced personalizing bias, and one that presented with a "self-blaming" attributional style. This led authors to conclude that attribution style in

Sz-spectrum disorders may be more of a state, rather than trait-driven characteristic, which requires further investigation. Additionally, there has been some suggestion that the presence of comorbid social anxiety may be associated with Sz-spectrum patients having a lower externalizing bias in comparison with controls (Achim et al., 2016). Of interest, a recent large study ($N=170$ Sz-spectrum patients) failed to find any moderating effects of age, sex, or race on attributional style (Pinkham, Kelsven, Kouros, Harvey, & Penn, 2017).

As with social perception, the validity and reliability of currently available measures have been criticized, and only the Ambiguous Intentions Hostility Questionnaire (AIHQ) was recommended for further evaluation by the previously-mentioned panel of experts (Pinkham et al., 2014b). Sz-spectrum patients have been found to score higher than controls on the subscales AIHQ hostility bias and AIHQ blame score (Pinkham, Penn, et al., 2016), with paranoid Sz patients displaying a greater tendency to appraise ambiguous social situations as hostile, and to more often attribute blame to other individuals, compared with nonparanoid patients (Pinkham, Harvey, & Penn, 2016). However, the AIHQ has weak psychometric properties, with internal consistency for both the aggression bias and blame score well below industry standards (Pinkham, Penn, et al., 2016). Further, factor analysis conducted on several social cognitive measures identified a two-factor structure in a Sz-spectrum sample, where hostile attributional style was identified as being independent to the social cognition skill factor (Buck, Healey, Gagen, Roberts, & Penn, 2016). In summary, the findings regarding attributional bias in established Sz-spectrum disorders is extremely mixed, and development of new, more reliable measures is needed before conclusions can be drawn.

## Pediatric and early course BD

While early course Sz-spectrum disorder research has very clearly defined criteria for classification into UHR or FEP groups, research on early course BD has not explicitly followed the same convention. In particular, there is no social cognitive research on prodromal BD akin to that on UHR for psychosis. Rather, groups considered at high risk for BD are comprised of youths (children or adolescents) with a genetic predisposition for the illness by proxy of having a parent or sibling with a BD diagnosis. Available research on social cognition in early course BD focuses on at-risk groups by this definition, as well as youth diagnosed with BD—henceforth referred to as pediatric BD (PBD)—and individuals assessed after seeking treatment for their first psychotic episode of BD. Early course BD research assessing social cognition comprehensively is sparse, with a single study reporting that offspring of BD patients were rated by their parents as having more difficulties with social reciprocity, social cognition, social awareness, social communication, social motivation, and autistic mannerisms than their control counterparts (Whitney et al., 2013). Other research, albeit limited, has mainly been studied behaviorally in the context of emotion perception, and to a lesser extent, theory of mind. In the following section we provide an overview of the extant literature in these areas.

## Emotion processing

To our knowledge, there have been only three behavioral studies of emotion processing published in youth at risk for BD. Whitney et al. (2013) failed to find an emotion recognition deficit in at-risk offspring of BD patients assessed on two emotion perception tasks. However, Brotman et al. (2008) reported that compared with their healthy counterparts, a group of unaffected siblings and offspring of BD patients were significantly less accurate at labeling the expressions of both child and adult faces. Importantly, the performance of this at-risk group did not differ from that of a PBD group. Similar findings were evident in another study, in which both PBD and at-risk youth were shown to require greater emotional intensity to correctly recognize facial expressions than healthy controls (Brotman et al., 2008). This abnormality was evident across several emotions, including happiness, surprise, fear, sadness, anger, and disgust.

Few studies have examined facial processing in PBD samples exclusively, but there is some further evidence of poor recognition of happy, sad, and fearful faces in adolescents with the illness (McClure, Pope, Hoberman, Pine, & Leibenluft, 2003). Impaired recognition of disgust, fear, and sadness has been partially attributed to the individual's current mood state, as some studies of manic adult patients show specific BD difficulties in recognizing fear and disgust (Rocca, Heuvel, Caetano, & Lafer, 2009). Children with BD may display an overall tendency to rate neutral faces as being more negative than healthy controls (Rich et al., 2006), and are likely to label angry facial expression as sad (Wegbreit et al., 2015). Even during the euthymic phase, PBD patients are prone to perceive extremely sad, and to a lesser extent, extremely happy facial expressions, as more moderate in intensity than their healthy counterparts (Rich et al., 2006; Schenkel, Pavuluri, Herbener, Harral, & Sweeney, 2007). This finding suggests that PBD patients may have learned to "tone down" the evaluation of intense affective stimuli to attenuate their emotional impact. Further, there appears to be an interaction between age and diagnosis in PBD, with younger PBD patients making more facial emotion recognition errors than healthy controls (Carter, Mundo, Parikh, & Kennedy, 2003; Wegbreit et al., 2015). As the experimental designs and statistical analyses that generated these findings accounted for medications, comorbidities, substance use, mood state, and global functioning, they support a hypothesis of disruption to the development of social cognitive abilities in PBD. First-episode BD patients also appear to have the same difficulties in recognizing mild expressions of happiness and sadness during both acute episodes, and after stabilization on pharmacological treatment (Daros, Ruocco, Reilly, Harris, & Sweeney, 2014). Thus, emotion recognition deficits in BD may be trait markers of the disease that may not immediately improve with acute pharmacologic treatment and clinical stabilization.

## Theory of mind

To date, few studies have directly investigated theory of mind in samples of at-risk for BD youth or PBD. In the single study of theory of mind in an at-risk for BD group, no deficit was apparent on a task that tapped into aspects of theory of mind, ranging from basic to more complex (Whitney et al., 2013). However, evidence shows that patients with PBD, specifically those with BD I, encounter greater difficulties in understanding other individuals' feelings and intentions when compared with their age-matched peers (Schenkel, Chamberlain, & Towne, 2014). One study linked manic symptoms to poorer performance on theory of mind tasks, specifically when it involved negative emotional stimuli. The authors argued that manic patients find it difficult to process negative stimuli, and are more likely to focus on themselves, rather than understand another person's perspective (Kerr, Dunbar, & Bentall, 2003). Another study revealed that PBD patients encounter difficulties in formulating socially appropriate responses during interpersonal situations (McClure et al., 2005). Schenkel, Marlow-O'Connor, Moss, Sweeney, and Pavuluri (2008) found a positive association between early theory of mind deficits and early BD onset, suggesting that BD interferes with the development of social cognitive skills. Some evidence suggests that the impaired social-cognitive skills observed in PBD may also be due to deficits in processing contextual cues to infer others' states of mind, and difficulties in understanding expressive aspects of pragmatic linguistic skills (appropriate use and interpretation of verbal/nonverbal language) (McClure et al., 2005).

A source of potential limitations in this domain is related to the type of tests administered. Some theory of mind tests reflected the real world, and included verbal and visual cues, and tapped into attentional, executive functions, and facial emotion processing systems (e.g., hinting task, eyes test). Others primarily included vignettes to examine intentions and thoughts, and may not provide a broad overview of the individual's cognitive functions (Tonelli, 2009). Given the limited number of studies in theory of mind and PBD, and the role of theory of mind in the development of social cognitive skills, additional studies in this field are urgently needed. Some thought should also be given to the type of theory of mind measures, and a more comprehensive battery should be used.

## Established BD

While the literature on social cognition in adult BD is significantly smaller than that of Sz-spectrum disorders, there has been an increased interest in this domain in recent years. While only a few studies have focused on attributional style in BD cohorts, much more focus has been dedicated to the areas of theory of mind and emotion processing. The majority of work on the latter addresses basic emotion perception, and there are a small number of studies on higher-level emotional intelligence. In the following section we provide an overview of findings related to these three areas.

### Emotion processing

Undoubtedly, the area of most focus in the behavioral emotion processing BD literature concerns the lower-level perception of facial emotions. Although there are exceptions, several studies provide evidence of a modest, but consistent deficit in the overall recognition of facial expressions in patients with BD that is present irrespective of current mood state, and not influenced by current mood symptoms (Mercer & Becerra, 2013; Samamé et al., 2015; Samamé, Martino, & Strejilevich, 2012; Van Rheenen & Rossell, 2013a). There is some evidence that this general impairment is more heavily driven by difficulties in the recognition of specific emotions, including anger and fear, and not others, such as happiness or sadness (Samamé et al., 2015). Some studies also suggest that patients with BD misattribute emotion to neutral facial expressions more frequently than controls (Thaler et al., 2013). A recent comprehensive study of facial emotion recognition in BD patients showed evidence of a consistent, albeit subtle impairment that is evident regardless of specific experimental stimuli/tasks, emotions, or expressive intensities (Van Rheenen & Rossell, 2014a). The precise cognitive mechanisms underlying this are not currently clear, but recent work has implicated the contribution of abnormal processing of spatial relationships between facial features, as well as poor general cognitive ability (Joshua et al., 2016; Van Rheenen, Joshua, et al., 2017; Van Rheenen, Meyer, & Rossell, 2014; Van Rheenen & Rossell, 2016).

Facial emotion recognition impairment in BD does not appear to represent an endophenotypic marker of the disorder, given limited evidence of impairment in first-degree relatives in the handful of studies examining the contribution of familial heritability (Bora & Özerdem, 2017). Thus, the impairment may represent a factor secondary to the disorder's expression. Additional rigorous studies in unaffected relatives are needed to confirm this, given that illness-related factors, including duration of illness or subclinical mood symptomatology, have not been associated with facial emotion recognition in metaanalyses (Samamé et al., 2012, 2015). Relevantly, there is some evidence indicating that facial emotion recognition impairment may be selective to only those BD patients who have experienced psychotic symptoms (Thaler et al., 2013). While the majority of studies have not reported separate analyses for those with or without a psychosis history, this does raise the potential that the deficit arises as a function of vulnerability to psychosis.

A recent metaanalysis showed that facial emotion recognition impairment in BD, although more subtle, is similar to that evident in Sz-spectrum disorders (Bora & Pantelis, 2016). However, empirical studies suggest that BD patients show less consistent impairments across different facial emotion recognition tasks (Rossell, Van Rheenen, Joshua, O'Regan, & Gogos, 2014), and a smaller magnitude of patient-control differences in terms of general accuracy than those with Sz-spectrum disorders (Bora & Pantelis, 2016; Goghari & Sponheim, 2013; Ruocco et al., 2014). A large study from the BSNIP consortium (Ruocco et al., 2014) comparing Sz, schizoaffective disorder, and BD patients with a psychosis history, showed that all patient groups were impaired at recognizing angry and neutral expressions compared with controls. Sz patients were also less accurate at recognizing fear and sadness than BD patients, and were the only patient group to show deficits in the recognition of happy faces compared with controls. Here, neutral stimuli were misinterpreted as either sad or happy, partially supporting previous findings showing misattribution of neutral faces to negatively valenced emotions in patients with Sz and BD with a psychosis history, but not BD patients without a psychosis history.

Preliminary research suggests that the observed deficits in facial emotion recognition in BD cannot be offset by the concurrent processing of emotional prosody (Van Rheenen & Rossell, 2014b). However, the limited number of studies examining prosodic emotion processing in samples with BD makes it difficult to determine whether this is due to the presence of a concurrent prosodic emotion recognition impairment, or a specific deficit in the higher order integration of stimuli. Indeed, the few studies investigating emotional prosody recognition in BD show mixed results (Van Rheenen & Rossell, 2013a); with slightly more reports of the presence of an impairment (Bozikas et al., 2007; Hoertnagl et al., 2014; Hofer et al., 2010; Murphy & Cutting, 1990; Rossell, Van Rheenen, Groot, Gogos, & Joshua, 2013; Van Rheenen & Rossell, 2013b) than not (Edwards et al., 2001; Vaskinn et al., 2007; Vederman et al., 2012). The extent to which this inconsistency is a function of the use of different tasks across studies is not yet clear. Mixed reports of impairments in male but not female patients (Van Rheenen & Rossell, 2013b) and vice versa (Bozikas et al., 2007), also calls into question the relevance of gender, which has not been examined as a covariate in all studies.

The specificity or generalizability of an emotional prosody impairment across emotions is also not clear, owing to study differences in the specific emotions examined, and the limited number of studies providing data about the identification of each emotion separately (Van Rheenen & Rossell, 2013a). Nonetheless, one study comparing patients with both BD and Sz indicated comparable impairment in the identification of anger, with both clinical groups misinterpreting anger as neutral prosody more frequently than controls (Hoertnagl et al., 2014). In turn, the findings of another study comparing patients with these disorders showed that while decrements in the *general* identification of emotional prosody was comparable between them, BD patients did not have the same difficulties in tuning out irrelevant information when naming auditory emotional expressions as Sz-spectrum patients (Rossell et al., 2013).

Taken together, the mixed findings regarding whether this basic perceptual ability is intact or impaired in BD indicate that concrete inferences about its nature are currently unjustified. Further work addressing the current methodological and statistical inconsistencies in the relevant literature are critically required to provide a significant progression of our knowledge in this area.

Regarding higher-level aspects of emotion processing, a number of recent studies in BD have included performance-based assessments to understand components of emotional intelligence, most commonly using the MSCIET. The full version of this test measures four branches of emotional intelligence across a 141-item assessment. However, typically only one branch—pertaining to the ability to manage one's emotions, has been assessed in BD because of its inclusion in the increasingly widely used MCCB. Compared with Sz-spectrum disorders that consistently show substantial impairments, findings from BD studies using assessment of only this branch have largely revealed an absence of impairment on this domain (Burdick et al., 2011; Junghee et al., 2013; Nitzburg, Burdick, Malhotra, & DeRosse, 2015; Sperry et al., 2015).

However, using the full MSCIET assessment, Aparicio et al. (2017) found impairments of medium to large effect across all but one branch of the test in a group of euthymic BD patients compared with controls. Frajo-Apor, Kemmler, et al. (2016) also reported a small decrement in the total MSCIET score (reflecting an emotional IQ) when comparing the performance of BD patients with normative data. A subsequent study by Varo et al. (2017) segregated a large euthymic BD sample into three subgroups based on MSCIET total scores that were equivocal, above, or below normative means. Using this method, 69% of the sample was considered to have normal levels of emotional intelligence. Only 12% of BD patients were considered to have low-range emotional intelligence, with scores 20 points below the normative average. This group showed the lowest cognitive, clinical, and functional outcome scores of all emotional intelligence subgroups. While the absence of direct comparisons to a control group limit these latter studies, Varo et al.'s findings of variability in emotional intelligence on the managing emotions branch of the MSCIET(Burdick et al., 2014; Lewandowski, Baker, McCarthy, Norris, & Öngür, 2017; Van Rheenen, Lewandowski, et al., 2017) are consistent with recent work indicating impairment on this test in only a subgroup of patients who show the most substantial cognitive impairments (Lewandowski et al., 2017; Van Rheenen, Lewandowski, et al., 2017).

## Theory of mind

There has been an increasing focus on theory of mind in BD in recent years, with more than 35 papers published on the topic, and several of these in the past 5 years. Metaanalytic findings indicate patient-control differences of small to medium effect (Bora, Bartholomeusz, & Pantelis, 2015; Samamé et al., 2012, 2015), which, although smaller than that of Sz-spectrum (Bora & Pantelis, 2016), are generally comparable in magnitude across visual versus verbal, and cognitive versus affective aspects of theory of mind (Bora et al., 2015). The trait-relevance of the deficit is apparent, given its presence across all mood phases of the illness (Bora et al., 2015). However, it does appear to be exacerbated in symptomatic patients, with more severe manic symptomatology associated with worse performance (Bora et al., 2015). With exceptions (Wang, Roberts, Liang, Shi, & Wang, 2015), theory of mind deficits have also been found in adult first-degree relatives of people with BD (Reynolds, Van Rheenen, & Rossell, 2014; Santos et al., 2017; Yücel et al., 2016), highlighting its potential as an endophenotypic marker for the disorder. Indeed, a recent metaanalysis showed that these deficits, while modest, were comparable in size with those seen in first-degree relatives for neurocognitive performance (Bora & Özerdem, 2017).

Deficits elicited by the most commonly employed tasks—the "Reading the Mind in the Eyes Test" and the "Faux Pas Test"—are consistently evident, although they have also been reported on other tasks indexing both first and second-order theory of mind. Equivocal patient-control performance on control components of these tasks suggests that the deficit, which reflects abnormalities in both basic and complex components of theory of mind, is unlikely to be a function of a general inability to understand social relationships or causal connections (Van Rheenen & Rossell, 2013c). Nonetheless, some work does show that general cognitive impairment, especially in executive functioning, may contribute (Bodnar & Rybakowski, 2017; Ospina et al., 2018; Van Rheenen et al., 2014; Wang et al., 2018). Gender, age, education, illness duration, and age of onset, however, are not predictive of theory of mind in BD (Bora et al., 2015).

## Attributional style

The research on attributional bias in adult BD is extremely scarce. Preliminary work shows that symptomatic and remitted BD patients are more likely to interpret hostility in ambiguous situations (Lahera et al., 2015), and attribute negative events to either themselves (Knowles et al., 2007; Lyon, Startup, & Bentall, 1999) or others, rather than the situation (Lahera et al., 2015), than controls. Subthreshold depressive symptoms appear to play a role in this bias (Lahera et al., 2015), but the role of other clinical and demographic variables is not clear.

## Conclusions

In Sz-spectrum patients, facial and prosodic emotion processing deficits appear to be of large effect, and are stable and robust across illness phases. While deficits in the overall recognition of emotions are evident, misinterpretation of negative, rather than positive, emotions is more common across both facial and prosodic emotion processing. However, facial emotion processing deficits appear to be influenced by state factors more so than are deficits in the processing of emotional prosody. Nonetheless, the finding of impaired facial emotion processing in early illness phases and in first degree relatives of Sz-spectrum probands suggests that they are primarily trait-like persistent features of the illness.

In contrast, facial emotion processing deficits in adult BD, while mood state-independent, are of much smaller effect than that of Sz-spectrum patients. However, consistent with Sz-spectrum findings, a generalized deficit in the overall processing of facial emotions appears to be replicable, although the extent to which deficits are driven by patients with a psychosis history is unclear. As in Sz-spectrum studies, there is evidence to suggest that neutral facial expressions are ascribed emotional significance more frequently by BD patients than controls. This effect has also been shown in PBD patients, which, alongside other preliminary evidence, indicates that facial emotion recognition deficits are present in the early course of BD. The paucity of rigorous research on emotion processing in at-risk, pediatric, and first-episode BD samples, however, limits conclusions about the stability, magnitude, and robustness of the effect, as well as the time-point at which the deficit is detectable, and whether it represents developmental delay, developmental arrest, or performance decline.

As preliminary evidence does suggest that facial emotion processing performance lacks specificity across Sz-spectrum disorders and BD in early-stage illness, it is therefore unlikely to serve as a phenotypic marker of transition to psychosis in UHR for psychosis populations. However, facial emotion processing deficits have been found in unaffected relatives of Sz-spectrum patients, but this is generally not reflected in studies of unaffected relatives of probands with BD. Thus, facial emotion processing performance may be valuable as an intermediate phenotype linking behavior to underlying genetics in Sz-spectrum disorders, whereas deficits in BD may represent an outcome secondary to the illness course.

Unlike the substantial literature showing prosody processing deficits across illness phases in Sz-spectrum cohorts, emotional prosody in BD has only been studied in the context of adults/established illness, where evidence for impairment is inconsistent, at best. Studies reporting deficits in some BD cohorts suggest a role for gender, which does not appear to be the case for Sz-spectrum patients. Likewise, only a proportion of adult BD patients appear to have impairments on emotional intelligence measures, while impairment in this higher-level aspect of emotion processing is evident in both early- and late-stage Sz-spectrum illness. It is possible that the low emotional intelligence subgroup of BD patients presents with a similar symptom profile to Sz-spectrum patients, but further work directly addressing this is necessary.

Theory of mind deficits are apparent in both Sz-spectrum and BD patients, with verbal and nonverbal theory of mind effect sizes generally comparable in established/adult illness. Although deficits are of a larger magnitude in Sz-spectrum disorders, deficits are evident in remitted illness phases for both disorders, as well as in unaffected relatives of probands. Thus, theory of mind performance appears to represent an endophenotype that cuts across diagnostic boundaries.

An increase in the magnitude of case-control effects of *verbal* theory of mind deficits from the UHR stage to established Sz-spectrum illness suggests a possible progressive decline in performance in these patients with increasing illness length. However, this is inconsistent with cross-sectional data in established Sz-spectrum patients showing that illness duration does not significantly correlate with theory of mind. It is also inconsistent with longitudinal research in FEP and established Sz-spectrum disorder. The inconsistent finding of impairment in UHR samples is likely related to the heterogeneity of the population and the dilution effect (given that not all UHR individuals will go on to develop a Sz-spectrum disorder). Impairments on *nonverbal* theory of mind tasks are also evident in FEP, but not as consistently as in established illness. In contrast, there is limited research on theory of mind in early-course BD, with a single study finding no deficits in individuals at genetic risk for the illness, and another showing impairment in PBD patients. In the absence of replication and longitudinal analysis, it is unclear whether the early theory of mind deficit found in PBD is stable, but findings in adult BD patients show that neither illness duration, nor age of onset, contribute to theory of mind performance.

In both Sz-spectrum and BD patients, the impact of neurocognition on facial emotion processing and theory of mind is not entirely clear, because many studies (particularly in BD) do not adequately control for concurrent neurocognitive ability. Intact social cognition is unusual in the context of impaired neurocognition, at least in Sz-spectrum disorders. Several correlational studies in Sz-spectrum and BD patients do also indicate that neurocognition predicts variance in both emotion recognition and theory of mind (Addington et al., 2006a; Brekke et al., 2005; Van Rheenen et al., 2014), albeit generally to only a small degree. However, in the context of null effects of education and IQ scores, deficits on theory of mind tasks, but generally not on cognitively relevant control tasks, suggests at least some independence between neurocognition and theory of mind in both disorders.

Discordantly, evidence of impairments on face processing control tasks in patients with established Sz-spectrum disorders suggest that facial emotion recognition deficits arise as part of a global cognitive deficit. Two recent studies provide evidence for the influence of face processing itself on facial emotion recognition; the first indicating impaired face processing performance on a task assessing the use of featural *and* configural face processing strategies(Joshua et al., 2016), and the second indicating impaired recognition of emotional faces in contexts in which both featural and configural face information was manipulated (Van Rheenen, Joshua, et al., 2017). In both of these studies, a selective impairment in only configural face processing of emotional *and* neutral faces was evident in BD patients. This discrepancy in the results of each cohort may explain the greater magnitude of facial emotion recognition impairments in Sz-spectrum compared with BD, given that BD patients appear to have some residual capacity for compensation using another face processing strategy when processing emotional faces (Van Rheenen, Joshua, et al., 2017). This is consistent with findings in BD showing a smaller magnitude of neurocognitive deficits relative to Sz-spectrum patients, and suggests that differences in the factors that mediate/moderate social cognition between disorders can partially explain differences in the magnitude of social cognitive deficits as well. This hypothesis is supported by statistical evidence from a recent metaanalytic in which greater neurocognitive impairment in Sz explicitly contributed to Sz-BD group differences in social cognition (Bora & Pantelis, 2016).

Given available evidence, it is clear that the literature on theory of mind and emotion processing in both Sz-spectrum and BD patients is much more comprehensive than that of social perception and attributional style. However, metaanalyses in Sz-spectrum disorders do show large patient-control effect size deficits for social perception that are evident in both FEP and established samples, are independent of age and illness duration, and remain stable for at least 5 years following illness onset. Similarly, hostile and personalizing biases are evident in both FEP and established Sz-spectrum samples. These findings are consistent with the limited research on attributional style in adult BD. However, as both depressive and psychotic symptoms are suggested to play a role in perpetuating these biases in BD and Sz-spectrum patients respectively, it appears that abnormal attributional style may represent a state, rather than a trait-like feature, in these disorders.

In summary, current evidence suggests that behavioral deficits in emotion processing, theory of mind, and attributional style cut across diagnostic boundaries of Sz-spectrum disorders and BD. At a surface level, differences appear to be quantitative rather than qualitative in nature, with both theory of mind and emotion processing deficits representing primarily trait-like illness features. The large overlap in social cognitive deficits between patients on this illness spectrum provides further support for a shift away from nosological categorization toward a transdiagnostic approach (Cuthbert & Insel, 2010), where specific phenomenology, such as emotion processing and theory of mind, may be utilized to predict illness outcomes and aid in treatment planning, irrespective of clinical diagnosis. Nonetheless, it is clear that further work covering all four identified domains of social cognition in these disorders will be important for progressing this approach and generating knowledge regarding the extent to which moderating/mediating factors contributing to social cognitive deficits overlap across disorders.

Such research should aim to overcome methodological assessment concerns in order to more clearly probe these constructs and rule out confounds. In particular, further research emphasis on social cognition in the early course of BD is crucial for comparison purposes, but also to provide insight into whether social cognitive impairments in established/adult BD are related to the expression of the illness, or are present prior to the manifestation of its descriptive mood symptoms. Further studies in well-characterized, unaffected, first-degree relative cohorts of probands with these disorders would complement this. Indeed, in Sz-spectrum research, the presence of lower-level emotion processing and theory of mind impairments in family members, patients in the UHR stage, those having recently experienced a first psychotic episode, and those with established illness suggests that social cognitive abnormalities follow a neurodevelopmental trajectory, and links more clearly to a primary genetic etiology than to secondary illness outcomes. It is difficult to draw the same conclusions in BD, given the paucity of research investigating the trajectory of social cognitive changes from early to late illness stages, and research involving first-degree relatives.

## References


Achim, A. M., Ouellet, R., Roy, M. A., & Jackson, P. L. (2012). Mentalizing in first-episode psychosis. *Psychiatry Research, 196*(2–3), 207–213.

Achim, A. M., Sutliff, S., Samson, C., Montreuil, T. C., & Lecomte, T. (2016). Attribution bias and social anxiety in schizophrenia. *Schizophrenia Research: Cognition, 4*, 1–3.

Addington, J., Girard, T. A., Christensen, B. K., & Addington, D. (2010). Social cognition mediates illness-related and cognitive influences on social function in patients with schizophrenia-spectrum disorders. *Journal of Psychiatry and Neuroscience, 35*(1), 49.

Addington, J., Saeedi, H., & Addington, D. (2006a). Facial affect recognition: a mediator between cognitive and social functioning in psychosis? *Schizophrenia Research, 85* (1–3), 142–150.

Addington, J., Saeedi, H., & Addington, D. (2006b). Influence of social perception and social knowledge on cognitive and social functioning in early psychosis. *The British Journal of Psychiatry, 189*(4), 373–378.

Adolphs, R., Tranel, D., & Damasio, A. R. (1998). The human amygdala in social judgment. *Nature, 393*(6684), 470–474.

Amminger, G. P., Schafer, M. R., Klier, C. M., Schlögelhofer, M., Mossaheb, N., Thompson, A., … Nelson, B. (2012). Facial and vocal affect perception in people at ultra-high risk of psychosis, first-episode schizophrenia and healthy controls. *Early Intervention in Psychiatry, 6*(4), 450–454.

Amminger, G. P., Schafer, M. R., Papageorgiou, K., Klier, C. M., Schlögelhofer, M., Mossaheb, N., … McGorry, P. D. (2012). Emotion recognition in individuals at clinical high-risk for schizophrenia. *Schizophrenia Bulletin, 38*(5), 1030–1039.

An, S. K., Kang, J. I., Park, J. Y., Kim, K. R., Lee, S. Y., & Lee, E. (2010). Attribution bias in ultra-high risk for psychosis and first-episode schizophrenia. *Schizophrenia Research, 118* (1–3), 54–61.

Aparicio, A., Santos, J. L., Jiménez-López, E., Bagney, A., Rodríguez-Jiménez, R., & Sánchez-Morla, E. M. (2017). Emotion processing and psychosocial functioning in euthymic bipolar disorder. *Acta Psychiatrica Scandinavica, 135*(4), 339–350.

Ay, R., Boke, O., Pazvantoglu, O., Şahin, A. R., Sarisoy, G., Arik, A. C., & Güz, H. (2016). Social cognition in schizophrenia patients and their first-degree relatives. *Noro Psikiyatri Arsivi, 53*(4), 338–343.

Bodnar, A., & Rybakowski, J. K. (2017). Mentalization deficit in bipolar patients during an acute depressive and manic episode: association with cognitive functions. *International Journal of Bipolar Disorders, 5*(1), 38.

Bora, E., Bartholomeusz, C., & Pantelis, C. (2015). Meta-analysis of Theory of Mind (ToM) impairment in bipolar disorder. *Psychological Medicine, 46*(2), 253–264.

Bora, E., & Özerdem, A. (2017). Social cognition in first-degree relatives of patients with bipolar disorder: a meta-analysis. *European Neuropsychopharmacology, 27*(4), 293–300.

Bora, E., & Pantelis, C. (2013). Theory of mind impairments in first-episode psychosis, individuals at ultra-high risk for psychosis and in first-degree relatives of schizophrenia: systematic review and meta-analysis. *Schizophrenia Research, 144*(1–3), 31–36.

Bora, E., & Pantelis, C. (2016). Social cognition in schizophrenia in comparison to bipolar disorder: a meta-analysis. *Schizophrenia Research, 175*(1), 72–78.

Bora, E., Yucel, M., & Pantelis, C. (2009). Theory of mind impairment in schizophrenia: meta-analysis. *Schizophrenia Research, 109*(1–3), 1–9.

Bozikas, V. P., Kosmidis, M. H., Tonia, T., Andreou, C., Focas, K., & Karavatos, A. (2007). Impaired perception of affective prosody in remitted patients with bipolar disorder. *The Journal of Neuropsychiatry and Clinical Neurosciences, 19*(4), 436–440.

Brekke, J., Kay, D. D., Lee, K. S., & Green, M. F. (2005). Biosocial pathways to functional outcome in schizophrenia. *Schizophrenia Research, 80*(2), 213–225.

Brotman, M. A., Guyer, A. E., Lawson, E. S., Horsey, S. E., Rich, B. A., Dickstein, D. P., ... Leibenluft, E. (2008). Facial emotion labeling deficits in children and adolescents at risk for bipolar disorder. *The American Journal of Psychiatry, 165*(3), 385–389.

Brotman, M. A., Skup, M., Rich, B. A., Blair, K. S., Pine, D. S., Blair, J. R., & Leibenluft, E. (2008). Risk for bipolar disorder is associated with face processing deficits across emotions. *Journal of the American Academy of Child and Adolescent Psychiatry, 47*, 1455–1461.

Browne, J., Penn, D. L., Raykov, T., Pinkham, A. E., Kelsven, S., Buck, B., & Harvey, P. D. (2016). Social cognition in schizophrenia: factor structure of emotion processing and theory of mind. *Psychiatry Research, 242*, 150–156.

Brune, M., Ozgurdal, S., Ansorge, N., von Reventlow, H. G., Peters, S., Nicolas, V., ... Lissek, S. (2011). An fMRI study of "theory of mind" in at-risk states of psychosis: comparison with manifest schizophrenia and healthy controls. *NeuroImage, 55*(1), 329–337.

Buck, B. E., Healey, K. M., Gagen, E. C., Roberts, D. L., & Penn, D. L. (2016). Social cognition in schizophrenia: factor structure, clinical and functional correlates. *Journal of Mental Health, 25*(4), 330–337.

Burdick, K. E., Goldberg, T. E., Cornblatt, B. A., Keefe, R. S., Gopin, C. B., DeRosse, P., & Malhotra, A. K. (2011). The MATRICS consensus cognitive battery in patients with bipolar I disorder. *Neuropsychopharmacology, 36*(8), 1587–1592.

Burdick, K. E., Russo, M., Frangou, S., Mahon, K., Braga, R. J., Shanahan, M., & Malhotra, A. K. (2014). Empirical evidence for discrete neurocognitive subgroups in bipolar disorder: clinical implications. *Psychological Medicine, 44*(14), 3083–3096.

Carter, T. D. C., Mundo, E., Parikh, S. V., & Kennedy, J. L. (2003). Early age at onset as a risk factor for poor outcome of bipolar disorder. *Journal of Psychiatric Research, 37*(4), 297–303.

Chan, R. C., Li, H., Cheung, E. F., & Gong, Q. Y. (2010). Impaired facial emotion perception in schizophrenia: a meta-analysis. *Psychiatry Research, 178*(2), 381–390.

Chung, Y. S., Barch, D., & Strube, M. (2014). A meta-analysis of mentalizing impairments in adults with schizophrenia and autism spectrum disorder. *Schizophrenia Bulletin, 40*(3), 602–616.

Chung, Y. S., Kang, D. H., Shin, N. Y., Yoo, S. Y., & Kwon, J. S. (2008). Deficit of theory of mind in individuals at ultra-high-risk for schizophrenia. *Schizophrenia Research, 99*(1–3), 111–118.

Comparelli, A., Corigliano, V., De Carolis, A., Mancinelli, I., Trovini, G., Ottavi, G., ... Girardi, P. (2013). Emotion recognition impairment is present early and is stable throughout the course of schizophrenia. *Schizophrenia Research, 143*(1), 65–69.

Corcoran, C. M., Keilp, J. G., Kayser, J., Klim, C., Butler, P. D., Bruder, G. E., ... Javitt, D. C. (2015). Emotion recognition deficits as predictors of transition in individuals at clinical high risk for schizophrenia: a neurodevelopmental perspective. *Psychological Medicine, 45*(14), 2959–2973.

Couture, S. M., Penn, D. L., Addington, J., Woods, S. W., & Perkins, D. O. (2008). Assessment of social judgments and complex mental states in the early phases of psychosis. *Schizophrenia Research, 100*(1–3), 237–241.

Cuthbert, B. N., & Insel, T. R. (2010). Toward new approaches to psychotic disorders: the NIMH research domain criteria project. *Schizophrenia Bulletin, 36*(6), 1061–1062.

Daros, A. R., Ruocco, A. C., Reilly, J. L., Harris, M. S., & Sweeney, J. A. (2014). Facial emotion recognition in first-episode schizophrenia and bipolar disorder with psychosis. *Schizophrenia Research, 153*(1), 32–37.

Dawson, S., Kettler, L., Burton, C., & Galletly, C. (2012). Do people with schizophrenia lack emotional intelligence? *Schizophr Research and Treatment, 2012*, 495174.

Dean, A. M., Goodby, E., Ooi, C., Nathan, P. J., Lennox, B. R., Scoriels, L., … Barnes, A. (2013). Speed of facial affect intensity recognition as an endophenotype of first-episode psychosis and associated limbic-cortical grey matter systems. *Psychological Medicine, 43*(3), 591–602.

Edwards, J., Jackson, H. J., & Pattison, P. E. (2002). Emotion recognition via facial expression and affective prosody in schizophrenia: a methodological review. *Clinical Psychology Review, 22*(6), 789–832.

Edwards, J., Pattison, P. E., Jackson, H. J., & Wales, R. J. (2001). Facial affect and affective prosody recognition in first-episode schizophrenia. *Schizophrenia Research, 48*(2–3), 235–253.

Fornells-Ambrojo, M., & Garety, P. A. (2009). Attributional biases in paranoia: the development and validation of the Achievement and Relationships Attributions Task (ARAT). *Cognitive Neuropsychiatry, 14*(2), 87–109.

Frajo-Apor, B., Kemmler, G., Pardeller, S., Plass, T., Mühlbacher, M., Welte, A. S., … Hofer, A. (2016). Emotional intelligence and non-social cognition in schizophrenia and bipolar I disorder. *Psychological Medicine*, 1–8.

Frajo-Apor, B., Pardeller, S., Kemmler, G., Welte, A. S., & Hofer, A. (Apr 2016). Emotional intelligence deficits in schizophrenia: the impact of non-social cognition. *Schizophrenia Research, 172*(1–3), 131–136.

Frith, C. D. (1992). *The cognitive neuropsychology of schizophrenia*. Hove: Lawrence Erlbaum.

Gabay, A. S., Kempton, M. J., & Mehta, M. A. (2015). Facial affect processing deficits in schizophrenia: a meta-analysis of antipsychotic treatment effects. *Journal of Psychopharmacology, 29*(2), 224–229.

Gee, D. G., & Cannon, T. D. (2011). Prediction of conversion to psychosis: review and future directions. *Revista Brasileira de Psiquiatria, 33*(Suppl. 2), s129–s142.

Gill, K. E., Cressman, V., Poe, S. L., Steinfeld, S., Ben-David, S., Keilp, J. G., … Corcoran, C. (2016). Social inference in individuals at clinical high risk for psychosis. *Early Intervention in Psychiatry, 10*(1), 77–80.

Gilleen, J., Xie, F., & Strelchuk, D. (2017). 38. Distinct theory of mind deficit profiles in schizophrenia and autism: a meta-analysis of published research. *Schizophrenia Bulletin, 43*(Suppl. 1), S22.

Goghari, V. M., & Sponheim, S. R. (2013). More pronounced deficits in facial emotion recognition for schizophrenia than bipolar disorder. *Comprehensive Psychiatry, 54*(4), 388–397.

Green, M. F., Bearden, C. E., Cannon, T. D., Fiske, A. P., Hellemann, G. S., Horan, W. P., … Nuechterlein, K. H. (2012). Social cognition in schizophrenia, Part 1: performance across phase of illness. *Schizophrenia Bulletin, 38*(4), 854–864.

Green, M. F., Penn, D. L., Bentall, R., Carpenter, W. T., Gaebel, W., Gur, R. C., … Heinssen, R. (2008). Social cognition in schizophrenia: an NIMH workshop on definitions, assessment, and research opportunities. *Schizophrenia Bulletin, 34*(6), 1211–1220.

Healey, K. M., Bartholomeusz, C. F., & Penn, D. L. (2016). Deficits in social cognition in first episode psychosis: a review of the literature. *Clinical Psychology Review, 50*, 108–137.

Healey, K. M., Penn, D. L., Perkins, D., Woods, S. W., & Addington, J. (2013). Theory of mind and social judgments in people at clinical high risk of psychosis. *Schizophrenia Research, 150*(2–3), 498–504.

Hill, S. K., Sweeney, J. A., Hamer, R. M., Keefe, R. S., Perkins, D. O., Gu, H., … Lieberman, J. A. (2008). Efficiency of the CATIE and BACS neuropsychological batteries in assessing cognitive effects of antipsychotic treatments in schizophrenia. *Journal of the International Neuropsychological Society, 14*(2), 209–221.

Hoekert, M., Kahn, R. S., Pijnenborg, M., & Aleman, A. (2007). Impaired recognition and expression of emotional prosody in schizophrenia: review and meta-analysis. *Schizophrenia Research, 96*(1–3), 135–145.

Hoertnagl, C. M., Yalcin-Siedentopf, N., Baumgartner, S., Biedermann, F., Deisenhammer, E. A., Hausmann, A., ... Hofer, A. (2014). Affective prosody perception in symptomatically remitted patients with schizophrenia and bipolar disorder. *Schizophrenia Research, 158*(1–3), 100–104.
Hofer, A., Baumgartner, S., Benecke, C., Biedermann, F., Deisenhammer, E. A., Hausmann, A., ... Yalcin, N. (2010). Perception of affective prosody in remitted patients with schizophrenia and bipolar disorder I. *Scizophrenia Research, 117*, 213.
Horan, W. P., Green, M. F., DeGroot, M., Fiske, A., Hellemann, G., Kee, K., ... Nuechterlein, K. H. (2012). Social cognition in schizophrenia, Part 2: 12-month stability and prediction of functional outcome in first-episode patients. *Schizophrenia Bulletin, 38*(4), 865–872.
Hur, J. W., Byun, M. S., Shin, N. Y., Shin, Y. S., Kim, S. N., Jang, J. H., & Kwon, J. S. (2013). General intellectual functioning as a buffer against theory-of-mind deficits in individuals at ultra-high risk for psychosis. *Schizophrenia Research, 149*(1–3), 83–87.
International Schizophrenia Consortium, Purcell, S. M., Wray, N. R., Stone, J. L., Visscher, P. M., O'Donovan, M. C., ... Sklar, P. (2009). Common polygenic variation contributes to risk of schizophrenia and bipolar disorder. *Nature, 460*(7256), 748–752.
Joshua, N. R. (2010). *Face processing in schizophrenia: An investigation of configural processing and the relationship with facial emotion processing and neurocognition* [PhD dissertation].
Joshua, N., Van Rheenen, T., Castle, D., & Rossell, S. (2016). Taking it at "face value": the use of face processing strategies in bipolar disorder and schizophrenia. *Journal of the International Neuropsychological Society, 22*, 1–10.
Junghee, L., Altshuler, L., Glahn, D. C., Miklowitz, D. J., Ochsner, K., & Green, M. F. (2013). Social and nonsocial cognition in bipolar disorder and schizophrenia: relative levels of impairment. *The American Journal of Psychiatry, 170*(3), 334–341.
Kee, K. S., Green, M. F., Mintz, J., & Brekke, J. S. (2003). Is emotion processing a predictor of functional outcome in schizophrenia? *Schizophrenia Bulletin, 29*(3), 487–497.
Kerr, N., Dunbar, R. I., & Bentall, R. P. (2003). Theory of mind deficits in bipolar affective disorder. *Journal of Affective Disorders, 73*(3), 253–259.
Kim, H. S., Shin, N. Y., Jang, J. H., Kim, E., Shim, G., Park, H. Y., ... Kwon, J. S. (2011). Social cognition and neurocognition as predictors of conversion to psychosis in individuals at ultra-high risk. *Schizophrenia Research, 130*(1–3), 170–175.
Knowles, R., Tai, S., Jones, S. H., Highfield, J., Morriss, R., & Bentall, R. P. (2007). Stability of self-esteem in bipolar disorder: comparisons among remitted bipolar patients, remitted unipolar patients and healthy controls. *Bipolar Disorders, 9*(5), 490–495.
Kohler, C. G., Turner, T. H., Bilker, W. B., Brensinger, C. M., Siegel, S. J., Kanes, S. J., ... Gur, R. C. (2003). Facial emotion recognition in schizophrenia: intensity effects and error pattern. *The American Journal of Psychiatry, 160*(10), 1768–1774.
Kohler, C. G., Walker, J. B., Martin, E. A., Healey, K. M., & Moberg, P. J. (2010). Facial emotion perception in schizophrenia: a meta-analytic review. *Schizophrenia Bulletin, 36* (5), 1009–1019.
Kucharska-Pietura, K., David, A. S., Masiak, M., & Phillips, M. L. (2005). Perception of facial and vocal affect by people with schizophrenia in early and late stages of illness. *The British Journal of Psychiatry, 187*, 523–528.
Lahera, G., Herrera, S., Reinares, M., Benito, A., Rullas, M., González-Cases, J., & Vieta, E. (2015). Hostile attributions in bipolar disorder and schizophrenia contribute to poor social functioning. *Acta Psychiatrica Scandinavica, 131*(6), 472–482.
Laursen, T. M., Agerbo, E., & Pedersen, C. B. (2009). Bipolar disorder, schizoaffective disorder, and schizophrenia overlap: a new comorbidity index. *The Journal of Clinical Psychiatry, 70*(10), 1432–1438.
Lavoie, M. A., Plana, I., Bedard Lacroix, J., Godmaire-Duhaime, F., Jackson, P. L., & Achim, A. M. (2013). Social cognition in first-degree relatives of people with schizophrenia: a meta-analysis. *Psychiatry Research, 209*(2), 129–135.

Lee, T. Y., Hong, S. B., Shin, N. Y., & Kwon, J. S. (2015). Social cognitive functioning in prodromal psychosis: a meta-analysis. *Schizophrenia Research, 164*(1–3), 28–34.
Lewandowski, K. E., Baker, J. T., McCarthy, J. M., Norris, L. A., & Öngür, D. (2017). Reproducibility of cognitive profiles in psychosis using cluster analysis. *Journal of the International Neuropsychological Society, 24*(4), 382–390.
Lincoln, S. H., Norkett, E. M., Frost, K. H., Gonzalez-Heydrich, J., & D'Angelo, E. J. (2017). A developmental perspective on social-cognition difficulties in youth at clinical high risk for psychosis. *Harvard Review of Psychiatry, 25*(1), 4–14.
Lyon, H. M., Startup, M., & Bentall, R. P. (1999). Social cognition and the manic defense: attributions, selective attention, and self-schema in bipolar affective disorder. *Journal of Abnormal Psychology, 108*, 273–282.
Mazza, M., Catalucci, A., Pino, M. C., Giusti, L., Nigri, A., Pollice, R., … Gallucci, M. (2013). Dysfunctional neural networks associated with impaired social interactions in early psychosis: an ICA analysis. *Brain Imaging and Behavior*, 7(3), 248–259.
McCleery, A., Lee, J., Fiske, A. P., Ghermezi, L., Hayata, J. N., Hellemann, G. S., … Green, M. F. (2016). Longitudinal stability of social cognition in schizophrenia: a 5-year follow-up of social perception and emotion processing. *Schizophrenia Research, 176*(2–3), 467–472.
McClure, E. B., Pope, K., Hoberman, A. J., Pine, D. S., & Leibenluft, E. (2003). Facial expression recognition in adolescents with mood and anxiety disorders. *American Journal of Psychiatry, 160*(6), 1172–1174.
McClure, E. B., Treland, J. E., Snow, J., Schmajuk, M., Dickstein, D. P., Towbin, K. E., … Leibenluft, E. (2005). Deficits in social cognition and response flexibility in pediatric bipolar disorder. *American Journal of Psychiatry, 162*(9), 1644–1651.
Mehl, S., Landsberg, M. W., Schmidt, A. C., Cabanis, M., Bechdolf, A., Herrlich, J., … Wagner, M. (2014). Why do bad things happen to me? Attributional style, depressed mood, and persecutory delusions in patients with schizophrenia. *Schizophrenia Bulletin, 40*(6), 1338–1346.
Mercer, L., & Becerra, R. (2013). A unique emotional processing profile of euthymic bipolar disorder? A critical review. *Journal of Affective Disorders, 146*(3), 295–309.
Miller, T. J., McGlashan, T. H., Rosen, J. L., Somjee, L., Markovich, P. J., Stein, K., & Woods, S. W. (2002). Prospective diagnosis of the initial prodrome for schizophrenia based on the Structured Interview for Prodromal Syndromes: preliminary evidence of interrater reliability and predictive validity. *The American Journal of Psychiatry, 159*(5), 863–865.
Minor, K. S., & Lysaker, P. H. (2014). Necessary, but not sufficient: links between neurocognition, social cognition, and metacognition in schizophrenia are moderated by disorganized symptoms. *Schizophrenia Research, 159*(1), 198–204.
Mote, J., & Kring, A. M. (2016). Facial emotion perception in schizophrenia: does sex matter? *World Journal of Psychiatry, 6*(2), 257–268.
Murphy, D., & Cutting, J. (1990). Prosodic comprehension and expression in schizophrenia. *Journal of Neurology, Neurosurgery, and Psychiatry, 53*, 727–730.
Nelson, B., Yuen, H. P., Wood, S. J., Lin, A., Spiliotacopoulos, D., Bruxner, A., … Yung, A. R. (2013). Long-term follow-up of a group at ultra high risk ("prodromal") for psychosis: the PACE 400 study. *JAMA Psychiatry, 70*(8), 793–802.
Nitzburg, G. C., Burdick, K. E., Malhotra, A. K., & DeRosse, P. (2015). Social cognition in patients with schizophrenia spectrum and bipolar disorders with and without psychotic features. *Schizophrenia Research: Cognition, 2*(1), 2–7.
Ospina, L. H., Nitzburg, G. C., Shanahan, M., Perez-Rodriguez, M. M., Larsen, E., Latifoglu, A., & Burdick, K. E. (2018). Social cognition moderates the relationship between neurocognition and community functioning in bipolar disorder. *Journal of Affective Disorders, 235*, 7–14. 2018/08/01.

Penn, D. L., Sanna, L. J., & Roberts, D. L. (2008). Social cognition in schizophrenia: an overview. *Schizophrenia Bulletin, 34*(3), 408–411.
Petkova, E., Lu, F., Kantrowitz, J., Sanchez, J. L., Lehrfeld, J., Scaramello, N., ... Butler, P. D. (2014). Auditory tasks for assessment of sensory function and affective prosody in schizophrenia. *Comprehensive Psychiatry, 55*(8), 1862–1874.
Pinkham, A. E., Harvey, P. D., & Penn, D. L. (2016). Paranoid individuals with schizophrenia show greater social cognitive bias and worse social functioning than non-paranoid individuals with schizophrenia. *Schizophrenia Research: Cognition, 3*, 33–38.
Pinkham, A. E., Kelsven, S., Kouros, C., Harvey, P. D., & Penn, D. L. (2017). The effect of age, race, and sex on social cognitive performance in individuals with schizophrenia. *The Journal of Nervous and Mental Disease, 205*(5), 346–352.
Pinkham, A. E., Penn, D. L., Green, M. F., Buck, B., Healey, K., & Harvey, P. D. (2014). The social cognition psychometric evaluation study: results of the expert survey and RAND panel. *Schizophrenia Bulletin, 40*(4), 813–823.
Pinkham, A. E., Penn, D. L., Green, M. F., & Harvey, P. D. (2016). Social cognition psychometric evaluation: results of the initial psychometric study. *Schizophrenia Bulletin, 42*(2), 494–504.
Premack, D., & Woodruff, G. (1978). Does the chimpanzee have a theory of mind? *The Behavioral and Brain Sciences, 04*, 515–526.
Reske, M., Habel, U., Kellermann, T., Backes, V., Jon Shah, N., von Wilmsdorff, M., ... Schneider, F. (2009). Differential brain activation during facial emotion discrimination in first-episode schizophrenia. *Journal of Psychiatric Research, 43*(6), 592–599.
Reynolds, M. T., Van Rheenen, T. E., & Rossell, S. L. (2014). Theory of mind in first degree relatives of individuals with bipolar disorder. *Psychiatry Research, 219*, 400–402.
Rich, B. A., Vinton, D. T., Roberson-Nay, R., Hommer, R. E., Berghorst, L. H., McClure, E. B., ... Leibenluft, E. (2006). Limbic hyperactivation during processing of neutral facial expressions in children with bipolar disorder. *Proceedings of the National Academy of Sciences, 103*(23), 8900–8905.
Rocca, C. C. A., Heuvel, E., Caetano, S. C., & Lafer, B. (2009). Facial emotion recognition in bipolar disorder: a critical review. *Revista Brasileira de Psiquiatria, 31*(2), 171–180.
Rossell, S. L., Van Rheenen, T. E., Groot, C., Gogos, A., & Joshua, N. R. (2013). Investigating affective prosody in psychosis: a study using the Comprehensive Affective Testing System. *Psychiatry Research, 210*(3), 896–900.
Rossell, S. L., Van Rheenen, T. E., Joshua, N. R., O'Regan, A., & Gogos, A. (2014). Investigating facial affect processing in psychosis: a study using the Comprehensive Affective Testing System. *Schizophrenia Research, 157*, 55–59.
Ruocco, A. C., Reilly, J. L., Rubin, L. H., Daros, A. R., Gershon, E. S., Tamminga, C. A., ... Sweeney, J. A. (2014). Emotion recognition deficits in schizophrenia-spectrum disorders and psychotic bipolar disorder: findings from the Bipolar-Schizophrenia Network on Intermediate Phenotypes (B-SNIP) study. *Schizophrenia Research, 158*(1–3), 105–112.
Samamé, C., Martino, D. J., & Strejilevich, S. A. (2012). Social cognition in euthymic bipolar disorder: systematic review and meta-analytic approach. *Acta Psychiatrica Scandinavica, 125*(4), 266–280.
Samamé, C., Martino, D. J., & Strejilevich, S. A. (2015). An individual task meta-analysis of social cognition in euthymic bipolar disorders. *Journal of Affective Disorders, 173*, 146–153.
Santos, J. M., Pousa, E., Soto, E., Comes, A., Roura, P., Arrufat, F. X., & Obiols, J. E. (2017). Theory of mind in euthymic bipolar patients and first-degree relatives. *The Journal of Nervous and Mental Disease, 205*(3), 207–212.
Savla, G. N., Vella, L., Armstrong, C. C., Penn, D. L., & Twamley, E. W. (2013). Deficits in domains of social cognition in schizophrenia: a meta-analysis of the empirical evidence. *Schizophrenia Bulletin, 39*(5), 979–992.

Schenkel, L. S., Chamberlain, T. F., & Towne, T. L. (2014). Impaired theory of mind and psychosocial functioning among pediatric patients with type I versus type II bipolar disorder. *Psychiatry Research*, *215*(3), 740–746.
Schenkel, L., Marlow-O'Connor, M., Moss, M., Sweeney, J., & Pavuluri, M. (2008). Theory of mind and social inference in children and adolescents with bipolar disorder. *Psychological Medicine*, *38*(6), 791–800.
Schenkel, L. S., Pavuluri, M. N., Herbener, E. S., Harral, E. M., & Sweeney, J. A. (2007). Facial emotion processing in acutely ill and euthymic patients with pediatric bipolar disorder. *Journal of the American Academy of Child and Adolescent Psychiatry*, *46*(8), 1070–1079.
Seiferth, N. Y., Pauly, K., Kellermann, T., Shah, N. J., Ott, G., Herpertz-Dahlmann, B., ... Habel, U. (2009). Neuronal correlates of facial emotion discrimination in early onset schizophrenia. *Neuropsychopharmacology*, *34*(2), 477–487.
Sergi, M. J., Rassovsky, Y., Nuechterlein, K. H., & Green, M. F. (2006). Social perception as a mediator of the influence of early visual processing on functional status in schizophrenia. *The American Journal of Psychiatry*, *163*(3), 448.
Simon, A. E., Borgwardt, S., Riecher-Rössler, A., Velthorst, E., de Haan, L., & Fusar-Poli, P. (2013). Moving beyond transition outcomes: meta-analysis of remission rates in individuals at high clinical risk for psychosis. *Psychiatry Research*, *209*(3), 266–272.
So, S. H., Tang, V., & Leung, P. W. (2015). Dimensions of delusions and attribution biases along the continuum of psychosis. *PLoS One*, *10*(12), e0144558.
Sperry, S. H., O'Connor, L. K., Öngür, D., Cohen, B. M., Keshavan, M. S., & Lewandowski, K. E. (2015). Measuring cognition in bipolar disorder with psychosis using the MATRICS consensus cognitive battery. *Journal of the International Neuropsychological Society*, *21*(06), 468–472.
Sprong, M., Schothorst, P., Vos, E., Hox, J., & van Engeland, H. (2007). Theory of mind in schizophrenia: meta-analysis. *The British Journal of Psychiatry*, *191*, 5–13.
Sullivan, S., Lewis, G., Mohr, C., Herzig, D., Corcoran, R., Drake, R., & Evans, J. (2014). The longitudinal association between social functioning and theory of mind in first-episode psychosis. *Cognitive Neuropsychiatry*, *19*(1), 58–80.
Thaler, N. S., Strauss, G. P., Sutton, G. P., Vertinski, M., Ringdahl, E. N., Snyder, J. S., & Allen, D. N. (2013). Emotion perception abnormalities across sensory modalities in bipolar disorder with psychotic features and schizophrenia. *Schizophrenia Research*, *147*(2–3), 287–292.
Thompson, A., Papas, A., Bartholomeusz, C., Allott, K., Amminger, G. P., Nelson, B., ... Yung, A. (2012). Social cognition in clinical "at risk" for psychosis and first episode psychosis populations. *Schizophrenia Research*, *141*(2–3), 204–209.
Tonelli, H. A. (2009). Cognitive "Theory of Mind" processing in bipolar disorder. *Revista Brasileira de Psiquiatria*, *31*(4), 369–374.
Tseng, H. H., Chen, S. H., Liu, C. M., Howes, O., Huang, Y. L., Hsieh, M. H., ... Hwu, H. G. (2013). Facial and prosodic emotion recognition deficits associate with specific clusters of psychotic symptoms in schizophrenia. *PLoS One*, *8*(6), e66571.
van Donkersgoed, R. J., Wunderink, L., Nieboer, R., Aleman, A., & Pijnenborg, G. H. (2015). Social cognition in individuals at ultra-high risk for psychosis: a meta-analysis. *PLoS One*, *10*(10), e0141075.
Van Rheenen, T. E., Bryce, S., Tan, E. J., Neill, E., Gurvich, C., Louise, S., & Rossell, S. L. (2016). Does cognitive performance map to categorical diagnoses of schizophrenia, schizoaffective disorder and bipolar disorder? A discriminant functions analysis. *Journal of Affective Disorders*, *192*, 109–115.
Van Rheenen, T. E., Joshua, N., Castle, D. J., & Rossell, S. L. (2017). Configural and featural face processing influences on emotion recognition in schizophrenia and bipolar disorder. *Journal of the International Neuropsychological Society*, *23*(3), 287–291.

Van Rheenen, T. E., Lewandowski, K. E., Tan, E. J., Ospina, L. H., Ongur, D., Neill, E., … Burdick, K. E. (2017). Characterizing cognitive heterogeneity on the schizophrenia—bipolar disorder spectrum. *Psychological Medicine*, *47*, 1848–1864.
Van Rheenen, T. E., Meyer, D., & Rossell, S. L. (2014). Pathways between neurocognition, social cognition and emotion regulation in bipolar disorder. *Acta Psychiatrica Scandinavica*, *30*(5), 397–405.
Van Rheenen, T. E., & Rossell, S. L. (2013a). Is the non-verbal behavioural emotion-processing profile of bipolar disorder impaired? A critical review. *Acta Psychiatrica Scandinavica*, *128*, 163–178.
Van Rheenen, T. E., & Rossell, S. L. (2013b). Auditory-prosodic processing in bipolar disorder; from sensory perception to emotion. *Journal of Affective Disorders*, *151*(3), 1102–1107.
Van Rheenen, T. E., & Rossell, S. L. (2013c). Picture sequencing task performance indicates theory of mind deficit in bipolar disorder. *Journal of Affective Disorders*, *151*(3), 1132–1134.
Van Rheenen, T. E., & Rossell, S. L. (2014a). Let's face it: facial emotion processing is impaired in bipolar disorder. *Journal of the International Neuropsychological Society*, *20* (02), 200–208.
Van Rheenen, T., & Rossell, S. (2014b). Multimodal emotion integration in bipolar disorder: an investigation of involuntary cross-modal influences between facial and prosodic channels. *Journal of the International Neuropsychological Society*, *20*, 1–9.
Van Rheenen, T., & Rossell, S. (2016). Facial emotion recognition impairments in bipolar disorder. A cognitive problem? *Journal of the International Neuropsychological Society*, *22*, 1–3.
Varo, C., Jimenez, E., Solé, B., Bonnín, C. M., Torrent, C., Valls, E., … Reinares, M. (2017). Social cognition in bipolar disorder: focus on emotional intelligence. *Journal of Affective Disorders*, *217*, 210–217.
Vaskinn, A., Sundet, K., Friis, S., Simonsen, C., Birkenaes, A. B., Engh, J. A., … Andreassen, O. A. (2007). The effect of gender on emotion perception in schizophrenia and bipolar disorder. *Acta Psychiatrica Scandinavica*, *116*(4), 263–270.
Vederman, A. C., Weisenbach, S. L., Rapport, L. J., Leon, H. M., Haase, B. D., Franti, L. M., … McInnis, M. G. (2012). Modality-specific alterations in the perception of emotional stimuli in bipolar disorder compared to healthy controls and major depressive disorder. *Cortex*, *48*, 1027–1034.
Ventura, J., Ered, A., Gretchen-Doorly, D., Subotnik, K. L., Horan, W. P., Hellemann, G. S., & Nuechterlein, K. H. (2015). Theory of mind in the early course of schizophrenia: stability, symptom and neurocognitive correlates, and relationship with functioning. *Psychological Medicine*, *45*(10), 2031–2043.
Ventura, J., Wood, R. C., Jimenez, A. M., & Hellemann, G. S. (2013). Neurocognition and symptoms identify links between facial recognition and emotion processing in schizophrenia: meta-analytic findings. *Schizophrenia Research*, *151*(1–3), 78–84.
Wang, Y. -G., Roberts, D. L., Liang, Y., Shi, J. -F., & Wang, K. (2015). Theory-of-mind understanding and theory-of-mind use in unaffected first-degree relatives of schizophrenia and bipolar disorder. *Psychiatry Research*, *230*(2), 735–737.
Wang, Y. -Y., Wang, Y., Zou, Y. -M., Ni, K., Tian, X., Sun, H. -W., … Chan, R. C. K. (2018). Theory of mind impairment and its clinical correlates in patients with schizophrenia, major depressive disorder and bipolar disorder. *Schizophrenia Research*, *197*, 349–356.
Wegbreit, E., Weissman, A. B., Cushman, G. K., Puzia, M. E., Kim, K. L., Leibenluft, E., & Dickstein, D. P. (2015). Facial emotion recognition in childhood-onset bipolar I disorder: an evaluation of developmental differences between youths and adults. *Bipolar Disorders*, *17*(5), 471–485.

Whitney, J., Howe, M., Shoemaker, V., Li, S., Marie Sanders, E., Dijamco, C., ... Chang, K. (2013). Socio-emotional processing and functioning of youth at high risk for bipolar disorder. *Journal of Affective Disorders, 148*, 112–117.

Yücel, M. O., Özgüven, H. D., Sakarya, A., Baskak, B., Kizil, E. T. Ö., & Sakarya, D. (2016). The relationship of verbal working memory and theory of mind in first degree relatives of patients with schizophrenia and bipolar disorder. *Turk Psikiyatri Dergisi, 27*(1), 8–14.

Yung, A. R., Phillips, L. J., McGorry, P. D., McFarlane, C. A., Francey, S., Harrigan, S., ... Jackson, H. J. (1998). Prediction of psychosis. A step towards indicated prevention of schizophrenia. *The British Journal of Psychiatry. Supplement, 172*(33), 14–20.

CHAPTER TWO

# Cognitive and social cognitive deficits in paranoia

**Bertalan Polner*, Anna Pengue†, Szabolcs Kéri*,‡, Ahmed A. Moustafa†**

*Department of Cognitive Science, Budapest University of Technology and Economics, Budapest, Hungary
†School of Social Sciences and Psychology, Marcs Institute for Brain and Behaviour, Western Sydney University, Penrith, NSW, Australia
‡Nyírő Gyula Hospital, National Institute of Psychiatry and Addictions, Budapest, Hungary

## The definition of paranoia and persecutory ideation

Paranoia has been defined as "the exaggerated or unfounded fear that others are trying to hurt, upset, or annoy you" (Cheng, 2008, p. 321). However, paranoia has diverse meanings (Freeman & Garety, 2000), and there is still much confusion regarding its definition (Manschreck, 1992). In a similar manner to the widespread use and misuse of the term "bipolar," the word "paranoia" is now commonly used in everyday language to typically convey an individual's general suspiciousness. Within the fields of psychology and psychiatry, the terms "paranoia" and "persecutory ideas" are often used interchangeably, yet they are viewed as different concepts by leading researchers in the field (Freeman & Garety, 2000). Traditionally, paranoia has been operationalized as a key feature or specific symptom of a number of severe psychiatric disorders, such as schizophrenia, paranoid personality disorder, major depression, and bipolar disorder. According to Manschreck (1992), the paranoid patient was commonly viewed as someone who was "disruptive, disturbed, or disturbing to others" (p. 242). However, according to Freeman and Garety (2000), paranoia is a much broader psychological construct, and its focus extends beyond the persecutory themes of ongoing threat and harm to self, to include such presentations as general suspiciousness, ideas of reference, grandiosity, and unsubstantiated jealousy (Freeman & Garety, 2000; McKay, Langdon, & Coltheart, 2007). These authors argue that paranoia is a clinical topic worthy of investigation in its "own right," especially in nonclinical individuals (Freeman & Freeman, 2009, p. 585).

*Social Cognition in Psychosis*
https://doi.org/10.1016/B978-0-12-815315-4.00002-1

The term "paranoia" (or "*paranoeo*") was first introduced in the writings of the ancient Greeks, and at the time, the concept held little diagnostic purpose, despite being synonymous with a person who was "crazy" or "out of their mind" (Lewis, 1970, p. 2). However, even dating back to the ancient Greeks, there appeared to be some indication that paranoid beliefs existed on a continuum with normal beliefs, with the paranoia definition ranging from terms such as "crazy" (a clinical disorder) to "thinking amiss" (a nonclinical experience) (Lewis, 1970, p. 2). In the latter half of the 19th century, paranoia began to be perceived as a clinically significant construct, and was first linked with delusional beliefs, specifically a person demonstrating "abnormal judgment about the relationship between the self and outer world" (Lewis, 1970, p. 4).

Paranoia (and other symptoms of psychosis) can be conceptualized along a continuum, with normality, in that experiences resembling psychosis are prevalent in a significant amount of individuals who do not meet diagnostic criteria for mental disorders (Strauss, 1969; Van Os, Hanssen, Bijl, & Ravelli, 2000). Paranoid ideation is frequently reported in the general population, in both adolescents and adults, with conservative estimates ranging between 10% and 24% (Altman, Collins, & Mundy, 1997; Freeman et al., 2011). Theoretically, a spectrum of paranoid beliefs ("The paranoia hierarchy") is hypothesized to exist, with social evaluative concerns and minor suspicions at a mild end, whereas severe paranoid delusions are at the extreme end (Freeman, Garety, et al., 2005). Implicating the social nature of paranoia in a large English representative sample, various levels of paranoia were associated with diverse social factors, such as being single or divorced, suffering from poverty, living in densely populated areas, less social support, finding the neighborhood unsafe, not feeling connected to others in the area, and also experiencing stress at work (Freeman et al., 2011). A taxometric analysis of scores on a self-report paranoia questionnaire was performed recently in a heterogeneous sample of almost 3000 participants, comprised of individuals from the general population, individuals at risk for psychosis, and also patients diagnosed with psychosis (Elahi, Algorta, Varese, McIntyre, & Bentall, 2017). The results supported the dimensional latent structure of paranoia, implicating a continuum between clinical and nonclinical manifestations of paranoia. Yet, it should be kept in mind that qualitative differences between clinical and nonclinical forms of paranoia might exist, such as the degree of control over the paranoid thoughts, whether the individual thinks that persecution is deserved, and to what degree the paranoid thoughts are associated with stressful social situations (Bentall & Udachina, 2013).

Overall, these research findings portray a picture of increased risks and numerous negative consequences linked with the presence of psychotic-like experiences in nonclinical individuals, and therefore it is very surprising that paranoia is rarely researched in nonclinical adolescents who are not psychotic. Both Tone and Davis (2012) and Freeman (2007) argue that more research with children and adolescents is urgently required to better understand the processes that are associated with the emergence of paranoid thinking in the young. Conversely, studying "normal-range" paranoid thinking (Tone & Davis, 2012, p. 1033) in younger-age samples may be "the ground out of which the clinical phenomena emerge" (Freeman, Dunn, et al., 2005, p. 307). In other words, studying nonclinical paranoia may advance both the understanding and treatment of clinically significant delusional and paranoid beliefs.

## The development of paranoid thinking

A limitation of the paranoia literature is that a developmental approach has been largely overlooked (Tone & Davis, 2012), notably restricted to the study of individuals above 18 years of age (Ellett, Lopes, & Chadwick, 2003; Freeman, Garety, et al., 2005). To our knowledge, very few studies have focused exclusively on the investigation of paranoia in nonclinical adolescents. One such study, conducted by Belgian researchers (Raes & Van Gucht, 2009), investigated paranoia and self-esteem in a sample of 131 nonclinical adolescents (age range 15–19); the results indicated that instability of self-esteem was associated with paranoia, over and above the effect of depressive symptoms and reduced global self-esteem. Another study by Campbell and Morrison (2007) assessed the relationship between psychotic-like experiences (including paranoia) in a large sample of 14–16 year olds. It is likely that a reticence to study paranoia in the early years may partly arise from a fear that some children and adolescents may unnecessarily receive stigmatizing labels and treatments. Furthermore, researchers may consider the rate of future transition from subclinical paranoia to active psychosis to be quite low (Carpenter & van Os, 2011; Tone & Davis, 2012). Additionally, it has also been suggested that a mistrust of others during adolescence may be a natural part of growing up, particularly when adolescents are prone to an increased self-consciousness about their appearance, and are more likely to be uneasy about whether they are liked or disliked by those around them (Bailey, Whittle, Farnworth, & Smedley, 2007). Adolescence is a life stage in which peers are considered to be of central importance to their lives, and a major developmental task includes a renegotiation of parental relationships to achieve increased autonomy (Fuligni, Eccles, Barber, & Clements, 2001). In a similar manner to anxious thoughts, some researchers have indicated that a degree of paranoid or suspicious thinking can be considered an adaptive psychological process in some environments, and can become problematic once this normal thinking style becomes overgeneralized (Ellett et al., 2003; Freeman & Garety, 2004). Consequently, the presence of paranoia within adolescents should not automatically signify psychopathology or a mental disorder to clinicians, because for many, the experience of paranoid thinking is often mild, or transient, and is associated with minimal distress or disruption (Ziermans, Schothorst, Sprong, & van Engeland, 2011).

Conversely, for some adolescents, paranoia may become maladaptive with far-reaching consequences, and due to the paucity of research, may remain an undetected and untreated phenomenon in those adolescents experiencing the most distress. An assessment of paranoia is not customarily included in clinical interviews, and the central features of clinical paranoia include the themes of preoccupation, interference, distress (Freeman & Garety, 2004), and feelings of being under the control of the paranoid ideas (Campbell & Morrison, 2007). Adolescence is a critical period associated with the onset of many serious disorders, and research clearly demonstrates that psychosis typically begins in adolescence (Harrop & Trower, 2001). Early work by Kraepelin in 1896 described "dementia praecox," a "disease" known to have paranoid features, to be a disorder primarily seen in adolescence and early adulthood (Kraepelin, 1919). Paranoia can have a profound impact on adolescent development. Studies have suggested that the experience of psychosis-like symptoms (including paranoid beliefs) in childhood and adolescence may pose as a risk factor for the later development of severe psychopathology (Rössler et al., 2011) and psychosis (Wigman et al., 2009). An influential longitudinal study conducted in New Zealand demonstrated that more than 25% of children endorsing lower-grade psychotic symptoms at 11 years of age developed a psychotic disorder by age 26 (Poulton et al., 2000). However, it is important to note that some of these individuals also reported hallucinations.

There has been a recent unsuccessful attempt to include a new diagnostic category of "attenuated psychosis syndrome" in the Fifth Edition of the *Diagnostic and Statistical Manual of Mental Disorders* (DSM-5) (Ruhrmann, Schultze-Lutter, & Klosterkötter, 2010). It was believed that this category would assist in the early identification and treatment of those at high-risk of developing psychosis, and diminish the frequently devastating and disabling consequences of psychotic illnesses. However, some researchers (Carpenter & van Os, 2011) argued that the potential pitfalls for such a category were too high, especially due to a lack of specificity in diagnosing the disorder outside of research studies, and the temporary course of many of the symptoms: psychotic-like symptoms are often transitory, and approximately 50%–85% of those who have psychotic symptoms will not make the conversion to a full-blown psychotic illness (Addington et al., 2011; Cannon et al., 2008). Addington et al. (2011) recently performed a longitudinal study investigating the outcomes of a group of high-risk individuals (age range: 12–36) who did not go on to develop psychosis (the "nonconverters"). They found that the "nonconverters" had "persistent disability" in the areas of social and role functioning (p. 803). In line with these findings, Riggio and Kwong (2009, 2011) found that paranoid ideation in young adulthood was related to adverse social outcomes, and poorer parental and peer relationships, as well as increased social isolation, loneliness, and social avoidance. Social isolation has also been posited to play a role in the maintenance of paranoid belief systems (Garety, Kuipers, Fowler, Freeman, & Bebbington, 2001). Therefore, it is apparent that caution must be taken when assessing paranoid symptoms in adolescence, as the early manifestations of paranoia may differ from the symptom expression commonly seen in adults (e.g., social avoidance and poor peer relationships).

## Fundamental cognitive processes and social cognition in paranoia

Can one consider paranoia as a special way of navigating in the social world? If one could find typical patterns in how the mind models others in those who experience paranoia, it would provide a justification for such a position. Based on extensive research conducted in recent decades, clinically informed theories have been developed to understand the origin and the maintenance of paranoia in terms of cognitive and affective mechanisms.

The threat anticipation model offers a multifactorial and multidimensional understanding of persecutory ideation (Freeman, 2007). A key premise of the model is that the purpose of delusional beliefs is to explain anomalous subjective experiences, unclear social signals, negative events, and uncanny coincidences. Thus, persecutory beliefs emerge as explanations for these experiences, and reflect maladaptive schemas that represent the self as vulnerable, others as malicious, and the world as evil. Affective disturbances are central to the model: first, paranoid ideation often follows stressful interpersonal events (Bentall, Wickham, Shevlin, & Varese, 2012; Kesting, Bredenpohl, Klenke, Westermann, & Lincoln, 2013; McDonnell, Stahl, Day, McGuire, & Valmaggia, 2018; Shevlin, McAnee, Bentall, & Murphy, 2015), and second, threat anticipation and worrying promote both the formation and the maintenance of persecutory delusions. Reasoning biases—such as jumping to conclusions (JTC)—increase the likelihood that suspicious feelings evolve into persecutory delusions that are held with great conviction. The maintenance of persecutory delusions is further supported by safety behaviors that prevent the individual from experiences that may provide evidence against the persecutory beliefs.

The attributional model considers paranoia as a form of defense against negative self-images, and focuses on attributional biases and problematic and fluctuating self-esteem (see Bentall & Udachina, 2013). The early version of the model stated that individuals with paranoia make external attributions for negative events in order to prevent themselves from experiencing themselves as negative. Later, the model was updated, as paranoid patients were found to be especially likely to make external-personal (blaming other people), but not external-situational attributions (blaming the circumstances) for negative events, a bias probably related to impaired reasoning about others' mental states and hasty decision making. The latest version of the model took bidirectional causal relationships between attributions and self-esteem into account: self-blaming or other-blaming interpretations of events are likely to reduce or protect self-esteem, respectively. In turn, relative to higher self-esteem, lower levels of self-esteem make internalizing attributions of negative events more likely. Accordingly, unstable self-esteem is expected in people who experience paranoia. Importantly, Bentall and Udachina (2013) have argued that paranoia can be considered a form of social adaptation, and cite studies showing that insecure attachment relationships and experiences of victimization and discrimination are associated with paranoid thinking (also see Freeman et al., 2011).

In addition, it has been proposed that the major cognitive deficits associated with paranoia, such as JTC, attributional biases, and impaired mentalizing (i.e., inferring the mental state of others), are all related to projecting the self into the past and the future (Corcoran, 2010). This line of reasoning rests on the insight that the anticipation of future threat is conceptually central to paranoia. Cognitively projecting the self through time is associated with the default mode network, and indeed, studies have shown aberrations in this network in schizophrenia (Whitfield-Gabrieli & Ford, 2012). Intriguingly, the default mode network is also associated with theory of mind (ToM) (Spreng, Mar, & Kim, 2009). However, it seems unlikely that the complex and dynamic nature of paranoia can be explained by a single, overarching cognitive deficit—as we shall see in following sections, it seems more plausible that paranoia is associated with deficits and distortions in multiple cognitive domains (see Bentall & Udachina, 2013; Freeman, 2007, 2016).

In a recent article, Freeman (2016) has argued that paranoia reflects variation in cognitive processes underpinning decisions concerning trust. Thus, persecutory delusions are best seen as beliefs concerning threats related to malicious others who are seen as willing to harm the individual. Six key sets of factors that maintain irrational threat beliefs were identified, which are worry, sleep disturbances, reasoning biases, anomalous experiences, negative self-schemas, and safety behaviors. Importantly, preliminary data suggests that these factors are promising therapeutic targets (Freeman et al., 2016).

To sum up, it can be seen that several socially relevant cognitive-affective correlates of paranoid ideation have been identified, such as hasty decision making and attributional biases, elevated threat anticipation and enhanced threat processing, ToM impairment, low and highly unstable self-esteem, and anomalous experiences (for recent reviews see Bentall & Udachina, 2013; Freeman, 2016). In the rest of the chapter, we will first provide a brief overview of the seminal findings with respect to the preceding factors, and also discuss what more recent neuroimaging studies have revealed about the neural substrates of social cognitive processes in paranoia. In addition, we provide an overview of insights into the daily life dynamics of paranoia yielded by experience sampling studies. Experience sampling (also known as ecological momentary assessment) is a structured diary technique that involves asking participants to report their feelings, thoughts, and behaviors during everyday life, typically several times a day for several consecutive days (see Myin-Germeys et al., 2009). We conclude the chapter with a review of recent studies that adopted a game theoretic framework to gain insight into how paranoia correlates with social cognition during social interactions.

## Jumping to conclusions

JTC refers to a hasty decision-making style that is characterized by drawing conclusions on the basis of relatively little information (Dudley, Taylor, Wickham, & Hutton, 2016). The metaanalysis of Dudley et al. (2016) revealed that people with psychosis show JTC relative to healthy individuals, and to individuals suffering from nonpsychotic mental disorders as well, and that JTC is associated with delusions. However, from this metaanalysis, it is unclear whether JTC is related to paranoid delusions.

In one investigation, JTC was found to lack specificity, such that after controlling for intelligence, differences in JTC disappeared in a group of individuals suffering from paranoid delusions (Bentall et al., 2009). Van Dael et al. (2005) initially found a significant relationship between JTC and paranoia in an at-risk sample, yet the effect turned insignificant after controlling for general intelligence. Executive functioning processes may trigger the restricted information-gathering style in paranoid individuals, and limit these individuals' ability to accept more benign explanations for their experiences or interactions with others. Moutoussis, Bentall, El-Deredy, and Dayan (2011) modeled the processes underlying choices in Bayesian probabilistic reasoning tasks and found that paranoid participants' quick decision-making on these tasks could be more accurately defined by their general cognitive impairments, specifically in the domains of executive functioning, challenging the notion that these tasks provide a pure measure of JTC. On the other hand, another study has suggested that the JTC bias might be specific to paranoia, in that patients with delusions of reference did not demonstrate the JTC bias (Menon, Addington, & Remington, 2013). In addition, JTC, as assessed in a social perception scenario, predicted the severity of persecutory delusions.

The tendency to gather less evidence before forming a decision has been found to be at the lower end of the continuum in paranoia-prone individuals. The study of nonclinical populations offers several advantages; as cognition in nonclinical individuals is not confounded by medication and hospitalization, such research can shed light on processes that may be involved in the formation of delusions and paranoid thinking during adolescence. Moreover, similar risk and demographic factors have been identified in both nonclinical and clinical paranoia (Freeman, 2007; Freeman et al., 2011). In addition, those at an increased risk of developing psychosis may be more accurately identified through this kind of research (Ziermans et al., 2011). Delusional ideation in association with the JTC bias has been extensively examined by numerous researchers; however, little research has focused on paranoia. Moreover, a gap in the literature exists in studying the phenomenon of paranoia in adolescents (Bailey et al., 2007). There has also been much speculation in the literature regarding potential links between the JTC bias and other cognitive functions.

Recently, researchers have started to investigate changes in JTC over time. Accordingly, the correlates of temporal fluctuations of paranoid beliefs was investigated in a 2-day-long online experience sampling study where 35 patients participated who were diagnosed with schizophrenia-spectrum disorders (Lüdtke, Kriston, Schröder, Lincoln, & Moritz, 2017). Participants were presented with textual descriptions of fictive and ambiguous scenarios, and were asked to rate the probability of certain interpretations of each situation—JTC was defined as rating interpretations highly likely. The results indicated that JTC showed some intra-individual variation over the 2-day period. Furthermore, JTC predicted subsequent paranoia over and above the effect of negative affect, and importantly, JTC was not predicted by preceding negative affect. These results contrast with findings from a study that experimentally manipulated anxiety and assessed its effect on JTC and paranoia in 90 healthy individuals (Lincoln, Lange, Burau, Exner, & Moritz, 2010). Experimentally induced anxiety was associated with an increase in both JTC and paranoia, and the increase in JTC mediated the association between anxiety and paranoia.

According to Freeman (2011), new interventions are required to specifically target cognitive biases, that is, erroneous reasoning processes. Cognitive remediation therapy (CRT) may prove to be a novel approach that helps improve cognitive outcomes for those who are distressed or feel controlled by their paranoid thoughts. CRT is a behavioral skill training program that aims to rectify cognitive difficulties, and evidence is accumulating regarding its use in individuals with severe disorders (Galletly & Rigby, 2013). In schizophrenia patients, a metaanalysis recently found CRT significantly improved a broad range of cognitive functions, including reasoning (Wykes, Huddy, Cellard, McGurk, & Czobor, 2011).

## Aberrant salience and sensitivity to social threat

Aberrant salience, that is, increased attention to neutral, irrelevant information, is a central concept in neurocognitive and phenomenological theories of psychosis, and has been extensively reviewed elsewhere (Howes & Kapur, 2009; Nelson, Whitford, Lavoie, & Sass, 2014; Sass & Byrom, 2015). Here, we will focus on studies that investigated aberrant salience attribution to socially relevant stimuli in connection with paranoid thinking.

The association of paranoia with aberrant salience to neutral social stimuli has been implicated by a study that compared male inpatients with paranoid schizophrenia with healthy male controls (Haralanova, Haralanov, Beraldi, Möller, & Hennig-Fast, 2012). As compared with controls, patients responded with higher subjective emotional arousal to emotionally neutral social scenes, but did not differ significantly in terms of subjective emotional arousal evoked by aversive social scenes. The authors suggested that paranoia might be characterized by a context-independent overarousal, which is easily projected to the social environment. Alternatively, it is also possible that patients interpret or see neutral images as threatening, and this perception leads to increased subjective arousal. Finally, it should be pointed out that the Haralanova et al. study did not include a nonparanoid schizophrenia control group, thus, it cannot be judged to what extent is aberrant salience attribution to neutral social stimuli specific to paranoia in schizophrenia.

It has been argued that measuring paranoid ideation is problematic, in that the investigator cannot precisely determine to what extent a suspicious thought is delusional or realistic (Freeman et al., 2008). To overcome the limitation of "justified suspicions" (Freeman et al., 2008, p. 1122), virtual reality can be used to place individuals in controlled and neutral social environments, and examine their perceptions and experiences in a virtual world where real harmful intentions are totally absent, as the behavior of characters in the virtual reality is completely under the control of the researcher. A study using this method in a nonclinical sample found that the tendency for anomalous experiences differentially predicted social anxiety and paranoid ideation in virtual reality: those prone to anomalous experiences were more likely to report having paranoid thoughts during a virtual metro ride, and at the same time, they were less likely to report experiencing social anxiety (Freeman et al., 2008). It should be noted, however, that a general proneness to anomalous experiences was measured before participants entered virtual reality, while anomalous experiences were not assessed during the session. In addition, as there is a strong correlation between proneness to anomalous experiences and delusional ideation (e.g., Claridge et al., 1996; Peters, Joseph, & Garety, 1999), one may argue that delusion proneness could have been a more accurate predictor of paranoid ideation in a virtual environment. Relatedly, a recent study examined paranoia in a virtual reality experiment in individuals at clinical high risk for psychosis (McDonnell et al., 2018). The severity of bullying experienced as a child significantly predicted interpersonal sensitivity (that is, being overly sensitive and aware of the feelings and actions of others), which in turn was associated with paranoid ideation during a virtual metro ride. In other words, interpersonal sensitivity appeared to be a full mediator of the association between bullying and paranoia.

A recent experience sampling study has described the temporal dynamics of aberrant salience and paranoia in daily life (So et al., 2017). For 2 weeks, patients with persecutory delusions were asked seven times a day to provide subjective accounts of their experiences. Aberrant salience and negative emotions temporally predicted persecutory delusions in daily life, while they were not predicted by persecutory delusions at the preceding assessment point. Interestingly, negative emotions temporally predicted aberrant salience as well, suggesting a direct and an indirect association between negative emotionality and paranoia.

Now we turn to enhanced sensitivity to social threat, an aspect of paranoia that is closely related to aberrant salience. It has been argued that paranoia can be conceptualized as the extreme of normal variation in the adaptive sensitivity to threatening social stimuli (Green & Phillips, 2004). Indeed, several studies have shown that paranoia is associated with abnormal perception of threat-related social stimuli. For example, paranoia is associated with a bias toward perceiving anger on neutral facial expressions. One study classified patients with schizophrenia into two groups based on whether they were actively paranoid ($n=68$) or not ($n=64$) (Pinkham, Brensinger, Kohler, Gur, & Gur, 2011). Participants were shown pictures of sad, happy, angry, fearful, or neutral faces, and were asked to identify the emotion. No significant differences were found between the two patient groups in terms of overall recognition accuracy. However, paranoid patients were less accurate in the identification of neutral faces. Follow-up analyses revealed that that paranoid patients were significantly more likely to categorize a neutral face as angry (26% of errors), relative to nonparanoid patients (6% of errors). These findings support the notion that paranoia is characterized by an elevated perception of threat in uncertain and neutral stimuli that are socially relevant. Furthermore, the findings are in line with the previously cited studies that highlighted the association of aberrant salience with paranoid ideation (e.g., Haralanova et al., 2012).

What are the neural substrates of a bias toward perceiving threat in paranoia? In one early neuroimaging study, participants identified emotion on angry, fearful, and disgusted face stimuli. The results showed that neural activation in regions involved in higher level visual processing (e.g., the fusiform gyrus), the cerebellum, and the insular cortex correlates with paranoid symptoms (Phillips et al., 1999). However, this study was limited in that it only compared five paranoid patients with five nonparanoid patients. Another study investigated the perception of dynamic facial expressions of fear in seven paranoid patients and eight nonparanoid male patients diagnosed with schizophrenia (Russell et al., 2007). Marginally significant effects suggested that hypoactivation of the amygdala is associated with paranoid symptomatology (Russell et al., 2007). However, a later study suggested that reduced amygdala responses in people with paranoia might be explained by increased baseline activity of the amygdala (Pinkham et al., 2015). Resting cerebral blood flow measurements and functional neuroimaging during a trustworthiness rating task were performed in 25 healthy control participants, 16 acutely paranoid patients with schizophrenia, and 16 patients with schizophrenia who were not paranoid. The paranoid schizophrenia group demonstrated significantly higher resting activity in the left amygdala (the effect was similar although marginally significant in the right amygdala), relative to both the nonparanoid schizophrenia group and the controls. When anxiety was controlled for, the paranoid group had significantly higher resting activity in the bilateral amygdala, as compared with patients with nonparanoid schizophrenia. During the trustworthiness rating task, patients with paranoid schizophrenia showed decreased activation in the bilateral amygdala, relative to controls, and in the right amygdala, relative to the nonparanoid schizophrenia group. The authors argue that a higher resting activity in the amygdala is a potential confound in functional neuroimaging studies, which may resolve the controversy of reduced amygdala responses in paranoia reported in other studies.

A different explanation has been provided a another study, which involved 22 controls, and 13 paranoid patients and 14 nonparanoid patients, and combined neuroimaging methods with skin conductance response recordings to measure arousal (Williams et al., 2007). Participants were presented with facial expressions of fear, anger, and disgust. They were asked to identify the gender of each face during scanning, and after scanning they were instructed to identify the emotions expressed. Relative to nonparanoid patients, paranoid patients were only impaired at recognizing disgust, and showed elevated skin conductance responses to fear, and more frequent skin conductance responses to disgust. For fear stimuli that evoked arousal, paranoid patients had reduced activation in the medial prefrontal cortex and bilateral amygdala. Additionally, for angry faces that evoked arousal, paranoia was associated with reduced activation in the anterior cingulum and the medial prefrontal cortex, and for arousal-evoking disgusted faces, paranoia was associated with reduced activation in the insula. Intriguingly, skin conductance responses to fearful faces correlated with suspiciousness and persecution among the patients, and reduced neural responses to arousal-evoking fearful faces predicted the severity of delusions, suspiciousness, and persecution. These findings imply that paranoia is related to a disconnection between somatic and neural representations of threat, indicated by responding to threat-related social stimuli with enhanced physiological arousal, but reduced amygdala activity. Curiously, increased arousal or neural responses to neutral stimuli were not associated with paranoia, which conflicts with the view that paranoia is characterized by aberrant salience attribution to neutral social stimuli (Pinkham et al., 2011). The study of Williams et al. (2007) differs from the study of Pinkham et al. (2011) in that their sample was smaller, and they applied different criteria to define paranoid and nonparanoid subgroups.

How does perception of social threat and paranoid thinking relate to the social context at different levels of severity of paranoia? Collip et al. (2011) conducted an experience sampling study with paranoid and nonparanoid patients with psychotic disorders and healthy controls with average and high levels of psychotic-like traits. On the basis of their scores on the Paranoia Scale (Fenigstein & Vanable, 1992), participants were classified into low, medium, and high paranoia groups. For 6 consecutive days, participants were asked 10 times a day to report whether they were alone or with company (if the latter, whether they were with familiar or less familiar company), their actual paranoid feelings, whether they perceived the social environment threatening, and finally, the general distress they experienced. Trait paranoia had no relationship with the frequency of being alone versus with company, or being with familiar versus less familiar others. Intriguingly, participants with low and intermediate levels of trait paranoia were more likely to experience paranoia and perceive a social threat when they were with less familiar others, while in highly paranoid participants, there were no differences in state paranoia or perceived social threat with respect to social context, and their state paranoia was typically high. This pattern of results suggests that suspiciousness may be somewhat adaptive in low and intermediate levels of trait paranoia, while for highly paranoid participants, paranoid ideation and social threat perception is disconnected from key aspects of the social context, implicating a key qualitative difference along the paranoia continuum.

Is paranoia related to regulation of interpersonal space? In a study involving 64 patients with schizophrenia-spectrum disorders and 24 controls, participants set the minimum tolerable interpersonal distance, and were also asked to report their subjective level of comfort at given distances (Schoretsanitis, Kutynia, Stegmayer, Strik, & Walther, 2016). Minimum tolerable interpersonal distance was assessed with a modified stop-distance paradigm: either participants had to approach the experimenter, and stop at a distance where they started to feel discomfort, or the experimenter was approaching the participants, and they had to ask the experimenter to stop as they had started to feel uncomfortable. Patients were classified into paranoid threat, paranoid power (experiencing grandiosity), and neutral affect subgroups on the basis of the global affect score of the Bern Psychopathology Scale (BPS; Strik et al., 2010). The BPS is a research instrument whose items are rated by clinicians on the basis of a manualized semistructured interview. The minimum tolerable distance for paranoid threat patients was more than two times larger than for the other groups. Relative to controls, patients with paranoid threat and neutral affect reported lower levels of comfort at fixed distances, which was higher at larger interpersonal distances. Patients with paranoid power reported a relatively high level of comfort, which did not vary with interpersonal distance. The study identified a sensitive behavioral marker of feelings of threat associated with paranoia, which might be used to facilitate screening for paranoia.

Another study examined how paranoia in the general population relates to experiencing social exclusion in a virtual environment (Westermann, Kesting, & Lincoln, 2012). Social exclusion was modeled with the Cyberball task, where the participant is throwing a ball with two other players, who either include the participant in the game, or ignore and exclude him or her, which induces profound social stress. This study examined the moderating effect of habitual emotion regulation strategies as well. The analyses revealed a curious interaction: for intermediate and highly paranoid participants, more frequent use of reappraisal (vs. suppression) predicted more intense paranoid experiences after being excluded in the game. Given that paranoia is intimately related to negative self-schemas, among other schemas (Freeman, 2007), it may seem reasonable that when only such schemas are available for reappraisal, it will lead to increased paranoid ideation.

To sum up, paranoia is associated with a bias toward perceiving threat in neutral social stimuli (Pinkham et al., 2011). Aberrant amygdala activation is a potential neural substrate of this impairment (Russell et al., 2007). One study has suggested that paranoia is characterized by a disconnect between arousal responses and neural activation evoked by social stimuli implicating threat (Williams et al., 2007). In daily life, high trait paranoia seems to be associated with intense feelings of social threat and paranoid ideation, independent of objective aspects of the social context (Collip et al., 2011). Feelings of social threat in paranoia are mirrored by a preference for larger interpersonal distances (Schoretsanitis et al., 2016). Finally, the counterintuitive finding that the tendency to use reappraisal is associated with increased paranoia in highly paranoid individuals is worth further attention, and could have important clinical implications (Westermann et al., 2012). The latter finding is perhaps explicable in terms of negative other- and self-schemas—these topics will be the focus of the following section.

## Attachment and the self

As we have discussed herein, negative self-schemas and low and fluctuating self-esteem are central topics in psychological explanations of paranoia (Bentall & Udachina, 2013; Freeman, 2007). Recently, the relationship between paranoia and the self has been extensively reviewed in multiple studies (Freeman & Garety, 2014; Kesting & Lincoln, 2013; Tiernan, Tracey, & Shannon, 2014). In general, these reviews concluded that paranoia is associated with negative beliefs about the self, and low and unstable self-esteem, while lower perceived deservedness of persecution appears to be associated with increased self-esteem. In this section, we review some more recent studies that provide details about potential psychological mechanisms underpinning the relationship between self-esteem and paranoia. Furthermore, self- and other-schemas are in a close conceptual relationship with attachment style, which has been defined as "the systematic pattern of relational expectations, emotions, and behavior that results from internalization of a particular history of attachment experiences and consequent reliance on a particular attachment-related strategy of affect regulation" (Mikulincer, Shaver, & Pereg, 2003, p. 79). Accordingly, in this section we review studies that addressed the relationship between attachment and paranoia.

Some studies have attempted to unravel the specific psychological mechanisms connecting reduced self-esteem and paranoia. It has been argued that paranoia and social anxiety overlap, in that they both relate to anxiety about how the self is represented by others (Matos, Pinto-Gouveia, & Gilbert, 2013). In the case of social anxiety, the self is seen as unattractive and likely to be rejected by critical others, while in paranoia, the self is seen as vulnerable, and likely to be harmed by malevolent others. As shame concerns whether the self is represented by others as undesirable, unworthy, and inferior, its relationship with paranoia and social anxiety is worth investigating. In a general population study involving 328 participants, shame, aspects of shame memories, paranoia, and social anxiety were measured using self-reports (Matos et al., 2013). A distinction can be made between external and internal shame: external shame relates to the experience of believing that the self is represented negatively by others, while internal shame refers to the experience accompanying representing ourselves as negative.

Moreover, early and traumatic experiences of shame can become core autobiographical memories, which will organize an individual's narrative identity. Paranoia was strongly related to external shame and to the traumatic impact of a childhood shame memory, and it was weakly related to internal shame. On the other hand, social anxiety was strongly associated with internal shame, and weakly with external shame. Centrality of the shame memory showed a specific association with paranoia, but it was unrelated to social anxiety. The authors suggest that early, traumatic shame memories (of harm, abuse, or neglect in close relationships) can become the source of persistent feelings of social threat, render the self vulnerable, and make others appear hateful and hostile. In addition, if traumatic shame memories are central to one's identity, then one will be more likely to interpret social events to be driven by negative representations of the self in others' minds, and perhaps their harmful intentions. Obviously, these schemas will be mirrored in expectations about future social scenarios as well (Corcoran, 2010).

Another study explored verbal self-attacking in 15 individuals experiencing persecutory delusions, 15 controls diagnosed with depression, and 19 nonclinical control participants (Hutton, Kelly, Lowens, Taylor, & Tai, 2013). Self-attacking is a form of self-criticism that involves feeling hate and disgust toward the self, and it is putatively rooted in experiences of neglect and abuse in early relationships with significant others. Hateful self-attacking and reduced self-reassurance distinguished between the group with persecutory delusions and healthy controls, while reduced constructive self-criticism was specific to individuals experiencing persecutory delusions, while healthy controls and participants with depression were more likely to criticize themselves for a corrective purpose. This study suggests that paranoia is not only related to expecting harm to the self from others, but also to mental self-harming. A similar pattern has been reported for nonclinical paranoia: in a sample of 131 university students, self-hating was linked to paranoid ideation, over and above the effect of depression (Mills, Gilbert, Bellew, McEwan, & Gale, 2007).

A longitudinal investigation provided insight into the dynamics between positive and negative beliefs about the self and symptoms of psychosis, including paranoia (Palmier-Claus, Dunn, Drake, & Lewis, 2011). A relatively large sample of first-episode psychosis patients ($N=256$) completed assessments four times over the course of one and a half years. Analyses have shown that initial decreases in positive beliefs about the self, and increases in negative beliefs about the self predicted increases in negative symptoms, but they were not related to changes in paranoia. However, variability of positive beliefs about the self over the study period predicted average paranoia levels, plus the overall mean of negative self-beliefs also predicted paranoia (although the latter effect was nonsignificant after controlling for depression), implicating that a fluctuating positive self-image and a constant negative self-image are both associated with increased likelihood of paranoid thinking.

Experience sampling studies can provide insight into temporal dynamics of psychopathology on a smaller scale (Myin-Germeys et al., 2009; Oorschot, Kwapil, Delespaul, & Myin-Germeys, 2009). Thewissen et al. (2011) conducted a 1-week long experience sampling study to investigate the temporal dynamics of paranoia, negative emotions, and self-esteem. Their sample ($N=158$) was comprised of paranoid and nonparanoid patients with psychotic disorders, and participants from the general population (who were preselected for average or high paranoid traits) as well. The study showed that increased anxiety and decreased self-esteem preceded paranoid episodes, and paranoid episodes were characterized by high levels of anxiety, depression, and anger, and profoundly reduced self-esteem. Thus, it seems unlikely that paranoia functions as a defense mechanism against low self-esteem (Bentall & Udachina, 2013), and it is more likely that negative emotions and reduced self-esteem are driving and maintaining paranoid ideation. The persistence of paranoia was positively predicted by paranoia and depression at the beginning of a given episode, while it was negatively predicted by anger and irritability at the onset of the episode, suggesting that various negative emotions might trigger paranoid episodes with different courses.

A subsequent experience sampling study followed 20 patients with psychotic disorders and 20 healthy controls for 6 days (Sitko, Varese, Sellwood, Hammond, & Bentall, 2016). Beyond paranoia, stress, and self-esteem, the measurements covered insecure attachment states and auditory hallucinations as well. Importantly, attachment insecurity was significantly higher and more variable in the patient group, relative to controls. This result is in line with previous studies that have shown a specific relationship between insecure attachment and paranoia in both clinical (Korver-Nieberg et al., 2013; Wickham, Sitko, & Bentall, 2015) and nonclinical samples (Pickering, Simpson, & Bentall, 2008). Furthermore, attachment insecurity was significantly predicted by preceding activity-related stress, while the effect of social stress was marginally significant (Sitko et al., 2016). Attachment insecurity, in turn, predicted subsequent paranoid thoughts, over and above the nonsignificant effect of self-esteem. Finally, attachment insecurity specifically predicted paranoia, but not auditory hallucinations. These findings underscore the importance of fluctuating attachment models to the occurrence of paranoia in daily life, and tentatively suggest that attachment might be more relevant to paranoia than self-esteem.

## Theory of mind and attribution

ToM refers to the ability to represent mental states (intentions, beliefs, and feelings) of the self and others (Brüne, 2005). There is compelling evidence indicating that ToM is impaired in patients with schizophrenia (see Bora, Yucel, & Pantelis, 2009; Brüne, 2005); yet, although theoretically appealing (Frith, 2004), it is far from evident that poor ToM would specifically predict paranoia. Instead, ToM impairment seems to be associated with negative symptoms (Freeman, 2016). In line with this, several studies have failed to find a significant association between ToM and paranoia in samples recruited from the general population (Fernyhough, Jones, Whittle, Waterhouse, & Bentall, 2008; McKay, Langdon, & Coltheart, 2005), or in adolescent patients with early psychosis (Korver-Nieberg et al., 2013).

Still, some studies have suggested that the ToM impairment in schizophrenia is indeed specific to paranoia. For example, this account is supported by a study that involved 25 individuals diagnosed with schizophrenia (some of whom were experiencing persecutory delusions) and 38 controls (Langdon, Siegert, McClure, & Harrington, 2005). Participants completed a battery of verbal and nonverbal tasks measuring ToM. Patients with persecutory delusions had a remarkable deficit on verbal, but not on nonverbal ToM tasks, independent of intelligence or general cognitive capacities. Importantly, poor performance on the more complex verbal ToM tasks correlated with persecutory delusions, independent of illness length, intelligence, and general cognitive capacity.

One study has compared ToM, attributional style, and paranoia in patients with schizophrenia who were experiencing persecutory delusions, and in patients with Asperger's syndrome (Craig, Hatton, Craig, & Bentall, 2004). The Hinting Task and the Reading the Mind in the Eyes task were used to assess ToM capacities. The Hinting Task probes the ability to understand indirect speech: participants are read short stories about interactions between two characters, and their task is to figure out the hidden message in the conversation. The Reading the Mind in the Eyes task requires participants to recognize complex cognitive emotions (such as "jealous" or "hateful") in pictures that show only the eye region of a face. Patients with persecutory delusions showed the greatest level of paranoia, and patients with Asperger's syndrome had intermediate levels of paranoia. Interestingly, although both patient groups had poor ToM, only the paranoid patients showed a bias toward external-personal attributions for negative events. These findings suggest that persecutory delusions are associated both with ToM impairment and an attributional bias; however, it should be noted that the analyses were not controlled for cognitive functions (although the groups were matched for intelligence) or other psychopathology.

How do ToM abilities relate to different levels of severity of paranoia? Versmissen et al. (2008) recruited patients with psychotic disorders, healthy first-degree relatives of patients with psychosis (familial high risk), and healthy people expressing high and intermediate levels of positive psychotic-like traits. ToM was measured with the Hinting Task (described herein). Controlling for age, gender, and intelligence, impaired ToM was found in the patient group, whereas ToM in the familial high-risk group was impaired relative to controls at the trend level, and healthy participants demonstrating high levels of psychotic-like traits did not differ significantly from the controls. Curiously, poor ToM performance correlated with auditory hallucinations, but not with past or current persecutory delusions among the patients, whereas in the group with high family risk for psychosis, poor ToM was associated with lifetime paranoid experiences, but not with auditory hallucinations. Relatedly, some studies have shown that impaired ToM is associated with subtle forms of suspiciousness in the general population. Studies involving university students have revealed a relationship between poor ToM and external-personal attributions (blaming others) for negative events, but no relationship was found with the complexity of explanations (Kinderman, Dunbar, & Bentall, 1998; Taylor & Kinderman, 2002).

How does ToM functioning relate to other psychological correlates of paranoia? Two studies have simultaneously investigated JTC and ToM in relation to paranoia. Bentall et al. (2009) tested 173 patients diagnosed with schizophrenia-spectrum disorders or major depression, plus 64 healthy controls. They measured ToM, JTC, general cognitive functions, threat anticipation, emotions, self-esteem, and attributional styles. Their results revealed that paranoia is predicted by reduced self-esteem, pessimism, prevalent negative emotions, and also by impaired cognition, as indexed by executive functions, JTC, and poor ToM. Furthermore, Corcoran and colleagues (Corcoran et al., 2008) tested a sample of 115 patients and 33 healthy controls. The patient group included currently paranoid and remitted paranoid patients with schizophrenia, currently paranoid patients with depression, and nonpsychotic patients with depression. Patients with persecutory delusions, irrespective of their diagnosis, demonstrated a JTC bias, and had poor performance on a verbal ToM task in which participants had to recognize false beliefs and deception in stories about social interactions.

Conflicting findings with respect to the relationship between paranoia and ToM might be related to measurement issues, as performance on ToM tasks can be influenced by multiple cognitive abilities. One study investigating the latent structure of ToM might clarify this point (Shryane et al., 2008). The authors recruited a relatively large and heterogeneous sample ($N=237$): among participants were patients with schizophrenia-spectrum disorders and patients with unipolar depression, and healthy controls as well. They completed the stories ToM task, where they were read short stories involving social interactions, and their task was to infer the mental representation of one person (first-order ToM) or of two persons (second-order ToM) involved in the story. Some stories involved false beliefs, while others involved deception. A three-factor model had the best fit to the data, which included a first-order and a second-order deception factor, and a single false-belief factor. Importantly, the latent factors representing first-order deception and false belief were both negatively associated with paranoia, over and above the effects of intelligence and gender.

Another study explored qualitative aspects of mentalizing with an ecologically more valid task in a sample of 80 patients with paranoid schizophrenia and 80 healthy controls (Montag et al., 2011). The authors applied a video-based assessment, the "Movie for the Assessment of Social Cognition" task, in which participants are shown a short movie about the interactions of four characters. Their task is to try to model the characters' mental states, and their understanding is probed with 48 questions at certain points during the movie. When the responses are scored, a crucial distinction is made between overmentalizing (excess representation of mental states) and undermentalizing (insufficient representation of mental states). After controlling for verbal memory, age, and gender, patients with paranoid schizophrenia had significantly more undermentalizing errors, while they did not differ from controls in terms of overmentalizing, or absence of ToM errors. Moreover, patients had more errors when they were to infer cognitive and affective mental states, as compared with the controls. Finally, positive and delusional symptoms demonstrated a specific correlation with overmentalizing errors, again highlighting the importance of taking qualitative aspects of ToM performance into account.

In a recent study, comprehensive assessment of social cognitive abilities and biases was performed in a sample of 123 students (Klein, Kelsven, & Pinkham, 2018). Paranoid ideation positively correlated with a blaming attributional style, negatively with recognizing emotions in pictures and detecting lies and sarcasm in videos of social situations, and paranoia was also negatively associated with trustworthiness ratings of faces. Importantly, paranoia was not significantly related to cognitive capacities. When anxiety and depression were controlled for, the preceding associations remained significant (except for the association with trustworthiness), and a significant negative correlation emerged between paranoia and indirect speech understanding (as measured by the Hinting Task). Although most effects were weak, and no significant correlations were found between paranoia and three additional social cognition tasks (that probed emotion recognition in pictures and videos and social knowledge related to various relational models), the results still indicate that subclinical paranoia is associated with an external-personal attributional bias and difficulties with recognizing emotions and inferring complex mental states.

These findings are somewhat paralleled by a study that applied the same social cognitive measures in a sample of 220 patients diagnosed with schizophrenia, or schizoaffective disorder (Buck, Pinkham, Harvey, & Penn, 2016): multiple regressions showed that paranoia was significantly and independently associated with a blaming attributional style and reduced trustworthiness. However, beyond the effect of the preceding two factors, paranoia was not significantly associated with social cognitive abilities (i.e., emotion recognition and ToM). On the other hand, the informant-rated tendency for hostile and malevolent interpretation of social situations was positively associated with emotion recognition in pictures, but negatively with emotion recognition and ToM assessed with videos.

Social cognitive capabilities are not only putatively related to paranoid symptomatology, but can also be crucial to social functioning. This notion is supported by a recent study that assessed emotion recognition with the Reading the Mind in the Eyes task and ToM with the Hinting Task in a sample of 88 patients (76 males) diagnosed with schizophrenia or schizoaffective disorder (Phalen, Dimaggio, Popolo, & Lysaker, 2017). The association between persecutory delusions and reduced social functioning was moderated by ToM (but not by emotion recognition): for patients who performed well on the Hinting Task (c. top 20%), persecutory delusions were not related to social functioning; on the contrary, in patients who had poorer ToM, the severity of persecutory delusions negatively predicted social functioning. The results suggest that intact social cognitive capabilities can protect against functionally impairing aspects of persecutory delusions, possibly by fostering alternative, nonmalevolent interpretations of intentions of others during daily social interactions.

A recent study undertook a novel, naturalistic approach to characterize social cognition in paranoia at the neural and the linguistic levels (Finn, Corlett, Chen, Bandettini, & Constable, 2018). During functional magnetic resonance imaging (fMRI), 22 healthy participants with varying degrees of paranoid ideation listened to a story that involved some uncertainty about the intentions of the characters. Voxels showing similar functional activations across subjects were identified with an intersubject correlation (ISC) analysis. In the whole sample, neural responses in the primary auditory cortex and in the superior temporal pole were largely similar, and several other association cortex areas also demonstrated similar responses. Importantly, in highly paranoid participants, activation patterns in brain areas typically associated with ToM (the left temporal pole and the right medial prefrontal cortex) (e.g., Carrington & Bailey, 2009) were more synchronized, relative to less paranoid participants. Follow-up analyses explored the specific effects of events in the narrative that were potentially suspicious and/or involved explicit inference about mental states. Strikingly, responses to these events in the left temporal pole and the right medial prefrontal cortex were positively correlated with paranoia, suggesting that socially ambiguous stimuli is more likely to activate ToM-related neural processes in highly paranoid participants. Immediately after the experiment, participants freely recalled the story. Key semantic and syntactic features of participants' speech could explain 72% of variance in paranoia; in particular, paranoia had a strong positive association with words related to affiliation, anxiety, and risk, while it was strongly and negatively linked to the presence of words referring to males, anger, and function words and conjunctions.

## Game theoretic approaches to social cognition and behavior in paranoia

Although evidence indicates that patients with schizophrenia have impaired ToM that is independent of general cognitive impairment, the deficit appears to be associated with negative and disorganized symptoms, while findings are controversial with respect to paranoid symptoms (Chan & Chen, 2011). It has been suggested that the inconsistent findings might be due to methodological and psychometric limitations (c.f. Shryane et al., 2008). In addition, Chan and Chen (2011) have argued that tasks used to assess ToM demand reasoning about the mental states of another person, and thus lack an essential component of real life social interactions: direct and continuous engagement with others. The authors suggested that adopting a game theoretical approach could remedy the preceding limitations of ToM research. And indeed, experiments rooted in game theory allow experimental investigation of social behavior, by putting participants in social interactions in which they have to decide whether to cooperate or compete with others (Raihani & Bell, 2017a).

For example, the Minnesota trust game includes the interaction of two players, and has separate conditions for the assessment of risk aversion, rational mistrust, and suspiciousness (Johnson, Rustichini, & MacDonald, 2009). In each condition, participants have to choose between two options. The first option always means a certain, but small, payoff for both players, while conditions differ in the second option. In the risk aversion condition, a coin flip determines whether both players are provided a larger payoff, or the participant is paid nothing, and his or her partner receives an even larger payoff; in this condition, a preference for the first option indicates risk aversion. In the rational mistrust condition, if a participant chooses the uncertain second option, his or her partner can choose between a less beneficial but cooperative outcome (larger payoff for both of them), or a tempting noncooperative outcome (even larger payoff for himself and nothing for the participant). As the partner is economically motivated to defect, choosing the first option in this condition indicates rational mistrust. Finally, in the suspiciousness condition, choosing the second option implicates that the participant's partner will have to choose between a beneficial cooperative outcome (large payoff for both players) and an economically irrational noncooperative outcome (smaller payoff for himself, nothing for the participant). Thus, preferring the first option mirrors suspiciousness in this condition. In a study involving university students, a score on a personality scale assessing beliefs about harmful intentions of others predicted suspiciousness in the game, but was not associated with risk aversion or rational mistrust. In contrast, scores on a self-report measure of risk aversion specifically predicted risk aversive preferences in the game. Thus, the study implicates that suspiciousness is associated with irrational, but not rational, mistrust.

Another study applied the prisoner's dilemma game to explore the correlates of paranoid ideation in the general population (Ellett, Allen-Crooks, Stevens, Wildschut, & Chadwick, 2013). In the prisoner's dilemma, two players choose between cooperation and defection. If both players choose to cooperate, they receive a relatively large payoff. The highest payoff can be received if one defects when the other cooperates, and the lowest payoff is received for cooperation when the partner defects. However, mutual defection results in lower payoffs than mutual cooperation. Trusting the other player and expecting him or her to cooperate is required for mutual cooperation, while distrust leads to defection. Ellett et al. (2013) have argued that the prisoner's dilemma is suitable for studying paranoia, as it is interpersonal, the game involves potential harm to the self by the other, and the other's intentions are unknown. Choosing to defect in the prisoner's dilemma was found to correlate with paranoid thoughts about the opponent's mental state during the game. Importantly, paranoid thoughts did not predict defection if the opponent was a computer, underscoring the inherently social nature of paranoia. A follow-up experiment revealed a specific relationship between paranoia and defection motivated by distrust (expecting the opponent to defect), but no significant association between paranoia and defecting due to greed (expecting the other player to cooperate).

One study has provided initial insight into the neural correlates of trust and paranoia (Gromann et al., 2013). Right-handed male patients with non-affective psychotic disorders and controls performed the trust game while they underwent fMRI. The trust game involves the interaction of two players: the first player has to decide how to split a given amount of money with the second player, who will receive the triple of the sum invested. Then, the second player has to decide how to split this amount with the first player. Although mutual cooperation is beneficial for both players, the second player can benefit from defection as well. In this study, the trust game was modified, in that participants played with a deceptive and a cooperative computer algorithm, but they were informed that they were playing with other humans. During cooperative games, patients made significantly lower investments. Relative to patients, in controls, stronger activation was observed in the right caudate nucleus during cooperative interactions, and in the right temporo-parietal junction during deceptive and cooperative interactions as well. In the patient group, the severity of paranoid symptoms negatively correlated with activation of the nucleus caudatus during cooperative interactions (potentially suggesting impaired reward processing), whereas more severe positive symptoms were associated with increased activation of the temporo-parietal junction during deceptive interactions (putatively associated with abnormal mentalizing). On the other hand, in healthy controls, the caudate nucleus activation showed a positive correlation with initial investments (a putative indicator of baseline trust), while this correlation was not significant in patients.

What are the mechanisms underlying reduced cooperation associated with paranoia? In a large-scale online study, individuals from the general population completed a self-report questionnaire measure of paranoia, and two game theoretical experiments with real partners: a Dictator Game and an Ultimatum Game (Raihani & Bell, 2017a). In the Dictator Game, the "dictator" decides how to split a given amount of money with the "receiver," who has to accept any offer. Thus, behavior in the Dictator Game indicates to what extent an individual is willing to share his or her resources with others in an asymmetric situation when no strategic concerns are to be considered. In contrast, in the Ultimatum Game, the partner can either accept or refuse the proposed split—in the latter case, none of the players receive any reward at all. The potential reaction of the partner has to be considered when making an offer, as it is more costly for the proposer than the responder if the responder refuses an offer that would be disproportionally beneficial for the proposer. The authors argued that if reduced cooperation is driven by distrust in paranoia, then highly paranoid participants should give more generous offers. Both games provide an indicator of punishment: in the Ultimatum Game, refusal of offers provides an opportunity to punish to proposer, whereas an option for the "receiver" to punish the "dictator" after offers was added to the Dictator Game (of which the "dictator" was unaware in advance). Paranoia was associated with punitive response tendencies, and with making lower offers in both games. Intriguingly, the association of paranoia with punitive decisions was partially mediated by attributing harmful intentions to others. From the unexpected finding that paranoia predicted lower offers, the authors inferred that reduced cooperation in paranoia is possibly better explained by increased self-interest (maximizing resources) than by lack of trust. Increased self-interest might be related to reduced subjective value of social relationships and/or hostility toward others. However, it is also possible that highly paranoid individuals were strongly convinced that their partner would try to harm them, regardless of the generosity of their offer; thus, it cannot be excluded that their lower offers were driven by rigid distrust.

Another large online study investigated the relationship between the social representation of others and paranoid thinking (Raihani & Bell, 2017b). Participants either played the Dictator Game as "receivers" or simply observed others playing the game. They were asked to rate the intentions of the "dictator" in terms of self-interest and harm. The dictator's choices were restricted in that he or she could either make a fair (splitting the money in half) or an unfair decision (keeping the whole amount). Paranoid ideation positively predicted attribution of harmful intent, whereas attribution of self-interest was not related to paranoia. Importantly, whether participants played as "receiver" or merely observed did not influence these associations, suggesting that paranoid individuals were more likely to attribute harmful intentions even when they were not directly involved in a social interaction. This latter finding implies that paranoia is related to representing others as harmful, even when they impose no direct threat to the self. This elegant study is in line with the previous literature that highlighted the role of negative other-schemas in paranoid ideation.

In summary, game theoretic studies have provided valuable insight into the social dynamics associated with paranoia in the general population. They suggest that paranoid thinking is not an extreme form of economically rational distrust (Ellett et al., 2013; Johnson et al., 2009), and might be better explained in terms of negative other-schemas (Ellett et al., 2013; Raihani & Bell, 2017b) or increased self-interest (Raihani & Bell, 2017a). An explorative neuroimaging study of patients with psychotic disorders has found that aberrant activation in the right caudate nucleus (brain region implicated in reward processing) during cooperative games was associated with paranoid symptoms, suggesting altered neural processing of rewarding social interactions in paranoia (Gromann et al., 2013).

## Acknowledgments

Bertalan Polner was supported by the ÚNKP-17-4-I New National Excellence Program of the Ministry of Human Capacities. This work was supported by the BME-Biotechnology FIKP grant of EMMI (BME FIKP-BIO).

## References

Addington, J., Cornblatt, B. A., Cadenhead, K. S., Cannon, T. D., McGlashan, T. H., Perkins, D. O., … Woods, S. W. (2011). At clinical high risk for psychosis: outcome for nonconverters. *American Journal of Psychiatry*, *168*(8), 800–805.

Altman, H., Collins, M., & Mundy, P. (1997). Subclinical hallucinations and delusions in nonpsychotic adolescents. *Journal of Child Psychology and Psychiatry*, *38*(4), 413–420.

Bailey, S., Whittle, N., Farnworth, P., & Smedley, K. (2007). A developmental approach to violence, hostile attributions, and paranoid thinking in adolescence. *Behavioral Sciences & the Law*, *25*(6), 913–929.

Bentall, R. P., Rowse, G., Shryane, N., Kinderman, P., Howard, R., Blackwood, N., … Corcoran, R. (2009). The cognitive and affective structure of paranoid delusions: a transdiagnostic investigation of patients with schizophrenia spectrum disorders and depression. *Archives of General Psychiatry*, *66*(3), 236–247. https://doi.org/10.1001/archgenpsychiatry.2009.1.

Bentall, R. P., & Udachina, A. (2013). Social cognition and the dynamics of paranoid ideation. In D. L. Roberts & D. L. Penn (Eds.), *Social cognition and schizophrenia: From evidence to treatment* (pp. 215–244). New York, NY: Oxford University Press.

Bentall, R. P., Wickham, S., Shevlin, M., & Varese, F. (2012). Do specific early-life adversities lead to specific symptoms of psychosis? A study from the 2007 the Adult Psychiatric Morbidity Survey. *Schizophrenia Bulletin*, *38*(4), 734–740. https://doi.org/10.1093/schbul/sbs049.

Bora, E., Yucel, M., & Pantelis, C. (2009). Theory of mind impairment in schizophrenia: meta-analysis. *Schizophrenia Research*, *109*(1), 1–9. https://doi.org/10.1016/j.schres.2008.12.020.

Brüne, M. (2005). "Theory of mind" in schizophrenia: a review of the literature. *Schizophrenia Bulletin*, *31*(1), 21–42. https://doi.org/10.1093/schbul/sbi002.

Buck, B. E., Pinkham, A. E., Harvey, P. D., & Penn, D. L. (2016). Revisiting the validity of measures of social cognitive bias in schizophrenia: additional results from the Social Cognition Psychometric Evaluation (SCOPE) study. *British Journal of Clinical Psychology*, *55*(4), 441–454. https://doi.org/10.1111/bjc.12113.

Campbell, M. L. C., & Morrison, A. P. (2007). The relationship between bullying, psychotic-like experiences and appraisals in 14–16-year olds. *Behaviour Research and Therapy*, *45*(7), 1579–1591. https://doi.org/10.1016/j.brat.2006.11.009.

Cannon, T. D., Cadenhead, K., Cornblatt, B., Woods, S. W., Addington, J., Walker, E., … McGlashan, T. (2008). Prediction of psychosis in youth at high clinical risk: a multisite longitudinal study in North America. *Archives of General Psychiatry*, *65* (1), 28–37.

Carpenter, W. T., & van Os, J. (2011). Should attenuated psychosis syndrome be a DSM-5 diagnosis? *American Journal of Psychiatry*, *168*(5), 460–463.

Carrington, S. J., & Bailey, A. J. (2009). Are there theory of mind regions in the brain? A review of the neuroimaging literature. *Human Brain Mapping*, *30*(8), 2313–2335. https://doi.org/10.1002/hbm.20671.

Chan, K. K. S., & Chen, E. Y. H. (2011). Theory of mind and paranoia in schizophrenia: a game theoretical investigation framework. *Cognitive Neuropsychiatry*, *16*(6), 505–529. https://doi.org/10.1080/13546805.2011.561576.

Cheng, M. (2008). Paranoia on the rise, experts say. *Bakersfield Californian*, *121*(321), A2.

Claridge, G., McCreery, C., Mason, O., Bentall, R., Boyle, G., Slade, P., & Popplewell, D. (1996). The factor structure of 'schizotypal' traits: a large replication study. *British Journal of Clinical Psychology*, *35*(1), 103–115.

Collip, D., Oorschot, M., Thewissen, V., Van Os, J., Bentall, R., & Myin-Germeys, I. (2011). Social world interactions: how company connects to paranoia. *Psychological Medicine*, *41*(05), 911–921. https://doi.org/10.1017/S0033291710001558.

Corcoran, R. (2010). The allusive cognitive deficit in paranoia: the case for mental time travel or cognitive self-projection. *Psychological Medicine*, *40*(8), 1233–1237. https://doi.org/10.1017/S003329170999211X.

Corcoran, R., Rowse, G., Moore, R., Blackwood, N., Kinderman, P., Howard, R., … Bentall, R. P. (2008). A transdiagnostic investigation of 'theory of mind' and 'jumping to conclusions' in patients with persecutory delusions. *Psychological Medicine*, *38*(11), 1577–1583. https://doi.org/10.1017/S0033291707002152.

Craig, J. S., Hatton, C., Craig, F. B., & Bentall, R. P. (2004). Persecutory beliefs, attributions and theory of mind: comparison of patients with paranoid delusions, Asperger's syndrome and healthy controls. *Schizophrenia Research*, *69*(1), 29–33. https://doi.org/10.1016/S0920-9964(03)00154-3.

Dudley, R., Taylor, P., Wickham, S., & Hutton, P. (2016). Psychosis, delusions and the "jumping to conclusions" reasoning bias: a systematic review and meta-analysis. *Schizophrenia Bulletin*, *42*(3), 652–665. https://doi.org/10.1093/schbul/sbv150.

Elahi, A., Algorta, G. P., Varese, F., McIntyre, J. C., & Bentall, R. P. (2017). Do paranoid delusions exist on a continuum with subclinical paranoia? A multi-method taxometric study. *Schizophrenia Research*, *190*, 77–81. https://doi.org/10.1016/j.schres.2017.03.022.

Ellett, L., Allen-Crooks, R., Stevens, A., Wildschut, T., & Chadwick, P. (2013). A paradigm for the study of paranoia in the general population: the Prisoner's Dilemma Game. *Cognition & Emotion*, *27*(1), 53–62. https://doi.org/10.1080/02699931.2012.689757.

Ellett, L., Lopes, B., & Chadwick, P. (2003). Paranoia in a nonclinical population of college students. *The Journal of Nervous and Mental Disease*, *191*(7), 425–430.

Fenigstein, A., & Vanable, P. A. (1992). Paranoia and self-consciousness. *Journal of Personality and Social Psychology*, *62*(1), 129.

Fernyhough, C., Jones, S. R., Whittle, C., Waterhouse, J., & Bentall, R. P. (2008). Theory of mind, schizotypy, and persecutory ideation in young adults. *Cognitive Neuropsychiatry*, *13*(3), 233–249. https://doi.org/10.1080/13546800801936516.

Finn, E. S., Corlett, P. R., Chen, G., Bandettini, P. A., & Constable, R. T. (2018). Trait paranoia shapes inter-subject synchrony in brain activity during an ambiguous social narrative. *Nature Communications*, *9*(1), 2043. https://doi.org/10.1038/s41467-018-04387-2.

Freeman, D. (2007). Suspicious minds: the psychology of persecutory delusions. *Clinical Psychology Review*, *27*(4), 425–457. https://doi.org/10.1016/j.cpr.2006.10.004.

Freeman, D. (2011). Improving cognitive treatments for delusions. *Schizophrenia Research*, *132*(2), 135–139.

Freeman, D. (2016). Persecutory delusions: a cognitive perspective on understanding and treatment. *The Lancet Psychiatry*, *3*(7), 685–692. https://doi.org/10.1016/S2215-0366(16)00066-3.

Freeman, D., Bradley, J., Waite, F., Sheaves, B., DeWeever, N., Bourke, E., … Dunn, G. (2016). Targeting recovery in persistent persecutory delusions: a proof of principle study of a new translational psychological treatment (the Feeling Safe Programme). *Behavioural and Cognitive Psychotherapy*, *44*(5), 539–552. https://doi.org/10.1017/S1352465816000060.

Freeman, D., Dunn, G., Garety, P. A., Bebbington, P., Slater, M., Kuipers, E., … Ray, K. (2005). The psychology of persecutory ideation I: a questionnaire survey. *Journal of Nervous and Mental Disease*, *193*(5), 302–308.

Freeman, D., & Freeman, J. (2009). Is paranoia increasing? *The Psychologist*, *22*(7), 582–585.

Freeman, D., & Garety, P. A. (2000). Comments on the content of persecutory delusions: does the definition need clarification? *British Journal of Clinical Psychology*, *39*(4), 407–414. https://doi.org/10.1348/014466500163400.

Freeman, D., & Garety, P. A. (2004). Bats amongst birds. *The Psychologist*, *17*(11), 642–645.

Freeman, D., & Garety, P. (2014). Advances in understanding and treating persecutory delusions: a review. *Social Psychiatry and Psychiatric Epidemiology*, *49*(8), 1179–1189. https://doi.org/10.1007/s00127-014-0928-7.

Freeman, D., Garety, P. A., Bebbington, P. E., Smith, B., Rollinson, R., Fowler, D., … Dunn, G. (2005). Psychological investigation of the structure of paranoia in a non-clinical population. *The British Journal of Psychiatry*, *186*(5), 427–435.

Freeman, D., Gittins, M., Pugh, K., Antley, A., Slater, M., & Dunn, G. (2008). What makes one person paranoid and another person anxious? The differential prediction of social anxiety and persecutory ideation in an experimental situation. *Psychological Medicine*, *38*(8), 1121–1132. https://doi.org/10.1017/S0033291708003589.

Freeman, D., McManus, S., Brugha, T., Meltzer, H., Jenkins, R., & Bebbington, P. (2011). Concomitants of paranoia in the general population. *Psychological Medicine*, *41*(05), 923–936. https://doi.org/10.1017/S0033291710001546.

Frith, C. D. (2004). Schizophrenia and theory of mind. *Psychological Medicine*, *34*(3), 385–389. https://doi.org/10.1017/S0033291703001326.

Fuligni, A. J., Eccles, J. S., Barber, B. L., & Clements, P. (2001). Early adolescent peer orientation and adjustment during high school. *Developmental Psychology*, *37*(1), 28.

Galletly, C., & Rigby, A. (2013). An overview of cognitive remediation therapy for people with severe mental illness. *ISRN Rehabilitation*, *2013*, 6. https://doi.org/10.1155/2013/984932.

Garety, P. A., Kuipers, E., Fowler, D., Freeman, D., & Bebbington, P. E. (2001). A cognitive model of the positive symptoms of psychosis. *Psychological Medicine, 31*(02), 189–195.

Green, M. J., & Phillips, M. L. (2004). Social threat perception and the evolution of paranoia. *Neuroscience & Biobehavioral Reviews, 28*(3), 333–342. https://doi.org/10.1016/j.neubiorev.2004.03.006.

Gromann, P. M., Heslenfeld, D. J., Fett, A. -K., Joyce, D. W., Shergill, S. S., & Krabbendam, L. (2013). Trust versus paranoia: abnormal response to social reward in psychotic illness. *Brain, 136*(6), 1968–1975. https://doi.org/10.1093/brain/awt076.

Haralanova, E., Haralanov, S., Beraldi, A., Möller, H. -J., & Hennig-Fast, K. (2012). Subjective emotional over-arousal to neutral social scenes in paranoid schizophrenia. *European Archives of Psychiatry and Clinical Neuroscience, 262*(1), 59–68. https://doi.org/10.1007/s00406-011-0227-1.

Harrop, C., & Trower, P. (2001). Why does schizophrenia develop at late adolescence? *Clinical Psychology Review, 21*(2), 241–265.

Howes, O. D., & Kapur, S. (2009). The dopamine hypothesis of schizophrenia: version III—the final common pathway. *Schizophrenia Bulletin, 35*(3), 549–562. https://doi.org/10.1093/schbul/sbp006.

Hutton, P., Kelly, J., Lowens, I., Taylor, P. J., & Tai, S. (2013). Self-attacking and self-reassurance in persecutory delusions: a comparison of healthy, depressed and paranoid individuals. *Psychiatry Research, 205*(1), 127–136. https://doi.org/10.1016/j.psychres.2012.08.010.

Johnson, M. K., Rustichini, A., & MacDonald, A. W. (2009). Suspicious personality predicts behavior on a social decision-making task. *Personality and Individual Differences, 47*(1), 30–35. https://doi.org/10.1016/j.paid.2009.01.050.

Kesting, M. -L., Bredenpohl, M., Klenke, J., Westermann, S., & Lincoln, T. M. (2013). The impact of social stress on self-esteem and paranoid ideation. *Journal of Behavior Therapy and Experimental Psychiatry, 44*(1), 122–128. https://doi.org/10.1016/j.jbtep.2012.07.010.

Kesting, M. -L., & Lincoln, T. M. (2013). The relevance of self-esteem and self-schemas to persecutory delusions: a systematic review. *Comprehensive Psychiatry, 54*(7), 766–789. https://doi.org/10.1016/j.comppsych.2013.03.002.

Kinderman, P., Dunbar, R., & Bentall, R. P. (1998). Theory-of-mind deficits and causal attributions. *British Journal of Psychology, 89*(2), 191–204. https://doi.org/10.1111/j.2044-8295.1998.tb02680.x.

Klein, H. S., Kelsven, S., & Pinkham, A. E. (2018). Increased social cognitive bias in subclinical paranoia. *Schizophrenia Research: Cognition, 12*, 74–76. https://doi.org/10.1016/j.scog.2018.05.002.

Korver-Nieberg, N., Fett, A. -K. J., Meijer, C. J., Koeter, M. W., Shergill, S. S., de Haan, L., & Krabbendam, L. (2013). Theory of mind, insecure attachment and paranoia in adolescents with early psychosis and healthy controls. *The Australian and New Zealand Journal of Psychiatry, 47*(8), 737–745. https://doi.org/10.1177/0004867413484370.

Kraepelin, E. (1919). *Dementia praecox and paraphrenia* [R.M. Barclay, Trans.] Edinburgh: Livingstone.

Langdon, R., Siegert, R. J., McClure, J., & Harrington, L. (2005). Schizophrenia, theory of mind, and persecutory delusions. *Cognitive Neuropsychiatry, 10*(2), 87–104. https://doi.org/10.1080/13546800344000327.

Lewis, A. (1970). Paranoia and paranoid: a historical perspective. *Psychological Medicine, 1*(1), 2–12. https://doi.org/10.1017/S0033291700039969.

Lincoln, T. M., Lange, J., Burau, J., Exner, C., & Moritz, S. (2010). The effect of state anxiety on paranoid ideation and jumping to conclusions. An experimental investigation. *Schizophrenia Bulletin, 36*(6), 1140–1148. https://doi.org/10.1093/schbul/sbp029.

Lüdtke, T., Kriston, L., Schröder, J., Lincoln, T. M., & Moritz, S. (2017). Negative affect and a fluctuating jumping to conclusions bias predict subsequent paranoia in daily life: an online experience sampling study. *Journal of Behavior Therapy and Experimental Psychiatry, 56*, 106–112. https://doi.org/10.1016/j.jbtep.2016.08.014.

Manschreck, T. C. (1992). Delusional disorders: clinical concepts and diagnostic strategies. *Psychiatric Annals, 22*(5), 241–251.

Matos, M., Pinto-Gouveia, J., & Gilbert, P. (2013). The effect of shame and shame memories on paranoid ideation and social anxiety: shame memories, paranoia and social anxiety. *Clinical Psychology & Psychotherapy, 20*(4), 334–349. https://doi.org/10.1002/cpp.1766.

McDonnell, J., Stahl, D., Day, F., McGuire, P., & Valmaggia, L. R. (2018). Interpersonal sensitivity in those at clinical high risk for psychosis mediates the association between childhood bullying victimisation and paranoid ideation: a virtual reality study. *Schizophrenia Research, 192*, 89–95. https://doi.org/10.1016/j.schres.2017.04.029.

McKay, R., Langdon, R., & Coltheart, M. (2005). Paranoia, persecutory delusions and attributional biases. *Psychiatry Research, 136*(2), 233–245. https://doi.org/10.1016/j.psychres.2005.06.004.

McKay, R., Langdon, R., & Coltheart, M. (2007). Jumping to delusions? Paranoia, probabilistic reasoning, and need for closure. *Cognitive Neuropsychiatry, 12*(4), 362–376. https://doi.org/10.1080/13546800701203769.

Menon, M., Addington, J., & Remington, G. (2013). Examining cognitive biases in patients with delusions of reference. *European Psychiatry, 28*(2), 71–73. https://doi.org/10.1016/j.eurpsy.2011.03.005.

Mikulincer, M., Shaver, P. R., & Pereg, D. (2003). Attachment theory and affect regulation: the dynamics, development, and cognitive consequences of attachment-related strategies. *Motivation and Emotion, 27*(2), 77–102.

Mills, A., Gilbert, P., Bellew, R., McEwan, K., & Gale, C. (2007). Paranoid beliefs and self-criticism in students. *Clinical Psychology & Psychotherapy, 14*(5), 358–364. https://doi.org/10.1002/cpp.537.

Montag, C., Dziobek, I., Richter, I. S., Neuhaus, K., Lehmann, A., Sylla, R., … Gallinat, J. (2011). Different aspects of theory of mind in paranoid schizophrenia: evidence from a video-based assessment. *Psychiatry Research, 186*(2), 203–209. https://doi.org/10.1016/j.psychres.2010.09.006.

Moutoussis, M., Bentall, R. P., El-Deredy, W., & Dayan, P. (2011). Bayesian modelling of jumping-to-conclusions bias in delusional patients. *Cognitive Neuropsychiatry, 16*(5), 422–447.

Myin-Germeys, I., Oorschot, M., Collip, D., Lataster, J., Delespaul, P., & van Os, J. (2009). Experience sampling research in psychopathology: opening the black box of daily life. *Psychological Medicine, 39*(09), 1533. https://doi.org/10.1017/S0033291708004947.

Nelson, B., Whitford, T. J., Lavoie, S., & Sass, L. A. (2014). What are the neurocognitive correlates of basic self-disturbance in schizophrenia?: Integrating phenomenology and neurocognition: Part 2 (aberrant salience). *Schizophrenia Research, 152*(1), 20–27. https://doi.org/10.1016/j.schres.2013.06.033.

Oorschot, M., Kwapil, T., Delespaul, P., & Myin-Germeys, I. (2009). Momentary assessment research in psychosis. *Psychological Assessment, 21*(4), 498–505. https://doi.org/10.1037/a0017077.

Palmier-Claus, J., Dunn, G., Drake, R., & Lewis, S. (2011). The negative and positive self: a longitudinal study examining self-esteem, paranoia and negative symptoms in individuals with first-episode psychosis: the negative and positive self. *Early Intervention in Psychiatry, 5*(2), 150–155. https://doi.org/10.1111/j.1751-7893.2010.00250.x.

Peters, E. R., Joseph, S. A., & Garety, P. A. (1999). Measurement of delusional ideation in the normal population: introducing the PDI (Peters et al. Delusions Inventory). *Schizophrenia Bulletin, 25*(3), 553–576.

Phalen, P. L., Dimaggio, G., Popolo, R., & Lysaker, P. H. (2017). Aspects of theory of mind that attenuate the relationship between persecutory delusions and social functioning in schizophrenia spectrum disorders. *Journal of Behavior Therapy and Experimental Psychiatry, 56*, 65–70. https://doi.org/10.1016/j.jbtep.2016.07.008.
Phillips, M. L., Williams, L., Senior, C., Bullmore, E. T., Brammer, M. J., Andrew, C., … David, A. S. (1999). A differential neural response to threatening and non-threatening negative facial expressions in paranoid and non-paranoid schizophrenics. *Psychiatry Research: Neuroimaging, 92*(1), 11–31. https://doi.org/10.1016/S0925-4927(99)00031-1.
Pickering, L., Simpson, J., & Bentall, R. P. (2008). Insecure attachment predicts proneness to paranoia but not hallucinations. *Personality and Individual Differences, 44*(5), 1212–1224. https://doi.org/10.1016/j.paid.2007.11.016.
Pinkham, A. E., Brensinger, C., Kohler, C., Gur, R. E., & Gur, R. C. (2011). Actively paranoid patients with schizophrenia over attribute anger to neutral faces. *Schizophrenia Research, 125*(2–3), 174–178. https://doi.org/10.1016/j.schres.2010.11.006.
Pinkham, A. E., Liu, P., Lu, H., Kriegsman, M., Simpson, C., & Tamminga, C. (2015). Amygdala hyperactivity at rest in paranoid individuals with schizophrenia. *American Journal of Psychiatry, 172*(8), 784–792. https://doi.org/10.1176/appi.ajp.2014.14081000.
Poulton, R., Caspi, A., Moffitt, T. E., Cannon, M., Murray, R., & Harrington, H. (2000). Children's self-reported psychotic symptoms and adult schizophreniform disorder: a 15-year longitudinal study. *Archives of General Psychiatry, 57*(11), 1053–1058.
Raes, F., & Van Gucht, D. (2009). Paranoia and instability of self-esteem in adolescents. *Personality and Individual Differences, 47*(8), 928–932. https://doi.org/10.1016/j.paid.2009.07.017.
Raihani, N. J., & Bell, V. (2017a). Conflict and cooperation in paranoia: a large-scale behavioural experiment. *Psychological Medicine*, 1–11. https://doi.org/10.1017/S0033291717003075.
Raihani, N. J., & Bell, V. (2017b). Paranoia and the social representation of others: a large-scale game theory approach. *Scientific Reports*. 7(1)https://doi.org/10.1038/s41598-017-04805-3.
Riggio, H. R., & Kwong, W. Y. (2009). Social skills, paranoid thinking, and social outcomes among young adults. *Personality and Individual Differences, 47*(5), 492–497.
Riggio, H. R., & Kwong, W. Y. (2011). Paranoid thinking, quality of relationships with parents, and social outcomes among young adults. *Journal of Family Issues, 32*(8), 1030–1049.
Rössler, W., Hengartner, M. P., Ajdacic-Gross, V., Haker, H., Gamma, A., & Angst, J. (2011). Sub-clinical psychosis symptoms in young adults are risk factors for subsequent common mental disorders. *Schizophrenia Research, 131*(1), 18–23.
Ruhrmann, S., Schultze-Lutter, F., & Klosterkötter, J. (2010). Probably at-risk, but certainly ill—advocating the introduction of a psychosis spectrum disorder in DSM-V. *Schizophrenia Research, 120*(1), 23–37. https://doi.org/10.1016/j.schres.2010.03.015.
Russell, T. A., Reynaud, E., Kucharska-Pietura, K., Ecker, C., Benson, P. J., Zelaya, F., … Phillips, M. L. (2007). Neural responses to dynamic expressions of fear in schizophrenia. *Neuropsychologia, 45*(1), 107–123. https://doi.org/10.1016/j.neuropsychologia.2006.04.026.
Sass, L., & Byrom, G. (2015). Phenomenological and neurocognitive perspectives on delusions: a critical overview. *World Psychiatry, 14*(2), 164–173. https://doi.org/10.1002/wps.20205.
Schoretsanitis, G., Kutynia, A., Stegmayer, K., Strik, W., & Walther, S. (2016). Keep at bay!—abnormal personal space regulation as marker of paranoia in schizophrenia. *European Psychiatry, 31*, 1–7. https://doi.org/10.1016/j.eurpsy.2015.10.001.

Shevlin, M., McAnee, G., Bentall, R. P., & Murphy, J. (2015). Specificity of association between adversities and the occurrence and co-occurrence paranoia and hallucinations: evaluating the stability of childhood risk in an adverse adult environment. *Psychosis*, 7(3), 206–216. https://doi.org/10.1080/17522439.2014.980308.

Shryane, N. M., Corcoran, R., Rowse, G., Moore, R., Cummins, S., Blackwood, N., ... Bentall, R. P. (2008). Deception and false belief in paranoia: modelling theory of mind stories. *Cognitive Neuropsychiatry*, *13*(1), 8–32. https://doi.org/10.1080/13546800701748805.

Sitko, K., Varese, F., Sellwood, W., Hammond, A., & Bentall, R. (2016). The dynamics of attachment insecurity and paranoid thoughts: an experience sampling study. *Psychiatry Research*, *246*, 32–38. https://doi.org/10.1016/j.psychres.2016.08.057.

So, S. H., Chau, A. K. C., Peters, E., Swendsen, J., Garety, P., & Kapur, S. (2017). Moment-to-moment associations between emotional disturbances, aberrant salience and persecutory delusions. *European Psychiatry*, *41*, S838. https://doi.org/10.1016/j.eurpsy.2017.01.1653.

Spreng, R. N., Mar, R. A., & Kim, A. S. N. (2009). The common neural basis of autobiographical memory, prospection, navigation, theory of mind, and the default mode: a quantitative meta-analysis. *Journal of Cognitive Neuroscience*, *21*(3), 489–510.

Strauss, J. S. (1969). Hallucinations and delusions as points on continua function: rating scale evidence. *Archives of General Psychiatry*, *21*(5), 581–586.

Strik, W., Wopfner, A., Horn, H., Koschorke, P., Razavi, N., Walther, S., & Wirtz, G. (2010). The Bern psychopathology scale for the assessment of system-specific psychotic symptoms. *Neuropsychobiology*, *61*(4), 197–209. https://doi.org/10.1159/000297737.

Taylor, J. L., & Kinderman, P. (2002). An analogue study of attributional complexity, theory of mind deficits and paranoia. *British Journal of Psychology*, *93*(1), 137–140. https://doi.org/10.1348/000712602162481.

Thewissen, V., Bentall, R. P., Oorschot, M., à Campo, J., van Lierop, T., van Os, J., & Myin-Germeys, I. (2011). Emotions, self-esteem, and paranoid episodes: an experience sampling study: emotions, self-esteem, and paranoia. *British Journal of Clinical Psychology*, *50*(2), 178–195. https://doi.org/10.1348/014466510X508677.

Tiernan, B., Tracey, R., & Shannon, C. (2014). Paranoia and self-concepts in psychosis: a systematic review of the literature. *Psychiatry Research*, *216*(3), 303–313. https://doi.org/10.1016/j.psychres.2014.02.003.

Tone, E. B., & Davis, J. S. (2012). Paranoid thinking, suspicion, and risk for aggression: a neurodevelopmental perspective. *Development and Psychopathology*, *24*(3), 1031–1046.

Van Dael, F., Versmissen, D., Janssen, I., Myin-Germeys, I., Van Os, J., & Krabbendam, L. (2005). Data gathering: biased in psychosis? *Schizophrenia Bulletin*, *32*(2), 341–351.

Van Os, J., Hanssen, M., Bijl, R. V., & Ravelli, A. (2000). Strauss (1969) revisited: a psychosis continuum in the general population? *Schizophrenia Research*, *45*(1), 11–20.

Versmissen, D., Janssen, I., Myin-Germeys, I., Mengelers, R., Campo, J. A., van Os, J., & Krabbendam, L. (2008). Evidence for a relationship between mentalising deficits and paranoia over the psychosis continuum. *Schizophrenia Research*, *99*(1), 103–110. https://doi.org/10.1016/j.schres.2007.09.024.

Westermann, S., Kesting, M. -L., & Lincoln, T. M. (2012). Being deluded after being excluded? How emotion regulation deficits in paranoia-prone individuals affect state paranoia during experimentally induced social stress. *Behavior Therapy*, *43*(2), 329–340. https://doi.org/10.1016/j.beth.2011.07.005.

Whitfield-Gabrieli, S., & Ford, J. M. (2012). Default mode network activity and connectivity in psychopathology. *Annual Review of Clinical Psychology*, *8*(1), 49–76. https://doi.org/10.1146/annurev-clinpsy-032511-143049.

Wickham, S., Sitko, K., & Bentall, R. P. (2015). Insecure attachment is associated with paranoia but not hallucinations in psychotic patients: the mediating role of negative self-esteem. *Psychological Medicine*, *45*(7), 1495–1507. https://doi.org/10.1017/S0033291714002633.

Wigman, J. T., Vollebergh, W. A., Raaijmakers, Q. A., Iedema, J., Van Dorsselaer, S., Ormel, J., ... van Os, J. (2009). The structure of the extended psychosis phenotype in early adolescence—a cross-sample replication. *Schizophrenia Bulletin*, *37*(4), 850–860.

Williams, L. M., Das, P., Liddell, B. J., Olivieri, G., Peduto, A. S., David, A. S., ... Harris, A. W. F. (2007). Fronto-limbic and autonomic disjunctions to negative emotion distinguish schizophrenia subtypes. *Psychiatry Research: Neuroimaging*, *155*(1), 29–44. https://doi.org/10.1016/j.pscychresns.2006.12.018.

Wykes, T., Huddy, V., Cellard, C., McGurk, S. R., & Czobor, P. (2011). A meta-analysis of cognitive remediation for schizophrenia: methodology and effect sizes. *American Journal of Psychiatry*, *168*(5), 472–485. https://doi.org/10.1176/appi.ajp.2010.10060855.

Ziermans, T. B., Schothorst, P. F., Sprong, M., & van Engeland, H. (2011). Transition and remission in adolescents at ultra-high risk for psychosis. *Schizophrenia Research*, *126*(1), 58–64.

CHAPTER THREE

# Social cognition and schizotypy

**Tovah Cowan, Thanh P. Le, Alex S. Cohen**
Psychology Department, Louisiana State University, Baton Rouge, LA, United States

Deficits in social cognitive functioning, involving cognitive processes allowing us to conceptualize and manage others and our relationships to them (Penn, Sanna, & Roberts, 2008), are generally considered ubiquitous in people with schizophrenia (e.g., Penn et al., 2008). These deficits are generally present throughout all phases of illness, are chronic, and treatment resilient. A modest research corpus has been devoted to understanding these social cognitive impairments as a neurodevelopmental anomaly that reflects a schizophrenia endophenotype. This chapter will review literature on social cognitive functioning and schizotypy. Schizotypy is defined as an underlying personality trait indicating a predisposition for schizophrenia spectrum disorders (e.g., schizophrenia or schizoaffective disorder, "cluster A" personality disorders). Although schizotypy is not pathological, there are particular reasons to explore the specific interaction of social cognition and schizotypy. Integrating these two concepts in a broad overview provides potential insight into how social cognitive processes are maintained or impacted by schizotypy. As summary by Cohen, Mohr, Ettinger, Chan, and Park (2015) studying people with schizotypy offers insight into potential mechanisms for adaptive traits, as well as an understanding of the schizophrenia spectrum less confounded by illness-related variables, such as medication, unemployment, hospitalization, and stigma (see also Barrantes-Vidal, Grant, & Kwapil, 2015). This chapter starts with a definitional section to situate readers and specify exactly what is meant by the concepts of schizotypy and social cognition. Subsequently, we will summarize what is known about schizotypy and social cognition from both a behavioral and psychophysiological perspective. The final section will integrate these findings and propose explanations for the patterns of social cognitive deficits in schizotypy.

## Definitions and grounding

### Schizotypy

The schizotypy construct has been used in empirical research for nearly six decades, and there is a general agreement that schizotypy reflects traits

*Social Cognition in Psychosis*
https://doi.org/10.1016/B978-0-12-815315-4.00003-3

associated with schizophrenia-spectrum pathology. Much work has been built on Paul Meehl's (1962) neurodevelopmental model of schizotypy, wherein schizotypy is a latent categorical construct reflecting a genetic predisposition to certain cognitive anomalies (schizotaxia), which, when combined with adverse learning and social influences (e.g., environmental factors) may result in observable "schizotypes" (e.g., people with high social anhedonia and/or psychosis-like experiences; Lenzenweger, 2015). While schizotypy as a latent construct in this model is categorical (which remains fiercely contested in the literature), the observable schizotypes are on a continuum, ranging from endophenotypes only observable in the laboratory to schizophrenia. This continuum also encompasses several of the Diagnostic and Statistical Manual of Mental Disorders 5th Edition (DSM-V; American Psychiatric Association, 2013) personality disorders. Importantly, an individual with schizotypy may not display any aspects of schizotypy, except in laboratory measures. We will focus our review on the less severe end of the illness continuum and will contextualize those populations in terms of more severe impairment or symptom presentations.

Despite consensus regarding many aspects of schizotypy, there is some disagreement about its phenotype and how it should be operationalized and defined. Common approaches include examining biological family members of individuals with schizophrenia, individuals with various clinical presentations (e.g., ultra-high risk studies), and individuals with self-reported schizotypal traits. The latter approach forms the lion's share of schizotypy research. This population is often identified by psychometric classification on one of several schizotypy questionnaires, such as the Schizotypal Personality Questionnaire (SPQ; Raine, 1991), the Chapman Schizotypy Scales (Chapman, Chapman, Kwapil, Eckblad, & Zinser, 1994), and the Oxford-Liverpool Inventory of Feelings and Experiences (Burch, Steel, & Hemsley, 1998). These measures differ somewhat in their conceptual scope. Collectively, however, it is clear that schizotypy is multidimensional, and comprised of a two- or three-factor structure: positive (e.g., magical thinking, unusual perceptual experiences, ideas of references, or paranoid ideation), negative (e.g., social anhedonia, constricted affect), and disorganized (e.g., odd speech and behavior) (Raine et al., 1994), which neatly connects to the three main facets of schizophrenia. Social anhedonia, in particular, is considered an important component of schizotypy as it has been shown to predict the development of "cluster A" schizotypal spectrum disorders and related pathology in longitudinal studies (Blanchard, Collins, Aghevli, Leung, & Cohen, 2009; Gooding, Tallent, & Matts, 2005; Kwapil, 1998).

## Social cognition

Social cognition is generally defined as a collection of processes by which we integrate information about others and use that information to have successful interactions (Penn et al., 2008). The social cognition literature is vast and spans a variety of methodologies and foci, from childhood development to clinical pathology. Importantly, social cognition and its associated processes has been integrated into the Research Domain Criteria framework (RDoC) (Insel et al., 2010)—a transdiagnostic approach to understanding psychopathology in terms of basic neuro-biologically coherent domains. As of the most recent version of the RDoC (National Institutes of Mental Health, 2017), the framework includes the domain of "social processes," encompassing the constructs of affiliation and attachment, social communication, perception and understanding of self, and perception and understanding of others. Each of these constructs contains subconstructs that then have research at the various levels of the RDoC framework, from genetics and molecules to self-report and paradigms.

Psychopathology researchers, including schizophrenia researchers specifically, have also generated their own frameworks of social cognition, though their primary scope has been studying schizophrenia. For example, Penn et al. (2008) categorize existing research in social cognition in schizophrenia according to three general constructs: emotion perception, theory of mind (ToM), and attributional style. Green et al. (2008) summarize the discussion at a National Institutes of Mental Health meeting on the subject by categorizing social cognition in schizophrenia according to five areas, adding social knowledge and emotional processing to the three described by Penn et al. (2008). This framework was then updated by Green, Horan, and Lee (2015), organizing social cognition over two levels: processes and subprocesses. These processes include the following: social cue perception (with subprocesses face and voice perception), mentalizing, experience and regulation of emotion (each of which alone is a subprocess), and experience sharing, or observation of another's behavior triggering neural activation similar to performing that behavior oneself (including the subprocesses of motor resonance and affect sharing). The process level clearly connects to the previous frameworks, with attributional biases and ToM having been combined. However, at the subprocess level, mechanisms for social cognition and the social cognitive processes themselves have been intermingled, including, for example, motor resonance, or the result of mirror neuron systems. Motor resonance allows more efficient use of information for social cognitive processes such as action or language understanding (Zwaan & Taylor, 2006), but is not a social cognitive process in itself.

There are three possible ways that social cognitive abnormalities could manifest in schizotypy. First, it could be that there are specific deficits, social cognitive and otherwise, inherent in schizotypy, which get worse with greater illness severity (and may be potentiating the increased severity). For example, it could be that social cue perception is increasingly impaired in individuals with schizotypy, whereas other domains, such as experience sharing and ToM, are generally preserved. If impairment in affected areas increases with illness severity, it primarily reflects motivational, and basic cognitive, social, or other "generalized" causes. Second, it could be that schizotypy is associated with a general social cognitive deficit that gets worse with greater severity. Finally, it is possible that there are different patterns of social cognitive performance along the schizotypy spectrum. This chapter will evaluate social cognition in schizotypy with a focus on understanding how social cognitive abilities vary as a function of severity of schizotype manifestation. We will present evidence of performance in people with schizotypy (defined through laboratory measures and questionnaires) in the domains of social perception, mentalizing and ToM, experience sharing, self-awareness, attributional style, and social knowledge, consistent with the framework established in the RDoC constructs, at least in part, as well as the majority of the schizophrenia frameworks. We will also interpret the findings in light of the three aforementioned possibilities. Understanding the specificity and generalizability of social cognitive abnormalities in schizotypy has important implications for our understanding, assessment, and treatment of schizophrenia-spectrum pathology.

## What is known about social cognition in schizotypy?

### Social perception

Social perception is typically partitioned into two categories, namely facial perception and vocal perception (Green et al., 2015). Vocal perception involves recognizing and discriminating acoustic properties of speech, and the affective information they convey, where facial perception is decoding affective information from others' facial expressions. Inherent in both categories is the ability to read others' emotional cues. Generally speaking, there is evidence that individuals with schizotypy show mildly impaired performance on tasks of social perception, though the nature of these impairments has not replicated well across studies. For example, van't Wout, Aleman, Kessels, Larøi, and Kahn (2004) administered both the SPQ and a facial affect recognition task, and found that positive schizotypy correlated with misclassifying angry faces as happy, and the subscale of unusual perceptual experiences correlated with misclassifying happy faces as angry or fearful. Other studies have extended these findings and observed that negative schizotypy is also related to deficits in facial affect (e.g., Abbott & Green, 2013; Williams, Henry, & Green, 2007). However, some studies do not replicate deficits in classifying facial emotion (e.g., Jahshan & Sergi, 2007; Shean, Bell, & Cameron, 2007), but these null findings are potentially limited by the emotion perception tasks used, which may be insufficiently sensitive. Brown and Cohen (2010) sought to use a more sensitive measure of facial emotion perception that included neutral stimuli, and found that people with schizotypy were worse at identifying emotions than controls, more likely to misclassify neutral faces as showing disgust, and that there was a relation between disorganized schizotypy and a systematic bias for classifying faces with a negative emotional valence. Individuals with schizotypy also show deficits in facial recognition, as demonstrated by Larøi, D'Argembeau, Brédart, and van der Linden (2007), who found that people higher in schizotypy traits endorsed greater difficulty recognizing themselves, and others they know, and also show perceptual distortion of faces, in comparison with people low in schizotypy. These difficulties and distortions were associated with positive and disorganized, but not negative, schizotypy. In sum, evidence suggests that there is a facial affect recognition negativity bias in people with schizotypy, and that kinds of schizotypy may differentially misread various emotions, which may be driven by a lack of ability to integrate facial cues more broadly.

These results are replicated in studies on emotion recognition from vocal information. Although there are significantly fewer studies overall, they consistently show deficits in vocal affect recognition, the ability to recognize emotions from speech, in schizotypy samples (Castro & Pearson, 2011), as well as in individuals diagnosed with schizotypal personality disorder (Baum & Nowicki, 1998; Wickline, Nowicki, Bollini, & Walker, 2012), and at ultra-high risk for psychosis (Amminger et al., 2012). One study partially extended this finding, with the specification that positive (particularly cognitive-perceptual) schizotypy was negatively related to vocal emotion recognition accuracy (Shean et al., 2007). Taken together, these studies and the studies of facial affect recognition show a deficit in social perception in schizotypy that mirrors the results of similar studies conducted in schizophrenia (Kohler et al., 2003; Tseng et al., 2013). Some researchers have reported a negative correlation between ability to identify happy emotions and both negative and positive symptoms (Tseng et al., 2013). In contrast, Kohler et al. (2003) observed that facial affect recognition in schizophrenia was worse relative to controls for fearful, disgusted, and neutral faces, but not for happy or angry faces, and worse performance was connected specifically to negative symptoms. Similar to facial emotion recognition, there is a pattern of general deficit in social perception, but mixed results at the level of what kinds of perceptions are most difficult, and which symptom profiles are most closely connected.

## Mentalizing and ToM

ToM involves the ability to attribute mental states to oneself and others (e.g., understanding others' actions, feelings, and intentions), and has been central to the schizotypy construct since its early formation (Langdon & Coltheart, 1999). Links between performance on ToM tasks, often developed for individuals with autism spectrum pathology, and schizotypal traits have provided evidence that only some aspects of ToM are affected in schizotypy. For example, schizotypy scores (reflecting positive, negative, and disorganization traits) have been associated with poor performance on a ToM task requiring recognition of social faux pas (Morrison, Brown, & Cohen, 2013), intact performance on measures of sarcasm (Jahshan & Sergi, 2007), and enhanced performance on a ToM task that involved detection of irony (McCleery et al., 2012). It is clear that behaviorally, high and low schizotypy cannot be reliably differentiated from each other based solely on ToM. However, there is substantial and consistent evidence for deficits in ToM in schizophrenia, in both populations in remission and those who are not, regardless of task (Bora, Yucel, & Pantelis, 2009). ToM seems to be an area where populations with schizophrenia are more severely impacted than populations with milder forms of schizotypy, who in some cases, may outperform controls.

## Experience sharing and empathy

Experience sharing is a broad term drawn from Green and colleague's framework (2015) to reference all ways in which another's experience is reflected in one's own senses. It can involve mimicking, consciously or subconsciously, another's emotions, movements, or visuospatial perspective. These are all components of various forms of empathy, the ability to take another's perspective and feel their feelings with the connotation of caring for them and using this information to help them. Empathy is closely related to ToM and mentalizing. However, unlike ToM and mentalizing, empathy also emphasizes understanding and feeling the emotions of others, cognitive and affective empathy, respectively. Of experience sharing, only visuospatial perspective taking has been explored in schizotypy (to our knowledge), where positive schizotypy was positively related to increased accuracy in perspective taking (Thakkar & Park, 2010). Overall, self-reported trait empathy is lower in schizotypy samples compared with normal controls (Henry, Bailey, & Rendell, 2008; Ripoll et al., 2013). When empathy is separated into cognitive and affective components, a different picture emerges. Positive schizotypy is associated with performance similar to controls, or even greater cognitive empathy (Henry et al., 2008; Wang et al., 2013), potentially connecting to the results on visuospatial perspective taking. Both negative and disorganized schizotypy are negatively related to affective and cognitive empathy (Henry et al., 2008; Thakkar & Park, 2010; Wang et al., 2013). In comparison, there are noted deficits in empathy among individuals with schizophrenia, though these are not preferentially related to specific syndromes or symptoms (e.g., negative, positive, or disorganized; Montag, Heinz, Kunz, & Gallinat, 2007; Smith et al., 2012). It seems that empathy is more globally impaired in individuals with schizophrenia than in individuals with schizotypy more generally.

## Self-awareness

Part of managing social relationships is the ability to understand one's self, and the boundaries between self and others (National Institutes of Mental Health, 2017). Consistently, there is a negative relationship between schizotypy and self-awareness. Schizotypy negatively relates to emotional identification, the ability to distinguish and describe one's own emotions (e.g., Seghers, McCleery, & Docherty, 2011, see Nicolò et al., 2011 for an exception), even though schizotypy is associated with increased attention to emotional stimuli (Kerns, 2005). Poor emotion identification is particularly associated with disorganized schizotypy (Kerns, 2006; Larøi, Van der Linden, & Aleman, 2008), and negative schizotypy (Larøi et al., 2008). Individuals with schizophrenia show a similar deficit, as seen in a metaanalysis by O'Driscoll, Laing, and Mason (2014). In both schizophrenia and schizotypy samples, negative traits and symptoms are associated with a difficulty in increasing and managing emotion (Henry et al., 2009; Morrison et al.,

2013; O'Driscoll et al., 2014). In fact, difficulties in upregulating positive emotion is a core feature of anhedonic negative symptoms in schizophrenia. Last, schizotypy is associated with decreased self-association, shown in weaker distinctions between the self and others (Larøi et al., 2007; Platek, Critton, Myers, & Gallup Jr., 2003), and decreased feelings of self-agency when judging what initiated a movement (Asai & Tanno, 2007). These findings replicate those in the schizophrenia literature (e.g., Irani et al., 2006; Knoblich, Stottmeister, & Kircher, 2004). As a whole, there are global deficits in emotional identification, emotion regulation, and self-boundaries in both schizotypy and schizophrenia.

## Attributional style

Attributional style is a set of "metacognitive" factors reflecting tendencies in inferring causes of external events (Abramson, Seligman, & Teasdale, 1978). Traditionally, attributional style is comprised of three dimensions, including internal-external, stable-unstable, and global-specific (Abramson et al., 1978). It was originally applied to understand enduring cognitive abnormalities in individuals with, or prone to, depression, but was soon applied to positive schizotypy, though research in this population has largely focused on the internal-external dimension. In contrast to depression, which tends to be associated with an internal attributional style, a number of studies have found that individuals with positive schizotypal traits evidence external attributions (Bentall, Corcoran, Howard, Blackwood, & Kinderman, 2001). Theoretical work expanding this external attributional style has focused on the tendency to infer/perceive hostility and deliberate threat from others (Morrison & Cohen, 2014). This is an important distinction from social anxiety, which is associated with perceived threat from social sources, but without the deliberate hostility motivation. Not surprisingly, positive schizotypal traits, notably those associated with suspicious beliefs, have been associated with increased attributions involving hostility and direct threat and aggressive reactions, particularly when situations are ambiguous with respect to the intentions of others, such as a group of teenagers who begin to laugh as you walk past them (Combs, Penn, Wicher, & Waldheter, 2007). Increased amygdala sensitivity for salient stimuli, particularly involving potential threat, is a proposed neurobiological mechanism (Galdos et al., 2010; Mohanty et al., 2005). Though less well-explicated, the attributional style of negative schizotypy is likely different, having been associated with "defeatist beliefs" that emphasize "internalizing" attributional style in a similar manner as in depression. Defeatist beliefs concern a "learned helplessness" and are a core feature of negative schizophrenia (Grant & Beck, 2008) and schizotypy (Luther et al., 2016) as well. Not surprisingly, negative schizotypy has not been associated with abnormal amygdala reactivity (Galdos et al., 2010). Paranoia in schizophrenia has specifically been associated with external attributional biases, though schizophrenia without paranoia may show attribution biases similar to controls (Moritz, Woodward, Burlon, Braus, & Andresen, 2007). Thus, attributional biases in schizotypy seem to be similar, albeit reduced in severity, compared with those in schizophrenia.

## Social knowledge

Very little information has been gathered in the domain of social knowledge and schizotypy. Social knowledge consists of two subskills: understanding social and cultural rules, and applying them appropriately in a given situation (Green et al., 2008). As such, social knowledge is distinct from general social functioning, as it is more closely related to social skills. Unfortunately, there is a lack of research on cultural social knowledge. However, there is consistent evidence for social knowledge deficits in people with schizotypy, especially as it relates to applying social rules, such as how to act during a job interview, to their own actions, but not in understanding the rules in the abstract (Waldeck & Miller, 2000, see also Morrison et al., 2013). Schizotypy has been related to less rejection of unfairness, and greater generosity, in Ultimatum and Dictator games (van't Wout & Sanfey, 2011). Individuals with higher negative schizotypy were more likely to accept unfair offers, and those with high positive schizotypy were more likely to make generous offers, even in the Dictator game when the offers could not be declined. These results can be interpreted as people with schizotypy insufficiently leveraging their position, and also that they are more pragmatic than controls in accepting unfair offers, because individuals with high negative schizotypy are less bothered by the social implications of unfair offers. This is only partially true in a schizophrenia population, where, relative to controls, individuals with schizophrenia were more generous, but were equally as likely to reject unfair offers (Agay, Kron, Carmel, Mendlovic, & Levkovitz, 2008). Other studies on schizophrenia populations show a similar ability to understand, but not apply, social rules (Bellack, Mueser, Wade, Sayers, & Morrison, 1992; Carini & Nevid, 1992). In the domain of social knowledge, it seems schizotypy results mirror those seen in the schizophrenia literature, except with even less bias toward fairness.

## What are potential mechanisms/causes of social cognitive abnormalities in schizotypy?

### Psychophysiological mechanisms

The neural mechanisms potentially underlying social cognition have recently been receiving attention. For example, Germine, Garrido, Bruce, and Hooker (2011) found that individuals with high social anhedonia, as compared with low social anhedonia, showed reduced neural activity in several face emotion processing regions during discrimination of emotional faces. In particular, reduced activity was most consistent in the anterior portion of the rostral medial prefrontal cortex (arMFC) during the facial affect perception. The arMFC region has been consistently associated with mentalizing and other aspects of social cognition (Amodio & Frith, 2006; Grossmann, 2013), and linked with abnormalities during an emotion face identity task, and other social-cognitive processing, in schizophrenia samples (Billeke & Aboitiz, 2013; Mukherjee et al., 2016). The deficient underactivation of the arMFC during facial affect perception may be related to differences in strategy during emotion discrimination (e.g., attending to low-level features to perform the task) or other differences in emotional information processing (Germine et al., 2011).

Similar to discrimination of emotional faces, studies have largely recruited patient populations to examine the neural mechanisms underlying mentalization, though some nonpatients studies exist. For example, performance across behavioral, self-report, and clinical interview measures of ToM was related to ventromedial prefrontal cortex gray matter volume in schizophrenia (Hooker, Bruce, Lincoln, Fisher, & Vinogradov, 2011), while activity in the medial prefrontal cortex (MPFC) during a task of predicting behavior based on mental states was related to self-reported social anhedonia in schizophrenia (Dodell-Feder, Tully, Lincoln, & Hooker, 2014). Wang et al. (2015) examined the relationship between positive and negative schizotypy and neural activity during ToM and empathy tasks. Performance was normal during the tasks, but schizotypal traits were associated with aberrant activity in the temporo-parietal junction (TPJ), a multisensory brain region that has been associated with self-processing in healthy controls (Amft et al., 2015), and autism (Lombardo, Chakrabarti, Bullmore, Baron-Cohen, & MRC AIMS Consortium, 2011), and in biological motion perception in schizophrenia (Matsumoto et al., 2018). Interestingly, negative schizotypy was associated with greater TPJ activity, suggesting the utilization of compensatory mechanisms. Furthermore, negative schizotypal traits have been associated with greater right TPJ thickness (Kühn, Schubert, &

Gallinat, 2012). Thus, the study of schizotypy provides insight on the influential role of the TPJ for successful mentalization. It may be that aberrant structural asymmetry and activity of the temporo-parietal region could reflect neural compensation resulting from protracted functional hyperactivity while engaged in self and other processing. These findings also highlight that the components of schizotypy may exhibit differential relationships with networks involved in social cognitive processing.

## Nonsocial cognition as a mechanism

Early investigators posited that social cognitive dysfunction may arise as a function of basic cognitive deficits (e.g., memory, attention, executive function). However, an accumulation of evidence from studies using structural equation modeling (SEM), confirmatory factor analysis (CFA), and exploratory factor analysis (EFA) have demonstrated that, while related, social cognition is distinct from basic cognition (Allen, Strauss, Donohue, & van Kammen, 2007; Pinkham & Penn, 2006). Furthermore, schizotypy is related to a host of social cognitive deficits as noted earlier, yet a recent metaanalysis indicates grossly normal cognitive abilities using traditional neuropsychological measures (Chun, Minor, & Cohen, 2013). However, it is still likely that basic cognitive abilities have an influential role in social cognitive processing within schizotypy. For example, mentalizing relies on schemas, or a set of ideas that organizes categories of information and the relationships among them (Penn et al., 2008). The ease or difficulty with which these schemas are activated is likely influenced by limited cognitive resources (e.g., working memory, attention deficits), which has been found in schizophrenia (Heinrichs & Zakzanis, 1998), and some studies of schizotypy (Cohen, Morrison, Brown, & Minor, 2012). Thus, a limited cognitive load may make the task of generating alternative thoughts or responses an extremely effortful process, thus facilitating attributional biases and deficits in other self-other processing. Furthermore, schizotypy has also been associated with both impaired (Mohr, Blanke, & Brugger, 2006) and enhanced (Thakkar & Park, 2010) egocentric mental imagery manipulation, which is notable, as mentalizing likely requires visuospatial perspective taking and simulation.

## Directionality of social cognition and schizotypy

From both a neurodevelopmental and cognitive "systems" approach (i.e., viewing cognition as a network of interdependent functions), the causal relationship between social cognition and schizotypy has yet to be explicated. It is tempting to assume that social cognitive deficits stem from basic central nervous system (CNS) disturbances (i.e., "schizotaxia"), and hence, reflect stable genetically-mediated markers of pathology. However, it may be the case that social cognitive impairments are related to the formation of deleterious daily experiences in schizotypy, so called "aversive drift," as described by Meehl (1962, 1990). From this perspective, aberrant social cognition would be proximal determinants of subclinical symptom presentation in schizotypy (e.g., magical ideation, ideas of reference, social anxiety). For example, deficits in mentalizing, which has been associated with reasoning biases such as the tendency to "jump to conclusions" or to seek "confirmation bias" (Penn et al., 2008), may lead to a wide range of social consequences, such as when to infer whether a stranger is friendly or seeking to cause harm. Investigators have postulated that deficits in mentalizing and social information processing may result in hostile attribution biases, which may lead to symptoms such as unusual perceptual experiences, ideas of reference, and suspicion (Lam, Raine, & Lee, 2016), each themselves correlates of positive schizotypy. While this is likely a mechanism for relatively brief emotion regulation and positive self-image management, it comes at the cost of negative perceptions of others.

Along similar lines, social cognitive deficits and aberrant experiences in schizotypy may emerge from comparable affective abnormalities. The characteristics of negative schizotypy, such as reduced pleasure and motivation to participant in social activities (Brown, Silvia, Myin-Germeys, & Kwapil, 2007), may contribute to impairment in social cognitive domains such as mentalizing and facial affect perception through reduced exposure and opportunity for practice. Alternatively, disrupted ToM and empathy may engender the extinction of social reinforcement, leading to increased negative schizotypal characteristics. Thus, it is unclear whether characteristics of schizotypy are consequences of social cognitive impairments, or whether subclinical symptoms impact social cognitive performance adversely. Parsing these relationships represents an avenue for future research.

## Summary and conclusion

In exploring social cognition in schizotypy, a pattern of impairment emerges that largely mimics the impairments shown in schizophrenia relative to controls. However, there are some areas with important differences: ToM is largely spared in schizotypy, and some studies show improved performance relative to controls in some ToM tasks, as well as for positive schizotypy in cognitive empathy and visual perspective taking. In contrast, in the manipulation of social rules around fairness, a schizotypy sample performed worse than both controls and individuals with schizophrenia. Based on this review of social cognition in schizotypy, it seems that areas of social cognition that revolve around emotional self-other boundaries and awareness are impacted (emotion perception, self-awareness, affective empathy, attributional styles, and social knowledge), whereas those that require mentalizing and cognitively empathizing are potentially spared. The impacted areas are heavily entangled with social anhedonia, with an unclear direction of causality, as discussed herein. The areas that are spared in schizotypy are not spared in schizophrenia. Acknowledging the magnitude of the deficits is a necessary endeavor to clearly understand the evolution of social cognition along the psychosis spectrum, though it is still possible to say that on the whole, milder forms of schizotypy show a specific pattern of social cognitive deficit, one distinct from the more severe form, schizophrenia. These differentiable patterns reinforce the idea that there are other processes involved in the development of schizophrenia beyond the latent trait of schizotypy, and these other processes either compensate for, or exacerbate social cognitive abnormalities. Given the particular pattern of social cognitive deficits evidenced in schizotypy, intact general cognitive abilities may be acting as a protective factor. Studying schizotypy allows us to explore social cognition in a population with fewer global deficits, with the eventual goal of identifying which differences lead to greater impairment. Schizotypy is an exciting and useful model in the gradual clarification of social cognition along the psychosis spectrum.

## References


Abbott, G. R., & Green, M. J. (2013). Facial affect recognition and schizotypal personality characteristics. *Early Intervention in Psychiatry*, 7(1), 58–63.

Abramson, L. Y., Seligman, M. E., & Teasdale, J. D. (1978). Learned helplessness in humans: critique and reformulation. *Journal of Abnormal Psychology*, *87*(1), 49.

Agay, N., Kron, S., Carmel, Z., Mendlovic, S., & Levkovitz, Y. (2008). Ultimatum bargaining behavior of people affected by schizophrenia. *Psychiatry Research*, *157*(1), 39–46.

Allen, D. N., Strauss, G. P., Donohue, B., & van Kammen, D. P. (2007). Factor analytic support for social cognition as a separable cognitive domain in schizophrenia. *Schizophrenia Research*, *93*(1–3), 325–333.

American Psychiatric Association. (2013). *Diagnostic and statistical manual of mental disorders* (5th ed.). Washington, DC: American Psychiatric Publishing.

Amft, M., Bzdok, D., Laird, A. R., Fox, P. T., Schilbach, L., & Eickhoff, S. B. (2015). Definition and characterization of an extended social-affective default network. *Brain Structure and Function*, *220*(2), 1031–1049.

Amminger, G. P., Schäfer, M. R., Klier, C. M., Schlögelhofer, M., Mossaheb, N., Thompson, A., ... Nelson, B. (2012). Facial and vocal affect perception in people at ultra-high risk of psychosis, first-episode schizophrenia and healthy controls. *Early Intervention in Psychiatry*, *6*(4), 450–454.

Amodio, D. M., & Frith, C. D. (2006). Meeting of minds: the medial frontal cortex and social cognition. *Nature Reviews Neuroscience*, 7(4), 268.

Asai, T., & Tanno, Y. (2007). The relationship between the sense of self-agency and schizotypal personality traits. *Journal of Motor Behavior*, *39*(3), 162–168.

Barrantes-Vidal, N., Grant, P., & Kwapil, T. R. (2015). The role of schizotypy in the study of the etiology of schizophrenia spectrum disorders. *Schizophrenia Bulletin*, *41*(Suppl. 2), S408–S416.

Baum, K. M., & Nowicki, S. (1998). Perception of emotion: measuring decoding accuracy of adult prosodic cues varying in intensity. *Journal of Nonverbal Behavior*, *22*(2), 89–107.

Bellack, A. S., Mueser, K. T., Wade, J., Sayers, S., & Morrison, R. L. (1992). The ability of schizophrenics to perceive and cope with negative affect. *The British Journal of Psychiatry*, *160*(4), 473–480.

Bentall, R. P., Corcoran, R., Howard, R., Blackwood, N., & Kinderman, P. (2001). Persecutory delusions: a review and theoretical integration. *Clinical Psychology Review*, *21*(8), 1143–1192.

Billeke, P., & Aboitiz, F. (2013). Social cognition in schizophrenia: from social stimuli processing to social engagement. *Frontiers in Psychiatry*, *4*, 4.

Blanchard, J. J., Collins, L. M., Aghevli, M., Leung, W. W., & Cohen, A. S. (2009). Social anhedonia and schizotypy in a community sample: the Maryland longitudinal study of schizotypy. *Schizophrenia Bulletin*, *37*(3), 587–602.

Bora, E., Yucel, M., & Pantelis, C. (2009). Theory of mind impairment in schizophrenia: meta-analysis. *Schizophrenia Research*, *109*(1), 1–9.

Brown, L. A., & Cohen, A. S. (2010). Facial emotion recognition in schizotypy: the role of accuracy and social cognitive bias. *Journal of the International Neuropsychological Society*, *16*(3), 474–483.

Brown, L. H., Silvia, P. J., Myin-Germeys, I., & Kwapil, T. R. (2007). When the need to belong goes wrong: The expression of social anhedonia and social anxiety in daily life. *Psychological Science*, *18*(9), 778–782.

Burch, G. S. J., Steel, C., & Hemsley, D. R. (1998). Oxford—Liverpool inventory of feelings and experiences: reliability in an experimental population. *British Journal of Clinical Psychology*, *37*(1), 107–108.

Carini, M. A., & Nevid, J. S. (1992). Social appropriateness and impaired perspective in schizophrenia. *Journal of Clinical Psychology*, *48*(2), 170–177.

Castro, A., & Pearson, R. (2011). Lateralisation of language and emotion in schizotypal personality: evidence from dichotic listening. *Personality and Individual Differences*, *51*(6), 726–731.

Chapman, L. J., Chapman, J. P., Kwapil, T. R., Eckblad, M., & Zinser, M. C. (1994). Putatively psychosis-prone subjects 10 years later. *Journal of Abnormal Psychology*, *103* (2), 171.

Chun, C. A., Minor, K. S., & Cohen, A. S. (2013). Neurocognition in psychometrically defined college schizotypy samples: we are not measuring the "right stuff". *Journal of the International Neuropsychological Society*, *19*(3), 324–337.

Cohen, A. S., Mohr, C., Ettinger, U., Chan, R. C., & Park, S. (2015). Schizotypy as an organizing framework for social and affective sciences. *Schizophrenia Bulletin*, *41*(Suppl. 2), S427–S435.

Cohen, A. S., Morrison, S. C., Brown, L. A., & Minor, K. S. (2012). Towards a cognitive resource limitations model of diminished expression in schizotypy. *Journal of Abnormal Psychology*, *121*(1), 109.

Combs, D. R., Penn, D. L., Wicher, M., & Waldheter, E. (2007). The Ambiguous Intentions Hostility Questionnaire (AIHQ): a new measure for evaluating hostile social-cognitive biases in paranoia. *Cognitive Neuropsychiatry*, *12*(2), 128–143.

Dodell-Feder, D., Tully, L. M., Lincoln, S. H., & Hooker, C. I. (2014). The neural basis of theory of mind and its relationship to social functioning and social anhedonia in individuals with schizophrenia. *NeuroImage: Clinical*, *4*, 154–163.

Galdos, M., Simons, C., Fernandez-Rivas, A., Wichers, M., Peralta, C., Lataster, T., … van Os, J. (2010). Affectively salient meaning in random noise: a task sensitive to psychosis liability. *Schizophrenia Bulletin*, *37*(6), 1179–1186.

Germine, L. T., Garrido, L., Bruce, L., & Hooker, C. (2011). Social anhedonia is associated with neural abnormalities during face emotion processing. *NeuroImage*, *58*(3), 935–945.

Gooding, D. C., Tallent, K. A., & Matts, C. W. (2005). Clinical status of at-risk individuals 5 years later: further validation of the psychometric high-risk strategy. *Journal of Abnormal Psychology*, *114*(1), 170.

Grant, P. M., & Beck, A. T. (2008). Defeatist beliefs as a mediator of cognitive impairment, negative symptoms, and functioning in schizophrenia. *Schizophrenia Bulletin*, *35*(4), 798–806.

Green, M. F., Horan, W. P., & Lee, J. (2015). Social cognition in schizophrenia. *Nature Reviews Neuroscience*, *16*(10), 620.

Green, M. F., Penn, D. L., Bentall, R., Carpenter, W. T., Gaebel, W., Gur, R. C., … Heinssen, R. (2008). Social cognition in schizophrenia: an NIMH workshop on definitions, assessment, and research opportunities. *Schizophrenia Bulletin*, *34*(6), 1211–1220.

Grossmann, T. (2013). The role of medial prefrontal cortex in early social cognition. *Frontiers in Human Neuroscience*, 7, 340.

Heinrichs, R. W., & Zakzanis, K. K. (1998). Neurocognitive deficit in schizophrenia: a quantitative review of the evidence. *Neuropsychology*, *12*(3), 426.

Henry, J. D., Bailey, P. E., & Rendell, P. G. (2008). Empathy, social functioning and schizotypy. *Psychiatry Research*, *160*(1), 15–22.

Henry, J. D., Green, M. J., Restuccia, C., de Lucia, A., Rendell, P. G., McDonald, S., & Grisham, J. R. (2009). Emotion dysregulation and schizotypy. *Psychiatry Research*, *166*(2), 116–124.

Hooker, C. I., Bruce, L., Lincoln, S. H., Fisher, M., & Vinogradov, S. (2011). Theory of mind skills are related to gray matter volume in the ventromedial prefrontal cortex in schizophrenia. *Biological Psychiatry*, *70*(12), 1169–1178.

Insel, T., Cuthbert, B., Garvey, M., Heinssen, R., Pine, D. S., Quinn, K., … Wang, P. (2010). Research domain criteria (RDoC): toward a new classification framework for research on mental disorders. *American Journal of Psychiatry*, *167*(7), 748–751. https://doi.org/10.1176/appi.ajp.2010.09091379.

Irani, F., Platek, S. M., Panyavin, I. S., Calkins, M. E., Kohler, C., Siegel, S. J., … Gur, R. C. (2006). Self-face recognition and theory of mind in patients with schizophrenia and first-degree relatives. *Schizophrenia Research*, *88*(1), 151–160.

Jahshan, C. S., & Sergi, M. J. (2007). Theory of mind, neurocognition, and functional status in schizotypy. *Schizophrenia Research*, *89*(1), 278–286.

Kerns, J. G. (2005). Positive Schizotypy and emotion processing. *Journal of Abnormal Psychology*, *114*(3), 392–401.

Kerns, J. G. (2006). Schizotypy facets, cognitive control, and emotion. *Journal of Abnormal Psychology*, *115*(3), 418.

Knoblich, G., Stottmeister, F., & Kircher, T. (2004). Self-monitoring in patients with schizophrenia. *Psychological Medicine*, *34*(8), 1561–1569.

Kohler, C. G., Turner, T. H., Bilker, W. B., Brensinger, C. M., Siegel, S. J., Kanes, S. J., … Gur, R. C. (2003). Facial emotion recognition in schizophrenia: intensity effects and error pattern. *American Journal of Psychiatry*, *160*(10), 1768–1774.

Kühn, S., Schubert, F., & Gallinat, J. (2012). Higher prefrontal cortical thickness in high schizotypal personality trait. *Journal of Psychiatric Research, 46*(7), 960–965.

Kwapil, T. R. (1998). Social anhedonia as a predictor of the development of schizophrenia-spectrum disorders. *Journal of Abnormal Psychology, 107*(4), 558–565.

Lam, B. Y., Raine, A., & Lee, T. M. (2016). Effect of theory of mind and peer victimization on the schizotypy–aggression relationship. *NPJ Schizophrenia, 2*, 16001.

Langdon, R., & Coltheart, M. (1999). Mentalizing, schizotypy, and schizophrenia. *Cognition, 71*(1), 43–71.

Larøi, F., D'Argembeau, A., Brédart, S., & van der Linden, M. (2007). Face recognition failures in schizotypy. *Cognitive Neuropsychiatry, 12*, 554–571.

Larøi, F., Van der Linden, M., & Aleman, A. (2008). Associations between dimensions of alexithymia and psychometric schizotypy in nonclinical participants. *The Journal of Nervous and Mental Disease, 196*(12), 927–930.

Lenzenweger, M. F. (2015). Thinking clearly about schizotypy: hewing to the schizophrenia liability core, considering interesting tangents, and avoiding conceptual quicksand. *Schizophrenia Bulletin, 41*(Suppl. 2), S483–S491.

Lombardo, M. V., Chakrabarti, B., Bullmore, E. T., Baron-Cohen, S., & MRC AIMS Consortium (2011). Specialization of right temporo-parietal junction for mentalizing and its relation to social impairments in autism. *Neuroimage, 56*(3), 1832–1838.

Luther, L., Salyers, M. P., Firmin, R. L., Marggraf, M. P., Davis, B., & Minor, K. S. (2016). Additional support for the cognitive model of schizophrenia: evidence of elevated defeatist beliefs in schizotypy. *Comprehensive Psychiatry, 68*, 40–47.

Matsumoto, Y., Takahashi, H., Miyata, J., Sugihara, G., Murai, T., & Takahashi, H. (2018). Neural basis of altered earlier attention and higher order biological motion processing in schizophrenia. *Social Neuroscience, 13*(5), 594–601.

McCleery, A., Divilbiss, M., St-Hilaire, A., Aakre, J. M., Seghers, J. P., Bell, E. K., & Docherty, N. M. (2012). Predicting social functioning in schizotypy: an investigation of the relative contributions of theory of mind and mood. *The Journal of Nervous and Mental Disease, 200*(2), 147.

Meehl, P. E. (1962). Schizotaxia, schizotypy, schizophrenia. *American Psychologist, 17*(12), 827.

Meehl, P. E. (1990). Toward an integrated theory of schizotaxia, schizotypy, and schizophrenia. *Journal of Personality Disorders, 4*(1), 1–99.

Mohanty, A., Herrington, J. D., Koven, N. S., Fisher, J. E., Wenzel, E. A., Webb, A. G., … Miller, G. A. (2005). Neural mechanisms of affective interference in schizotypy. *Journal of Abnormal Psychology, 114*(1), 16.

Mohr, C., Blanke, O., & Brugger, P. (2006). Perceptual aberrations impair mental own-body transformations. *Behavioral Neuroscience, 120*(3), 528.

Montag, C., Heinz, A., Kunz, D., & Gallinat, J. (2007). Self-reported empathic abilities in schizophrenia. *Schizophrenia Research, 92*(1), 85–89.

Moritz, S., Woodward, T. S., Burlon, M., Braus, D. F., & Andresen, B. (2007). Attributional style in schizophrenia: evidence for a decreased sense of self-causation in currently paranoid patients. *Cognitive Therapy and Research, 31*(3), 371–383.

Morrison, S. C., Brown, L. A., & Cohen, A. S. (2013). A multidimensional assessment of social cognition in psychometrically defined schizotypy. *Psychiatry Research, 210*(3), 1014–1019.

Morrison, S. C., & Cohen, A. S. (2014). The moderating effects of perceived intentionality: exploring the relationships between ideas of reference, paranoia and social anxiety in schizotypy. *Cognitive Neuropsychiatry, 19*(6), 527–539.

Mukherjee, P., Sabharwal, A., Kotov, R., Szekely, A., Parsey, R., Barch, D. M., & Mohanty, A. (2016). Disconnection between amygdala and medial prefrontal cortex in psychotic disorders. *Schizophrenia Bulletin, 42*(4), 1056–1067.

National Institutes of Mental Health. (2017, May 18). *RDoC snapshot: Version 3*. (saved 5/18/2017). Retrieved March 22, 2018, from *https://www.nimh.nih.gov/research-priorities/rdoc/constructs/rdoc-snapshot-version-3-saved-5-18-2017.shtml*.

Nicolò, G., Semerari, A., Lysaker, P. H., Dimaggio, G., Conti, L., D'Angerio, S., … Carcione, A. (2011). Alexithymia in personality disorders: correlations with symptoms and interpersonal functioning. *Psychiatry Research*, *190*(1), 37–42.

O'Driscoll, C., Laing, J., & Mason, O. (2014). Cognitive emotion regulation strategies, alexithymia and dissociation in schizophrenia, a review and meta-analysis. *Clinical Psychology Review*, *34*(6), 482–495.

Penn, D. L., Sanna, L. J., & Roberts, D. L. (2008). Social cognition in schizophrenia: an overview. *Schizophrenia Bulletin*, *34*(3), 408–411.

Pinkham, A. E., & Penn, D. L. (2006). Neurocognitive and social cognitive predictors of interpersonal skill in schizophrenia. *Psychiatry Research*, *143*(2–3), 167–178.

Platek, S. M., Critton, S. R., Myers, T. E., & Gallup, G. G., Jr. (2003). Contagious yawning: the role of self-awareness and mental state attribution. *Cognitive Brain Research*, *17*(2), 223–227.

Raine, A. (1991). The SPQ: a scale for the assessment of schizotypal personality based on DSM-III-R criteria. *Schizophrenia Bulletin*, *17*, 555–564.

Raine, A., Reynolds, C., Lencz, T., Scerbo, A., Triphon, N., & Kim, D. (1994). Cognitive-perceptual, interpersonal, and disorganized features of schizotypal personality. *Schizophrenia Bulletin*, *20*(1), 191–201.https://doi.org/10.1093/schbul/20.1.191.

Ripoll, L. H., Zaki, J., Perez-Rodriguez, M. M., Snyder, R., Strike, K. S., Boussi, A., … New, A. S. (2013). Empathic accuracy and cognition in schizotypal personality disorder. *Psychiatry Research*, *210*(1), 232–241.

Seghers, J. P., McCleery, A., & Docherty, N. M. (2011). Schizotypy, alexithymia, and socioemotional outcomes. *The Journal of Nervous and Mental Disease*, *199*(2), 117.

Shean, G., Bell, E., & Cameron, C. D. (2007). Recognition of nonverbal affect and schizotypy. *The Journal of Psychology*, *141*(3), 281–292.

Smith, M. J., Horan, W. P., Karpouzian, T. M., Abram, S. V., Cobia, D. J., & Csernansky, J. G. (2012). Self-reported empathy deficits are uniquely associated with poor functioning in schizophrenia. *Schizophrenia Research*, *137*(1), 196–202.

Thakkar, K. N., & Park, S. (2010). Empathy, schizotypy, and visuospatial transformations. *Cognitive Neuropsychiatry*, *15*(5), 477–500.

Tseng, H. H., Chen, S. H., Liu, C. M., Howes, O., Huang, Y. L., Hsieh, M. H., … Hwu, H. G. (2013). Facial and prosodic emotion recognition deficits associate with specific clusters of psychotic symptoms in schizophrenia. *PLoS One*, *8*(6), e66571.

van't Wout, M., Aleman, A., Kessels, R. P., Larøi, F., & Kahn, R. S. (2004). Emotional processing in a non-clinical psychosis-prone sample. *Schizophrenia Research*, *68*(2), 271–281.

van't Wout, M., & Sanfey, A. G. (2011). Interactive decision-making in people with schizotypal traits: a game theory approach. *Psychiatry Research*, *185*(1), 92–96.

Waldeck, T. L., & Miller, L. S. (2000). Social skills deficits in schizotypal personality disorder. *Psychiatry Research*, *93*(3), 237–246.

Wang, Y., Liu, W. H., Li, Z., Wei, X. H., Jiang, X. Q., Neumann, D. L., … Chan, R. C. (2015). Dimensional schizotypy and social cognition: an fMRI imaging study. *Frontiers in Behavioral Neuroscience*, *9*, 133.

Wang, Y., Neumann, D. L., Shum, D. H., Liu, W. H., Shi, H. S., Yan, C., … Chan, R. C. (2013). Cognitive empathy partially mediates the association between negative schizotypy traits and social functioning. *Psychiatry Research*, *210*(1), 62–68.

Wickline, V. B., Nowicki, S., Bollini, A. M., & Walker, E. F. (2012). Vocal and facial emotion decoding difficulties relating to social and thought problems: highlighting schizotypal personality disorder. *Journal of Nonverbal Behavior*, *36*(1), 59–77.

Williams, B. T., Henry, J. D., & Green, M. J. (2007). Facial affect recognition and schizotypy. *Early Intervention in Psychiatry, 1*(2), 177–182.

Zwaan, R. A., & Taylor, L. J. (2006). Seeing, acting, understanding: motor resonance in language comprehension. *Journal of Experimental Psychology: General, 135*(1), 1–11.

## Further Reading

Dickey, C. C., Panych, L. P., Voglmaier, M. M., Niznikiewicz, M. A., Terry, D. P., Murphy, C., … McCarley, R. W. (2011). Facial emotion recognition and facial affect display in schizotypal personality disorder. *Schizophrenia Research, 131*(1), 242–249.

Kerr, S. L., & Neale, J. M. (1993). Emotion perception in schizophrenia: specific deficit or further evidence of generalized poor performance? *Journal of Abnormal Psychology, 102*(2), 312.

Kohler, C. G., Walker, J. B., Martin, E. A., Healey, K. M., & Moberg, P. J. (2009). Facial emotion perception in schizophrenia: a meta-analytic review. *Schizophrenia Bulletin, 36*(5), 1009–1019.

CHAPTER FOUR

# Impact of social cognitive deficits on community functioning

**Margherita Bechi, Marco Spangaro**
Department of Clinical Neurosciences, IRCCS San Raffaele Scientific Institute, Milan, Italy

## Introduction

According to the World Health Organization (WHO), the lifetime prevalence of any mental disorder is around 25%, with 450 million people suffering from a psychiatric disease and its related psychosocial impairments. Despite the different clinical and psychopathological features, almost every severe mental illness (SMI) negatively affects functional outcome by impairing personal, work, and social functioning, thus further worsening the burden of the disease.

Research in this field has shown that, in addition to psychopathology, functional deficits are highly related to disruptions in other domains, such as neurocognition and social cognition.

Despite the development of effective antipsychotic treatments and the achievement of symptomatic remission, more than half of the subjects affected by schizophrenia show functional impairment overall. Indeed, clinically stable patients with schizophrenia show deficits in almost every area of daily functioning. This places a great social and financial burden on health systems, families, and society. Patients' level of functioning is reported as one of the main predictors of disease costs. Furthermore, severe functional impairments are associated with higher risk of physical morbidity and premature mortality. Deficits in daily functioning are, in fact, detectable even before the acute onset of illness, and persist after remission of positive symptomatology. In this context, recovery or functional remission has therefore become the ultimate treatment goal in schizophrenia, and different rehabilitative interventions have been developed in order to offer patients the opportunity to achieve a satisfying life.

For this reason, current research has focused on the identification of functional predictors and influencing factors, showing that the functioning of patients is mainly related to their neurocognitive and social cognitive

*Social Cognition in Psychosis*
https://doi.org/10.1016/B978-0-12-815315-4.00004-5

abilities, which are known to be impaired in schizophrenia. However, there is an ongoing debate on the specific relationship between the two domains of cognition and functional outcome.

In this chapter, comparing the results of different studies, these topics will be discussed with a particular focus on the role of social cognitive subdomains, which appear to be crucial for several real-world skills.

Moreover, given the observation of initial and mild functional impairments in the early phases of the illness, or even before acute clinical onset, in order to better understand the longitudinal evolution of the deficits, psychosocial functioning of ultra-high risk (UHR) for psychosis (Yung & McGorry, 2007) and first-episode psychosis (FEP) individuals will be analyzed.

In conclusion, this chapter will also include a section on the functioning of individuals affected by bipolar disorder (BD). Even during euthymic phases, bipolar patients show a consistent functional impairment, leading to social withdrawal, unemployment, and several other deficits in daily living. Despite having different clinical manifestations and course of illness, the functioning of bipolar and psychotic patients shows areas of overlap. The differences are mainly caused by distinct neurocognitive and social cognitive impairments, which translate in different functional outcomes.

These findings bring forward evidence that neurocognition, and especially social cognition, strongly determine functional outcome, and that in order to achieve recovery, effective and targeted psychosocial interventions are needed.

## Community functioning in schizophrenia

Schizophrenia represents a pervasive, persistent, and heterogeneous chronic disease associated with global functional impairment. Psychosocial decline starts before acute onset of illness, and persists even during periods of symptoms' remission (Reichenberg et al., 2009). Patients with schizophrenia show a widespread impairment of several domains of community functioning, including self-care activities and psychosocial and work functioning. However, the degree of neurocognitive, social cognitive, and functional impairment is characterized by great variability between subjects (Keefe et al., 2006). Therefore, research has focused on the individuation of outcome predictors and influencing factors, in order to achieve a thorough comprehension of the disease, and to develop targeted and effective interventions (Mariachiara Buonocore, Bosia, et al., 2018).

## Independent living

Approximately only one-third of patients with schizophrenia are able to live independently, managing medications, bills, shopping, cooking, and transportation without external support (Karagianis et al., 2009). Among these, more than 75% rely on public disability compensation (Harvey, 2009). Changing perspective, this data means that less than 1 patient out of 10 is able to live independently, without need of economic support. Moreover, a worldwide study including more than 17,000 psychotic patients showed marked impairments in everyday activities, such as cooking, travel, and shopping (Karagianis et al., 2009). Subjects with schizophrenia are usually not able to follow a balanced diet, to regularly practice sport, or to perform self-care activities (Breier, Schreiber, Dyer, & Pickar, 1991), which is reflected in greater prevalence of cardio-metabolic disease, further worsening the burden of the illness.

## Social functioning

Schizophrenia is characterized by multiple social impairments across domains of functioning, including overall reductions in quantity and quality of interpersonal interactions (Harvey, 2009) and poor social networks (Patterson et al., 1997). Moreover, patients frequently show severe difficulties in creating and maintaining connections outside the family and the assistance sphere. Subjects with schizophrenia manifest difficulties in sharing personal experiences, and often exhibit socially disruptive behavior, characterized by inappropriate comments and impairments in judging appropriate interpersonal distance. Impaired social functioning is only partially related to the disorganized behavior, and other symptoms of the illness (Patterson, Moscona, McKibbin, Davidson, & Jeste, 2001).

## Work functioning

Impairment of work performance precedes the onset of disease, and negatively impacts patients' daily functioning. Many patients experience job loss after disease onset, and even when clinically stabilized, only a small percentage of patients obtain and maintain paid employment. Research has found that difficulties with appropriate social behaviors at work may contribute to challenges obtaining and maintaining employment (Becker & Milke, 1998). Still, obtaining and maintaining a job represents a crucial step for functional recovery of subjects with schizophrenia (Saavedra, Lopez, Gonzales, &

Cubero, 2016). It is known that working activity and educational levels are associated with better psychosocial health among the general population. Work leads to increases in self-esteem and self-efficacy, and is associated with a reduction of hospital admissions. Moreover, work functioning facilitates overcoming social stigma and isolation, creating an active role for patients in society (Bevan et al., 2013).

## Factors affecting community functioning in schizophrenia

As mentioned herein, disability in multiple functional domains is common in schizophrenia. Even with regular pharmacological compliance and good response to antipsychotics, during the course of illness, most patients with schizophrenia exhibit clinical relapse and periodic exacerbations of positive symptomatology. However, clinical symptomatic remission only partially correlates with functional recovery (Foussias & Remington, 2010; Mueser & McGurk, 2004) such as obtaining and maintaining a meaningful role in the community. In order to achieve functional recovery, it is crucial to identify functional outcome predictors and influencing factors, which is also necessary for the development of more targeted and effective interventions.

Functional disability of patients with schizophrenia is the sum of multiple illness-related factors, individual resources, and environmental elements (Galderisi, Vignapiano, Mucci, & Boutros, 2014). Symptom factors, including lack of insight, and residual negative symptoms, such as amotivation and anhedonia, have been associated with functional outcome, even more consistently than positive symptoms (Lincoln, Wilhelm, & Nestoriuc, 2007; Milev, Ho, Arndt, & Andreasen, 2005; Rabinowitz et al., 2012; Ventura, Hellemann, Thames, Koellner, & Nuechterlein, 2009). Additionally, almost 40% of patients present persistent reductions in social motivation and pleasure-related activities, resulting in deficits in daily living (Kirkpatrick, Buchanan, Ross, & Carpenter, 2001; Leifker, Bowie, & Harvey, 2009). Together, these elements greatly contribute to the social withdrawal associated with schizophrenia, and represent an important barrier to functional recovery (Marwaha & Johnson, 2005).

Among the strongest predictors of functioning, however, are deficits in neurocognition and social cognition, which are common across the psychoses (Bowie et al., 2008; Brekke, Hoe, Long, & Green, 2007; Couture, Penn,

& Roberts, 2006; Green, 1996; Green, Kern, Braff, & Mintz, 2000; Lysaker et al., 2013; Vauth, Rusch, Wirtz, & Corrigan, 2004). Impairments in neurocognitive domains persist across the course of illness, and are only partially influenced by antipsychotic treatment (Bechi, Spangaro, et al., 2018; Harvey, Bowie, & Friedman, 2001). Despite their crucial role in community functioning, neurocognitive abilities alone are able to explain only 40% of patients' functional outcome variability, leaving the remaining 60% unexplained (Best, Gupta, Bowie, & Harvey, 2014). In this context, social cognition has emerged as one of the most promising predictors, showing consistent associations with both neurocognition and different functional domains (Schmidt, Mueller, & Roder, 2011).

Social cognition refers to the set of processes underlying social interactions, and includes several distinct abilities. Theory of Mind (ToM), emotion perception and processing, empathy, social perception (SP), and attributional style (AS) are the main and most studied sociocognitive domains. Patients with schizophrenia show impairments in multiple dimensions of social cognition, with large effect sizes (Savla, Vella, Armstrong, Penn, & Twamley, 2013). Social cognitive deficits are already detectable in the prodromal phase and during the early course of the illness, persisting during periods of symptom remission (de Achaval et al., 2010; Fett et al., 2011; Kohler, Walker, Martin, Healey, & Moberg, 2010). Disorganized and negative symptoms have been consistently associated with social cognition in schizophrenia, which is largely independent of positive symptomatology (Bora & Pantelis, 2013; Green et al., 2012). Studies investigating the relationship between neurocognition and social cognition and their impact on functioning report partial independence of the two cognitive domains (Fett & Maat, 2013; Ventura, Wood, & Hellemann, 2013). Studies have found that social cognition is able to explain over 20% of variance in functional outcome, and mediates the effect of neurocognition on functioning (Schmidt et al., 2011).

As indicated by a recent metaanalysis, among the different sociocognitive subdomains, ToM is considered the most strongly associated with functional outcome, strongly influencing community functioning (Velthorst et al., 2015). This study showed that social cognition was able to explain 16% of functional variance, with neurocognition explaining only the 6%. Given the high variability of functional outcomes, in order to individuate and analyze other possible influencing factors, two different methods can be adopted: the first specifically analyzes the contribution of every subdomain of social cognition, whereas the second investigates patients' actual functioning and quality of life, starting from functional domains. Here we will examine the association between social cognition and functioning, using both methods.

## Impact of social cognitive deficits on competence and real world behavior in schizophrenia

Overall functioning represents the sum of two different functional constructs: real-world behavior (or real-world functioning: what the patient actually does) and functional capacity (patient's ability under optimal conditions) (Best et al., 2014; Bowie et al., 2008; Brune, Schaub, Juckel, & Langdon, 2011). Real-world behavior can be assessed observing actual performance of real-world activities, whereas functional capacity can be evaluated in a neutral environment, using performance-based measures concerning the ability to accomplish real-world tasks (Harvey, Velligan, & Bellack, 2007). Consistent discrepancies have been reported between functional capacity and actual functional performance, due to different factors influencing everyday life behavior (Bowie, McGurk, Mausbach, Patterson, & Harvey, 2012; Gupta, Holshausen, Mausbach, Patterson, & Bowie, 2012).

A recent study investigated the influence of neurocognition, social cognition, and psychopathology on the two functional constructs, also analyzing the relationship between real-world behavior, measured with the Quality of Life Scale (QLS) (Heinrichs, Hanlon, & Carpenter, 1984), and functional capacity, measured with the UCSD Performance-based Skills Assessment (UPSA-B) (Bechi, Bosia, et al., 2017; Mausbach et al., 2010). Consistent with the literature (Leifker et al., 2009; Menendez-Miranda et al., 2015), no correlation was observed between the two constructs of functioning. The analysis of single contributors showed that IQ was the most influencing factor of functional capacity, confirming previous evidence (McLaughlin et al., 2016). Results concerning real-life behavior, on the other hand, showed that empathy was the most influencing factor, explaining 12% of overall variance. Previous studies reported that social dysfunction in schizophrenia is associated with empathy deficit. Indeed, empathy is crucial for social cognition, including abilities such as experience sharing and mentalizing (Michaels et al., 2014; Shamay-Tsoory et al., 2007; Smith et al., 2014). This data suggests that empathy has a negative impact on social capacity, and affects daily functioning as well. As an example, erroneous empathic judgments may lead to altered social inferences, resulting in impaired school and work performance (Dodell-Feder, Tully, & Hooker, 2015).

Cognitive aspects of empathy were previously directly associated with real-life functioning (Michaels et al., 2014; Smith et al., 2012).

The study also reported emotion recognition as a predictor of real-life behavior, also showing a minor, but significant, effect of psychopathology, particularly of negative symptoms, consistent with previous literature (Bowie et al., 2008, 2010; Bowie, Reichenberg, Patterson, Heaton, & Harvey, 2006; Galderisi et al., 2013; Leifker et al., 2009; Smith, Hull, Huppert, & Silverstein, 2002; Stefanopoulou et al., 2011).

Taken together, these data suggest that neurocognition and social cognition may affect social capacity and real-world functioning differently, with social cognition more strongly related to actual functioning in daily life (Tas et al., 2013; Velthorst et al., 2015). Consistent with these findings, a study by Maat and colleagues investigated the relationship between social cognition, neurocognition, and psychopathology with quality of life in a sample of 1032 patients with schizophrenia. The authors reported the major role of social cognition in determining quality of life, with a prevalent effect of ToM compared with other social cognitive subdomains (Maat, Fett, & Derks, 2012).

## Impact of social cognitive domains on community functioning in schizophrenia

As mentioned herein, in addition to the analysis of functional capacity and real-world behavior, the effect of social cognitive subdomains on functioning is critical to understanding the complex associations between social cognition and functional outcomes. The rationale for this approach is that every social cognitive construct could lead to a distinct mediating pathway, and differentially influence functional status during the course of illness.

### Theory of Mind

Several studies reported that ToM, defined as the ability to attribute mental states to others, is impaired in schizophrenia, significantly affecting functional outcome (Biedermann, Frajo-Apor, & Hofer, 2012; Couture et al., 2006; Schmidt et al., 2011; Sprong, Schothorst, Vos, Hox, & van Engeland, 2007; Vauth et al., 2004). It was shown that ToM deficit was able to explain up to 20%–30% of functional impairment, although a high variability characterizes this data (Ventura et al., 2015). Indeed, there is a great variability in ToM performance between patients with schizophrenia generally (Bechi, Bosia, Agostoni, et al., 2018). For example, studies have found that 20%–50% of patients show normal, or close to normal, ToM abilities, despite having difficulties in social interactions, whereas other patients show significant impairments (Brune & Schaub, 2012; Lavelle, Healey, & McCabe, 2014; McCabe, Leudar, & Antaki, 2004).

Psychosocial deficit precedes acute clinical onset, and the degree of ToM impairment represents one of the best predictors of community functioning, regardless of duration of illness (Brune, 2005; Green, 2016; Harrington, Siegert, & McClure, 2005). Actually, the ability to attribute mental states to others explains 50% of the variance in social abilities in schizophrenia, independent of IQ, neurocognition, and medication (Brune, Abdel-Hamid, Lehmkamper, & Sonntag, 2007). Moreover, as evidenced by the study of Rocca and colleagues, the degree of ToM impairment is directly associated with patients' functional outcome (Rocca et al., 2016). The authors identified three clusters of patients with schizophrenia, characterized by unimpaired, impaired, and very impaired performance in a ToM task, reporting a linear relationship between daily functioning and social cognition across the three clusters. Similarly, Brüne and colleagues grouped patients into "poor" and "fair" mentalizers, showing that "fair" mentalizers presented fewer social behavioral abnormalities, as well as less behavioral disorganization (Brune & Schaub, 2012). Interestingly, no difference in an executive functioning task between groups was reported, again suggesting that social cognition and neurocognition are at least partially dissociable constructs. Similar results were more recently reported by Bechi, Bosia, Agostoni, et al. (2018), showing differences of quality of life, assessed by QLS, between "poor" and "fair" mentalizers, particularly in self-directness (Bechi, Bosia, Agostoni, et al., 2018). Moreover, when compared with healthy controls, all patients showed impairments in both neurocognition and ToM, and these deficits were predictive of functional outcome in the whole sample.

Taken together, these studies strongly indicate that ToM is crucial for both social cognition and overall functioning. Consistently, other studies showed that ToM impairment was also associated with a deficit in personal independence and hygiene (Ventura et al., 2015), and levels of school and work functioning (Brune et al., 2011).

This last data is of particular interest, because having and maintaining work can determine a wide range of clinical advantages, including less severe symptoms of disease, improved sense of recovery, strengthened self-esteem, and better quality of life (Ventura et al., 2015). ToM abilities play a central role in interacting with colleagues, and are involved in the comprehension of work tasks and instructions (Lo & Siu, 2015). Horan and colleagues reported that lower levels of baseline ToM were associated with poorer work outcomes (Horan et al., 2012). Consistently, a subsequent study found that patients with higher ToM also showed higher occupational rates (Couture et al., 2006). More recently, Bechi et al. designed a ToM rehabilitative intervention that was effective in improving ToM, and this improvement was associated with better work functioning (Bechi, Spangaro, et al., 2017). In conclusion, among patients with schizophrenia, ToM is associated with functional outcomes in several real-life domains, and constitutes a critical ability for effective social interactions (Hoe, Nakagami, Green, & Brekke, 2012; Schmidt et al., 2011; Ventura et al., 2015).

## Emotion processing

Emotions represent mental and physiological conditions associated with psychophysiological changes, determined by natural or learned stimuli. Emotion processing (EP) in self and others plays a central role within social cognition, as it is a fundamental element of social interaction. The ability to decode facial expressions represents a necessary requirement for good social functioning (Adolphs, Tranel, Damasio, & Damasio, 1994), because it permits inference of the emotional state of others, and ability to select, in a rapid and appropriate way, a behavioral response. Several studies have reported a deficit of recognition of emotions in schizophrenia (Addington & Addington, 1998; Barkl et al., 2014; Dodell-Feder, Tully, & Hooker, 2015; Fett et al., 2011; Kee, Kern, & Green, 1998; Kohler et al., 2003), which seems to be permanently present during the course of illness (Green, Horan, & Lee, 2015; Irani, Seligman, Kamath, Kohler, & Gur, 2012; Penn et al., 2000).

Particularly, research has focused on the deficit in recognition of facial emotional expressions, as it is considered a crucial aspect for functional outcome (Fett et al., 2011; Green, 2016). Patients with schizophrenia show difficulty in understanding others' emotions, with a secondary deficit in reacting adequately to the social context, and thus eventually obtaining poor results in work, interpersonal relationships, independence, and self-care (Hoe et al., 2012; Poole, Tobias, & Vinogradov, 2000). The ability to adequately interpret and react to facial expressions is important for effective interaction with others, and for active participation in the social environment (Batty & Taylor, 2003). Indeed, misunderstanding facial and emotional expressions alters the inferences made about others' emotional states, thus limiting the ability to elaborate on social hypotheses, leading to maladaptive psychosocial consequences, and eventually increasing the tendency to socially withdraw (Hooker & Park, 2002; Izard, 2001; Schultz, Izard, Ackerman, & Youngstrom, 2001).

Thus, deficits in recognizing emotional expressions in patients with schizophrenia may contribute to challenges in social and interpersonal relationships (Combs et al., 2007; Green et al., 2000; Hoe et al., 2012).

Impairment in recognizing others' emotions can lead to emotional isolation and social withdrawal (Schneider et al., 2006), thus directly affecting social functioning (Gard, Kring, Gard, Horan, & Green, 2007). Kee and colleagues evaluated EP and work functioning in a sample of psychotic patients at baseline and after 1 year, reporting a stable, significant association (Kee, Green, Mintz, & Brekke, 2003). Therefore, better EP performance could also be predictive of higher work functioning levels. Consistently, a subsequent study reported a positive correlation between correct detection of disgust and global and work functioning (Hofer et al., 2009). Disgust recognition is important, because it could be a signal of criticism concerning inadequate working behaviors. Therefore, a correct detection of this emotion can lead to more appropriate conduct.

EP has also been associated with aspects of personal functioning, such as appropriate appearance and hygiene (Kee et al., 2009).

In summary, patients with schizophrenia show an impairment in EP that negatively affects their overall functioning. In fact, EP seems to be necessary for adequate interpersonal relationships, for daily living, and for social, work, and school interactions (Bechi, Bosia, Spangaro, et al., 2018; Hoe et al., 2012; Poole et al., 2000).

## Empathy

Empathy is one of the less studied social cognitive domains in schizophrenia, but can potentially highly influence daily functioning, as proposed by different studies (Green et al., 2015; Lee, Zaki, Harvey, Ochsner, & Green, 2011; Sparks, McDonald, Lino, O'Donnell, & Green, 2010). Empathy is defined as the ability to understand and share others' feelings, and to respond emotionally. Therefore, it has been hypothesized that its impairment could negatively affect social and relational life (Blair, 2005; Henry, Bailey, & Rendell, 2008; Peuskens & Gorwood, 2012; Sabbag et al., 2011). Altered empathic evaluations may lead to aberrant SP, and thus to inadequate social behaviors and impaired functioning (Lee et al., 2011). In this view, Michaels et al. (2014) analyzed empathy abilities in a sample of 52 patients with schizophrenia, compared with a healthy control group (Michaels et al., 2014). Authors showed impaired empathy levels among psychotic subjects, associated with worse social functioning. Similar results were also reported by other studies, further supporting the initial hypothesis of a significant effect of empathy on functional outcome of subjects with schizophrenia (Horan et al., 2015; Smith et al., 2012). Particularly, Smith et al. showed that perspective taking, a cognitive subcomponent of empathy, was able to explain 15% of variability of patients' overall functioning (Smith et al., 2012). However, further studies are needed in order to better understand and define the role of empathy in daily functioning in schizophrenia.

In addition to ToM, EP, and empathy, other social cognitive domains, including AS and SP, have been found to influence functioning of patients with schizophrenia. However, to date, few studies have been made available, and those show conflicting results.

## Attributional style

A few studies have investigated the relationship between AS and functioning in schizophrenia, showing contradictory results. Lysaker, Bryson, Marks, Greig, and Bell (2004) analyzed the influence of AS and negative symptoms on social functioning, evidencing that an altered AS could be predictive of impaired social interactions, worse participation in community life and, in general, of a lower quality of life (Lysaker et al., 2004). However, the study included only a small sample of male subjects. Similar results more recently reported by Lahera and colleagues showed that altered AS, together with residual psychotic symptoms and the tendency to blame others for negative situations, is associated with impaired functioning (Lahera et al., 2015). In contrast, Mancuso and colleagues previously analyzed the structure of social cognition and its relationship with functional outcome in psychosis (Mancuso, Horan, Kern, & Green, 2011). The authors individuated three main factors: hostile AS, lower-level social cue detection (including EP and social knowledge), and higher-level inferential and regulatory processes. Analyzing the effect of these factors on functional outcome, hostile AS did not correlate with any measure of functioning. Similar negative results were also reported by Buck, Healey, Gagen, Roberts, and Penn (2016) (Buck et al., 2016). Different from the study of Mancuso, however, social cognition was explained by a two-factor model composed of social cognitive skills (including ToM, EP, and SP) and hostile AS. As noted herein, the authors did not show any association between hostile AS and functional outcome.

## Social perception

Different from AS, the study of Mancuso et al. individuated SP as one of the main predictors of functional capacity (Mancuso et al., 2011). SP refers to the ability to decode and interpret social cues in others, including social context processing and social knowledge (Addington, Saeedi, & Addington, 2006). A review by Fett and colleagues also indicated that SP was an influencing factor on community functioning, the strongest predictor after ToM (Fett et al., 2011). The authors also reported a significant association of SP performance with neurocognition. In line with these data, more recently, Karpouzian showed that patients with an adequate level of community functioning did not show any impairment of SP, despite having a ToM deficit (Karpouzian, Alden, Reilly, & Smith, 2016). The authors hypothesized that perceptual aspects of social cognition could be more related to community functioning (Couture et al., 2006); whereas a ToM deficit, persistent also among remitted subjects, could be a trait-characteristic of the illness, persisting regardless of functioning.

## Conclusions

In addition to the severe difficulties secondary to psychopathology, patients with schizophrenia experience deficits in multiple aspects of social cognition, which are strongly associated with daily functioning. Research and rehabilitative treatments in schizophrenia have been focusing on improving functioning in order to give patients the opportunity for functional recovery (Buonocore, Bosia, et al., 2018). In this context, different effective rehabilitative interventions have been developed targeting cognition, metacognition, and social cognition, showing initial encouraging results, and thus lowering the burden associated with the disease (Bechi et al., 2012, 2013, 2015a; Buonocore et al., 2015).

# Patients at UHR for psychosis: Clinical features and community functioning

The first symptoms of psychosis can be prior to the acute onset of the illness, sometimes by years (van Donkersgoed, Wunderink, Nieboer, Aleman, & Pijnenborg, 2015). Some patients show an early social functioning impairment, progressive social withdrawal, and negative symptoms (MacBeth & Gumley, 2008). Moreover, other signs that can be present before onset are subclinical self-experienced disturbances in thought, speech, and perception processes, and mild positive symptoms, such as hallucinations or delusional ideas (Schultze-Lutter, Ruhrmann, Berning, Maier, & Klosterkotter, 2010). Given their subacute manifestation, these symptoms are usually defined as attenuated psychotic symptoms (APS). Subjects with functional impairment, APS, or brief limited intermittent psychotic symptoms (BLIPS) and family history of schizophrenia are considered UHR individuals (Yung & McGorry, 2007). UHR subjects are also recognizable due to the presence of basic symptoms (BS). BS represent mild alterations in affect, speech, perceptions, thought, or stress tolerance, and are usually detectable before ALS and BLIPS (Bora et al., 2014).

The study of UHR individuals allows identification of specific risk factors for the development of psychosis and mechanisms of symptom evolution, without the confounding effects of medication and acute and chronic illness factors. Data indicate that among UHR subjects, the transition rate to psychosis is 22% after 1 year, and 36% after 3 years follow-up (Fusar-Poli et al., 2012). However, there is huge outcome variability, also due to methodologic issues concerning diagnosis. Long-term clinical outcomes range from treatment-resistant psychosis, to full functional and symptomatic remission, and the degree of functional impairment represents one of the best outcome predictors.

Comparing UHR subjects with healthy controls, Fusar-Poli et al. (2012) showed that the two groups significantly differed on baseline unemployment and independent living. Moreover, fewer UHR individuals had previously lived alone, depending on family and needing support for instrumental role functioning (e.g., employment) and independent living (Fusar-Poli et al., 2012). Interestingly, living in a communal structure and baseline unemployment were associated with a higher risk of developing a psychotic episode within the following 2 years.

However, even in patients who did not convert to psychosis, Brandizzi and colleagues (Brandizzi et al., 2015) found that 43% of nonconverters reported poor functioning at 6 years follow-up. Interestingly, functional outcome was predicted by baseline functioning, employment status, and intensity of prodromal psychotic symptoms. The observation of a significant impairment in psychosocial functioning among UHR subjects, despite the absence of psychotic symptoms, further supports the need of early targeted interventions independent from longitudinal outcome (transition/nontransition).

## Predictors of community functioning in UHR subjects

It is well established that social cognitive deficits constitute the main functional predictors of patients with schizophrenia. Therefore, the role of social cognitive deficits in UHR and their impact on daily functioning and transition to psychosis is of great interest. Additionally, duration of untreated prodromal phase has been inversely associated with long-term prognosis. In order to optimize the detection of people at risk of psychosis, different markers have been studied, helping to better define clinical and biological features of the disease.

While a number of studies have examined neurocognition in UHR and found that it is a significant predictor of outcomes, as in patients with schizophrenia (Cotter et al., 2014), fewer studies have investigated social cognition among UHR subjects, reporting an intermediate degree of impairment between schizophrenia and healthy controls. Specifically, UHR subjects exhibit impairments in the AS domain, associated with the degree of paranoid symptoms (Thompson, Bartholomeusz, & Yung, 2011), and ToM deficits are detectable among UHR converters (Lee, Hong, Shin, & Kwon, 2015).

In addition to representing the most frequently impaired abilities among UHR individuals, ToM and AS are the best functional outcome predictors as well. Indeed, Cotter (Cotter et al., 2017), analyzing possible influences of social cognitive abilities previously linked to functioning in schizophrenia, found an association of ToM with functional outcome among UHR subjects as well. A subsequent study including 65 UHR subjects and 30 healthy controls analyzed social cognitive function, social skills, and a broad range of functioning measures. UHR patients showed significant decrements on a ToM task and on an EP task, which were associated with role functioning and social skill performance. Overall functioning was associated with the level of AS, whereas ToM was associated with self-reported functioning (Glenthoj et al., 2016). As previously mentioned, social cognitive abilities among UHR individuals are less impaired than those of patients with schizophrenia. In line with this evidence, Ohmuro and colleagues (Ohmuro et al., 2016) recently showed that, with respect to schizophrenia, the strength of the association between ToM deficit and functional outcome was less consistent among FEP and UHR subjects. As noted herein, in patients with schizophrenia, social cognition mediates the relationship between neurocognition and functioning. This association has been examined in an UHR sample. Barbato et al. (2013) reported that social cognition failed to represent a mediator between neurocognition and functioning, different from what is observed in schizophrenia (Barbato et al., 2013). However, in this model, both neurocognition and social cognition were associated with functional outcome. The authors hypothesized that this discrepancy could be due to the minor degree of impairment of social abilities in UHR subjects. This may be due to the fact that in this sample, only 30% of UHR individuals subsequently converted to psychosis (Fusar-Poli et al., 2012).

Given that the premorbid functional level seems to be a good predictor of future transition to psychosis, studies have investigated the role of social cognition and neurocognition as well. Kim et al. (2011), in a longitudinal study, found that UHR converters were more impaired in social cognition (SC) than nonconverters and healthy subjects (Kim et al., 2011) at baseline. Moreover, authors also developed a significant model of transition to psychosis, including social cognitive and neurocognitive variables as predictors. Similar data were subsequently reported by Zhang et al. (2017), who found that UHR converters showed a stronger association between social cognition and neurocognition than nonconverters (Zhang et al., 2017). Moreover, after conversion to psychosis, the authors found a stronger association between the two cognitive domains, thus suggesting that the degree of association of social and neurocognitive abilities could represent a marker of transition to psychosis.

Despite these intriguing findings, more studies are needed to clearly define the role of social cognitive domains in the functioning of UHR. Development in this research area, particularly through long-term, large-scale longitudinal studies, would allow researchers to optimize the detection of people at risk of psychosis, and to eventually reduce the rate of conversion by enhancing protective functional predictors.

## First-episode psychosis: Social cognitive and functional features

FEP patients represent one of the most interesting targets for investigating the impact of social cognition on daily functioning. The study of cognition-outcome interrelations in FEP may help us better understand the trajectories of cognition and psychosis in general, and may also be helpful in developing effective interventions at various stages of these disorders (Bora, Yucel, & Pantelis, 2010a, 2010b; McGorry et al., 2010).

FEP patients show psychopathological features similar to chronic psychosis. Impairment of social and functional abilities is one of the main early manifestations of the illness (Birchwood, Todd, & Jackson, 1998; Grant, Addington, Addington, & Konnert, 2001). FEP subjects often experience significant difficulties creating and maintaining close relationships, and need financial support from parents or public assistance (Gillberg, Hellgren, & Gillberg, 1993; Lay, Blanz, Hartmann, & Schmidt, 2000; Vyas, Hadjulis, Vourdas, Byrne, & Frangou, 2007). Disruption of social functioning predates the first hospitalization (Horan, Subotnik, Snyder, & Nuechterlein, 2006), and a large part of psychosocial deterioration occurs during the first 5 years of illness (Green, Kern, & Heaton, 2004). Therefore, given the crucial nature of this stage of illness, it is important to emphasize the need of early pharmacological and psychosocial treatments. Consistently, it was reported that early interventions have a stronger impact than treatments provided later in the course of illness (Miller et al., 2002).

As noted herein, neurocognitive and social cognitive deficits are detectable even before disease onset. However, despite a recent increase in the study of social cognition in FEP, it is still not clear how social cognitive deficits evolve during the course of psychotic illness. Specifically, it is not well established how social cognitive impairment progresses in the period of time before and after disease onset. A recent review including 48 studies showed consistent social cognitive deficits among FEP patients compared with healthy controls, particularly in EP (fear and sadness recognition) and ToM, rather than SP and AS (Healey, Bartholomeusz, & Penn, 2016), and that degree of impairment was similar between patients with chronic schizophrenia and FEP. Moreover, social cognitive impairment was stable over time, and associated with negative and positive symptoms.

## Impact of social cognitive deficits on community functioning in FEP patients

The study of community functioning of FEP subjects shows prognoses ranging from full functional recovery to progressive functional decline, consistent with FEP clinical course; whereas chronic patients with psychosis show a smaller range of poorer prognoses. Therefore, the analysis of FEP functional predictors may be useful in the prediction of functional decline and functional recovery and resilience (Menezes, Arenovich, & Zipursky, 2006).

Several studies have examined the influence of social cognition on the functioning of FEP subjects, with conflicting results. Addington proposed a model to explain the functioning of psychotic patients, including FEP, that explained 79.7% of the variance in social function, and demonstrated that the link between cognition and social function was fully mediated by social cognition (Addington & Piskulic, 2011). Despite the high heterogeneity of the sample, as in schizophrenia, this study further supported the concept of social cognition as a mediator between neurocognition and daily functioning. Similarly, Stouten and colleagues analyzed the influence of psychopathology, neurocognition, and social cognition on psychosocial functioning in 153 nonaffective FEP patients (Stouten, Veling, Laan, Van der Helm, & Van der Gaag, 2014). Assessments were performed at baseline and after 1 year. They reported that at baseline, psychosocial functioning was primarily associated with positive and negative symptoms. Conversely, greater ToM abilities at baseline were associated with a worsening of disturbing behavior. However, variation of social functioning was predicted by both psychopathology and social and neurocognitive deficits. Data suggested that cognitive impairment could represent a more accurate longitudinal predictor of psychosocial problems and functional recovery in the early course of psychosis. This hypothesis is also consistent with previous literature (Allott, Liu, Proffitt, & Killackey, 2011) indicating different neurocognitive (Leeson, Barnes, Hutton, Ron, & Joyce, 2009; Nuechterlein et al., 2011; Van Winkel et al., 2007) and social cognitive domains (Horan et al., 2012) as valid longitudinal predictors of psychosocial functioning and/or functional change in FEP patients.

Contrary to this hypothesis, a recent study of Woolverton (Woolverton, Bell, Moe, Harrison-Monroe, & Breitborde, 2017) found no longitudinal relationships between social cognition and psychosocial functioning course among 71 patients with FEP. In this study, the authors analyzed the possible influence of five domains of social cognition relevant to psychotic disorders on social functioning at baseline and after 1 year, reporting only modest cross-sectional relationships. These unexpected findings fail to align with previous research that has documented a more robust relationship between these two constructs (Couture et al., 2006), and raises critical questions with regard to the nature of the association between social cognition and social functioning among individuals with FEP (Fett et al., 2011). However, this was not the first study to suggest a more complicated relationship between social cognition and social functioning. In fact, Stouten and colleagues found that greater social knowledge at baseline among individuals with FEP was associated with improvements in self-care behaviors over a 12-month follow-up period. Conversely, greater ToM abilities were associated with a worsening of disturbing behavior over the same follow-up period (Stouten, Veling, Laan, van der Helm, & van der Gaag, 2014).

Taken together, these data suggest that the relationship between social cognition and social functioning is not as linear as previously assumed (Harvey & Penn, 2010), raising questions concerning the direction of this association among individuals with psychosis.

## Bipolar disorder: Clinical features and functional outcomes

Patients affected by BD experience moderate to severe functional impairment in several domains such as work, social activities, and family relationships (Coryell et al., 1993; Judd et al., 2008; Keck, 2006; MacQueen, Young, & Joffe, 2001; Sanchez-Moreno et al., 2009; Wingo, Harvey, & Baldessarini, 2009). Functional deficits are present even during periods of sustained and substantial remission (Jaeger & Vieta, 2007), and are associated with poor quality of life and reduced self-esteem, also contributing to an increased suicide risk.

There is a consistent gap between clinical remission and functionality. The old conception in which patients were able to return to their normal life after symptom remission is now abandoned. Given the persistence of psychosocial deterioration during euthymic phases, functional deficit is considered to be independent from depressive symptomatology and from clinical remission, with only 40% of patients recovering their premorbid levels (DelBello, Hanseman, Adler, Fleck, & Strakowski, 2007). Therefore, individuation of functional predictors represents one of the main goals of BD research.

Several clinical and psychopathological variables, such as comorbid substance abuse (Tohen, Greenfield, Weiss, Zarate, & Vagge, 1998), treatment adverse effects, presence of psychotic symptoms, low premorbid functioning (Zarate, Tohen, Land, & Cavanagh, 2000), recurrence of episodes (MacQueen et al., 2001), and early onset (Tohen, Jacobs, & Feldman, 2000) have been associated with worse psychosocial outcomes. In addition to associations with symptom and treatment-related factors, among BD subjects, functional impairment was directly related to cognitive deficits, particularly of executive functions, attention, and verbal memory (Dickerson et al., 2004; Martinez-Aran et al., 2007). In a recent study, Jiménez-López et al. (2018) investigated possible differences in psychosocial functioning between BD patients with or without a history of episodes with psychotic symptoms (Jiménez-López et al., 2018). The authors found no differences between groups, also reporting that neurocognitive dysfunction and subclinical depressive symptoms were the best functional predictors of the whole sample. In line with this data, a recent large-scale review analyzed 92 cross-sectional and longitudinal studies on functional impairment and its relationship to symptomatic, neurocognitive, personality, and stress variables in BD. Results suggested that functional recovery after a mood episode follows a different course than symptomatic and syndromal recovery. Longer term functional impairment is only partly explained by the number of manic/hypomanic episodes. Cognitive impairment and subsyndromal states were the strongest correlates of functional impairment in BD, with personality and psychosocial stressors playing secondary roles. In this study, social cognitive deficits are not included in the analyses of the possible contributors of functional impairment (Gitlin & Miklowitz, 2017).

## Social cognition and community outcomes in BD

Similar to schizophrenia, BD is characterized by functional and cognitive deficits (Bora, Yücel, & Pantelis, 2009; Tabarés-Seisdedos et al., 2008). However, the degree of impairment of social cognition in BD appears to be more modest relative to schizophrenia. Whereas social cognitive deficit represents a main feature of schizophrenia, its presence in BD is not widely recognized.

Lee et al. compared social cognitive (facial affect perception, emotional regulation, empathic accuracy, mental state attribution, and self-referential memory) and neurocognitive (speed of processing, attention/vigilance, working memory, verbal memory, visual memory, and reasoning/problem solving) performance of stabilized patients with BD, schizophrenia, and healthy controls (Lee et al., 2013). BD patients showed both neurocognitive and sociocognitive deficits compared with healthy controls; however, deficits were less severe in patients with BD than in patients with schizophrenia. Interestingly, the study reported that among BD subjects, neurocognitive deficit was higher than social cognitive deficit, observing an opposite pattern among patients with schizophrenia. This observation further indicates a partial independence of the two cognitive domains, also suggesting that neurocognitive deficit does not always implicate a disruption in social cognition. The authors hypothesized that BD patients could have compensatory mechanisms protecting social cognitive circuits, which are missing in schizophrenia. In a recent metaanalysis (Bora, Bartholomeusz, & Pantelis, 2016), the authors found a medium-level impairment of ToM among bipolar patients compared with healthy controls, with more pronounced deficits during acute episodes. Interestingly, ToM deficits were independent from psychotic features of manic episodes. The persistence of a social cognitive impairment during euthymia contrasts with the results of Lee and colleagues, and further suggests that social cognition should represent a main target of research in order to thoroughly comprehend functional outcome in BD.

Given the central role of social cognition in functional outcomes in schizophrenia and the conflicting findings regarding the nature of social cognitive functioning in BD, the relationship between social cognition and functional outcome among patients with BD has been examined in several social cognitive domains. Some evidence suggests that social cognition, particularly ToM, significantly predicts impaired functioning in euthymic bipolar patients (Bora et al., 2016). Even studies that failed to find a significant association reported that both ToM abilities and overall functioning were impaired among BP patients, when compared with healthy controls (Barrera, Vazquez, Tannenhaus, Lolich, & Herbst, 2013).

In sum, while a growing body of research has suggested ToM deficits across mood phases in BD, how these impairments may affect life functioning remains unclear. More research is needed to not only better define ToM deficits in BD, but also to understand the clinical importance of ToM performance on life functioning. Inconsistencies between studies could be due to the effect of other influencing factors, such as residual depressive symptoms (subsyndromal), as suggested by Konstantakopoulos and colleagues (Konstantakopoulos, Ioannidi, Typaldou, Sakkas, & Oulis, 2016). In this study, the authors showed that residual depressive symptoms and social cognitive deficits negatively affected psychosocial functioning of remitted BD patients, highlighting a central role of ToM as a mediator between clinical and cognitive variables and functioning. These data suggest that social cognitive deficit, although associated with different features according to the disease, affect functioning in both schizophrenia and BD in a similar way.

In addition to ToM, to date, few studies have investigated whether other components of social cognition are related to psychosocial functioning in BD. Bipolar patients are characterized by difficulties in identifying emotional expressions and conceptualizing others' mental states (Addington & Addington, 1998; Bozikas et al., 2006, 2007; Getz, Shear, & Strakowski, 2003; Lembke & Ketter, 2002). Two different studies among BD patients found that facial EP ability was related to psychosocial functioning

(Martino, Strejilevich, Fassi, Marengo, & Igoa, 2011), lower ratings of depression, and greater quality of life (Hoertnagl et al., 2011), also reporting a trend level relationship with general functioning. There were also reports of a trend level relationship between EP accuracy and higher rates of employment and general functioning. Although preliminary, and from small samples, these findings support the role of emotion regulation in influencing psychosocial function in BD, which is worthy of future research.

Another social cognitive domain of interest that has been poorly investigated is AS, which shows a potentially relevant association with psychosocial functioning in BD. Lahera et al. found that patients with schizophrenia and BD showed similar AS in ambiguous situations, with a tendency to interpret hostility (Lahera et al., 2015). The authors also reported similar EP impairments between patient groups, whereas ToM was more disrupted among subjects with schizophrenia. Interestingly, among BD subjects, AS was associated with subsyndromal depressive symptoms, suggesting that misperception of social situations is negatively influenced by psychopathology. Particularly, subthreshold depressive symptoms were associated with hostile intent attribution, anger, blame attribution, and aggressiveness, but not with other domains of social cognition. It is conceivable that in order to understand others' emotions, one's own emotional system needs to be preserved, without influences of attenuated depressive symptoms. This hypothesis is also consistent with recent studies showing a lack of awareness in patients with varying degrees of depression.

Taken together, evidence indicates that social cognition plays a significant role in global functioning in BD. Relative to schizophrenia, other influencing factors such as demographic, clinical, and neurocognitive variables seem to have a greater effect on functional outcome. Therefore, indicators of neurocognitive and sociocognitive abilities, and clinical variables, should be included in complex models in order to explain the clinical-functional gap of patients with BD.

## Conclusions

Patients affected by SMI show marked difficulties in achieving social and functional goals necessary to be fully integrated in society. Social cognitive deficits may further worsen social withdrawal associated with SMI. Therefore, social stigma results from people's prejudices and patients' loss of self-esteem, and apprehension regarding the external environment.

The resulting psychosocial deficits lead to low quality of life for both patients and their families, requiring considerable efforts and financial support from caregivers, and increasing the health costs related to the disease.

Given these premises, improving patients' functioning and quality of life represents a main target in treating SMI, especially considering that pharmacological treatments are not effective in this regard.

As previously described, functional impairment is only partially associated with psychopathology, but is strongly influenced by social cognitive and neurocognitive abilities, known to be impaired in SMI. Therefore, in order to better understand the disease, neurocognition and social cognition became main targets of research.

In particular, social cognition, especially ToM and EP, predicts a large part of functional impairment variability in schizophrenia. However, functioning is influenced by both cognition and environment, and the study of the effect of other sociocognitive subdomains can lead to a better comprehension of the processes that negatively affect functional outcome (Fett et al., 2011).

In addition to chronic schizophrenia, functional outcome plays a central role during the early and prodromal stages of the disease. Indeed, among UHR subjects, the level of functioning represents one of the best predictors of conversion to psychosis (Kim et al., 2011), and is considered a clinical outcome indicator in FEP as well. Both UHR and FEP individuals are characterized by neurocognitive and sociocognitive deficits influencing functional outcome, although less strongly than in schizophrenia (Ohmuro et al., 2016).

Functional deficits are detectable also in BD, even during euthymia and phases of symptomatic remission (Jaeger & Vieta, 2007). Moreover, impaired levels of functioning are associated sociocognitive disabilities, along with low self-esteem, social withdrawal, unemployment, and higher suicide risk (Bora et al., 2016).

Based on these data, rehabilitative interventions targeting neurocognition (i.e., cognitive remediation therapy—CRT) and social cognition (i.e., metalizing training) have recently been developed (Cavallaro et al., 2009; Bechi et al., 2015; Kurtz, Gagen, Rocha, Machado, & Penn, 2016), with the aim to improve patients' functional outcome. Improvement of neurocognitive abilities after CRT has been associated with better real-word functioning in schizophrenia (Medalia & Saperstein, 2013). Interestingly, these improvements were stable after 5 years when CRT was combined with standard rehabilitation (Buonocore, Bosia, et al., 2018; Buonocore, Spangaro, et al., 2018).

More recently, sociocognitive training has been developed, which is effective in improving both social cognition and functional domains such as work and interpersonal relationships (Buonocore et al., 2017). When delivered during the early or prodromal stages of illness, integrated neurocognitive and sociocognitive intervention can delay the disease onset and reduce the subsequent functional decline. However, complex relationships among social cognition, neurocognition, and clinical features appear to exist, and each on its own is only able to partially explain functional outcome variability in SMI. Therefore, a thorough study of functional outcome and its predictors is needed, in order to allow patients to achieve full functional recovery.

## References

Addington, J., & Addington, D. (1998). Facial affect recognition and information processing in schizophrenia and bipolar disorder. *Schizophrenia Research*, *32*(3), 171–181. S0920996498000425 [pii].

Addington, J., & Piskulic, D. (2011). Social cognition and functional outcome are separate domains in schizophrenia. *Schizophrenia Research*, *127*(1–3), 262–263. https://doi.org/10.1016/j.schres.2010.04.005. S0920-9964(10)01258-2 [pii].

Addington, J., Saeedi, H., & Addington, D. (2006). Influence of social perception and social knowledge on cognitive and social functioning in early psychosis. *The British Journal of Psychiatry*, *189*(4), 373–378.

Adolphs, R., Tranel, D., Damasio, H., & Damasio, A. (1994). Impaired recognition of emotion in facial expressions following bilateral damage to the human amygdala. *Nature*, *372* (6507), 669–672. https://doi.org/10.1038/372669a0.

Allott, K., Liu, P., Proffitt, T. M., & Killackey, E. (2011). Cognition at illness onset as a predictor of later functional outcome in early psychosis: systematic review and methodological critique. *Schizophrenia Research*, *125*(2–3), 221–235. https://doi.org/10.1016/j.schres.2010.11.001. S0920-9964(10)01630-0 [pii].

Barbato, M., Liu, L., Penn, D. L., Keefe, R. S., Perkins, D. O., Woods, S. W., & Addington, J. (2013). Social cognition as a mediator between neurocognition and functional outcome in individuals at clinical high risk for psychosis. *Schizophrenia Research*, *150*(2–3), 542–546. https://doi.org/10.1016/j.schres.2013.08.015. S0920-9964(13)00451-9 [pii].

Barkl, S. J., Lah, S., Starling, J., Hainsworth, C., Harris, A. W., & Williams, L. M. (2014). Facial emotion identification in early-onset psychosis. *Schizophrenia Research*, *160*(1–3), 150–156. https://doi.org/10.1016/j.schres.2014.10.035. S0920-9964(14)00604-5 [pii].

Barrera, A., Vazquez, G., Tannenhaus, L., Lolich, M., & Herbst, L. (2013). Theory of mind and functionality in bipolar patients with symptomatic remission. *Revista de Psiquiatría y Salud Mental*, *6*(2), 67–74. https://doi.org/10.1016/j.rpsm.2012.07.004. S1888-9891(12)00151-6 [pii].

Batty, M., & Taylor, M. J. (2003). Early processing of the six basic facial emotional expressions. *Cognitive Brain Research*, *17*(3), 613–620.

Bechi, M., Bosia, M., Agostoni, G., Spangaro, M., Buonocore, M., Bianchi, L., … Cavallaro, R. (2018). Can patients with schizophrenia have good mentalizing skills? Disentangling heterogeneity of theory of mind. *Neuropsychology*, *32*(6), 746–753. https://doi.org/10.1037/neu0000456.

Bechi, M., Bosia, M., Spangaro, M., Buonocore, M., Cavedoni, S., Agostoni, G., ... Smeraldi, E. (2017). Exploring functioning in schizophrenia: predictors of functional capacity and real-world behaviour. *Psychiatry Research, 251*, 118–124.

Bechi, M., Bosia, M., Spangaro, M., Buonocore, M., Cocchi, F., Pigoni, A., ... Smeraldi, E. (2015). Combined social cognitive and neurocognitive rehabilitation strategies in schizophrenia: neuropsychological and psychopathological influences on Theory of Mind improvement. *Psychological Medicine, 45*(15), 3147–3157. https://doi.org/10.1017/S0033291715001129. S0033291715001129 [pii].

Bechi, M., Bosia, M., Spangaro, M., Pigoni, A., Buonocore, M., Scrofani, D., ... Smeraldi, E. (2018). Visual and audio emotion processing training for outpatients with schizophrenia: an integrated multisensory approach. *Neuropsychological Rehabilitation, 28*(7), 1131–1144.

Bechi, M., Riccaboni, R., Ali, S., Fresi, F., Buonocore, M., Bosia, M., ... Cavallaro, R. (2012). Theory of mind and emotion processing training for patients with schizophrenia: preliminary findings. *Psychiatry Research, 198*(3), 371–377.

Bechi, M., Spangaro, M., Agostoni, G., Bosinelli, F., Buonocore, M., Bianchi, L., ... Cavallaro, R. (2018). Intellectual and cognitive profiles in patients affected by schizophrenia. *Journal of Neuropsychology*. https://doi.org/10.1111/jnp.12161.

Bechi, M., Spangaro, M., Bosia, M., Zanoletti, A., Fresi, F., Buonocore, M., ... Cavallaro, R. (2013). Theory of Mind intervention for outpatients with schizophrenia. *Neuropsychological Rehabilitation, 23*(3), 383–400.

Bechi, M., Spangaro, M., Pigoni, A., Ripamonti, E., Buonocore, M., Cocchi, F., ... Cavallaro, R. (2017). Exploring predictors of work competence in schizophrenia: the role of theory of mind. *Neuropsychological Rehabilitation*, 1–13.

Becker, J. T., & Milke, R. M. (1998). Cognition and aging in a complex work environment: relationships with performance among air traffic control specialists. *Aviation, Space, and Environmental Medicine, 69*(10), 944–951.

Best, M. W., Gupta, M., Bowie, C. R., & Harvey, P. D. (2014). A longitudinal examination of the moderating effects of symptoms on the relationship between functional competence and real world functional performance in schizophrenia. *Schizophrenia Research: Cognition, 1*(2), 90–95. https://doi.org/10.1016/j.scog.2014.03.002.

Bevan, S., Gulliford, J., Steadman, K., Taskila, T., Thomas, R., & Moise, A. (2013). *Working with schizophrenia: Pathways to employment, recovery and inclusion*. The Work Foundation.

Biedermann, F., Frajo-Apor, B., & Hofer, A. (2012). Theory of mind and its relevance in schizophrenia. *Current Opinion in Psychiatry, 25*(2), 71–75. https://doi.org/10.1097/YCO.0b013e3283503624.

Birchwood, M., Todd, P., & Jackson, C. (1998). Early intervention in psychosis. The critical period hypothesis. *The British Journal of Psychiatry. Supplement, 172*(33), 53–59.

Blair, R. J. (2005). Responding to the emotions of others: dissociating forms of empathy through the study of typical and psychiatric populations. *Consciousness and Cognition, 14*(4), 698–718. S1053-8100(05)00085-1 [pii] https://doi.org/10.1016/j.concog.2005.06.004.

Bora, E., Bartholomeusz, C., & Pantelis, C. (2016). Meta-analysis of Theory of Mind (ToM) impairment in bipolar disorder. *Psychological Medicine, 46*(2), 253–264.

Bora, E., Lin, A., Wood, S. J., Yung, A. R., McGorry, P. D., & Pantelis, C. (2014). Cognitive deficits in youth with familial and clinical high risk to psychosis: a systematic review and meta-analysis. *Acta Psychiatrica Scandinavica, 130*(1), 1–15. https://doi.org/10.1111/acps.12261.

Bora, E., & Pantelis, C. (2013). Theory of mind impairments in first-episode psychosis, individuals at ultra-high risk for psychosis and in first-degree relatives of schizophrenia: systematic review and meta-analysis. *Schizophrenia Research, 144*(1–3), 31–36. https://doi.org/10.1016/j.schres.2012.12.013. S0920-9964(12)00706-2 [pii].

Bora, E., Yucel, M., & Pantelis, C. (2010a). Cognitive impairment in schizophrenia and affective psychoses: implications for DSM-V criteria and beyond. *Schizophrenia Bulletin, 36*(1), 36–42. https://doi.org/10.1093/schbul/sbp094. sbp094 [pii].
Bora, E., Yucel, M., & Pantelis, C. (2010b). Neurocognitive markers of psychosis in bipolar disorder: a meta-analytic study. *Journal of Affective Disorders, 127*(1–3), 1–9. https://doi.org/10.1016/j.jad.2010.02.117. S0165-0327(10)00238-7 [pii].
Bora, E., Yücel, M., & Pantelis, C. (2009). Theory of mind impairment: a distinct trait-marker for schizophrenia spectrum disorders and bipolar disorder? *Acta Psychiatrica Scandinavica, 120*(4), 253–264.
Bowie, C. R., Depp, C., McGrath, J. A., Wolyniec, P., Mausbach, B. T., Thornquist, M. H., … Pulver, A. E. (2010). Prediction of real-world functional disability in chronic mental disorders: a comparison of schizophrenia and bipolar disorder. *The American Journal of Psychiatry, 167*(9), 1116–1124. https://doi.org/10.1176/appi.ajp.2010.09101406. appi.ajp.2010.09101406 [pii].
Bowie, C. R., Leung, W. W., Reichenberg, A., McClure, M. M., Patterson, T. L., Heaton, R. K., & Harvey, P. D. (2008). Predicting schizophrenia patients' real-world behavior with specific neuropsychological and functional capacity measures. *Biological Psychiatry, 63*(5), 505–511. S0006-3223(07)00506-9 [pii] https://doi.org/10.1016/j.biopsych.2007.05.022.
Bowie, C. R., McGurk, S. R., Mausbach, B., Patterson, T. L., & Harvey, P. D. (2012). Combined cognitive remediation and functional skills training for schizophrenia: effects on cognition, functional competence, and real-world behavior. *The American Journal of Psychiatry, 169*(7), 710–718. https://doi.org/10.1176/appi.ajp.2012.11091337. 1148700 [pii].
Bowie, C. R., Reichenberg, A., Patterson, T. L., Heaton, R. K., & Harvey, P. D. (2006). Determinants of real-world functional performance in schizophrenia subjects: correlations with cognition, functional capacity, and symptoms. *The American Journal of Psychiatry, 163*(3), 418–425. 163/3/418 [pii] https://doi.org/10.1176/appi.ajp.163.3.418.
Bozikas, V. P., Kosmidis, M. H., Tonia, T., Andreou, C., Focas, K., & Karavatos, A. (2007). Impaired perception of affective prosody in remitted patients with bipolar disorder. *Journal of Neuropsychiatry and Clinical Neuroscience, 19*(4), 436–440. https://doi.org/10.1176/jnp.2007.19.4.436.
Bozikas, V. P., Tonia, T., Fokas, K., Karavatos, A., & Kosmidis, M. H. (2006). Impaired emotion processing in remitted patients with bipolar disorder. *Journal of Affective Disorder, 91*(1), 53–56. https://doi.org/10.1016/j.jad.2005.11.013.
Brandizzi, M., Valmaggia, L., Byrne, M., Jones, C., Iwegbu, N., Badger, S., … Fusar-Poli, P. (2015). Predictors of functional outcome in individuals at high clinical risk for psychosis at six years follow-up. *Journal of Psychiatric Research, 65*, 115–123. https://doi.org/10.1016/j.jpsychires.2015.03.005. S0022-3956(15)00074-6 [pii].
Breier, A., Schreiber, J. L., Dyer, J., & Pickar, D. (1991). National Institute of Mental Health longitudinal study of chronic schizophrenia. Prognosis and predictors of outcome. *Archives of General Psychiatry, 48*(3), 239–246.
Brekke, J. S., Hoe, M., Long, J., & Green, M. F. (2007). How neurocognition and social cognition influence functional change during community-based psychosocial rehabilitation for individuals with schizophrenia. *Schizophrenia Bulletin, 33*(5), 1247–1256. sbl072 [pii] https://doi.org/10.1093/schbul/sbl072.
Brune, M. (2005). "Theory of mind" in schizophrenia: a review of the literature. *Schizophrenia Bulletin, 31*(1), 21–42. sbi002 [pii] https://doi.org/10.1093/schbul/sbi002.
Brune, M., Abdel-Hamid, M., Lehmkamper, C., & Sonntag, C. (2007). Mental state attribution, neurocognitive functioning, and psychopathology: what predicts poor social competence in schizophrenia best? *Schizophrenia Research, 92*(1–3), 151–159. S0920-9964(07)00058-8 [pii] https://doi.org/10.1016/j.schres.2007.01.006.

Brune, M., & Schaub, D. (2012). Mental state attribution in schizophrenia: what distinguishes patients with "poor" from patients with "fair" mentalising skills? *European Psychiatry, 27*(5), 358–364. https://doi.org/10.1016/j.eurpsy.2010.10.002. S0924-9338(10)00201-4 [pii].

Brune, M., Schaub, D., Juckel, G., & Langdon, R. (2011). Social skills and behavioral problems in schizophrenia: the role of mental state attribution, neurocognition and clinical symptomatology. *Psychiatry Research, 190*(1), 9–17. https://doi.org/10.1016/j.psychres.2010.03.015. S0165-1781(10)00114-9 [pii].

Buck, B. E., Healey, K. M., Gagen, E. C., Roberts, D. L., & Penn, D. L. (2016). Social cognition in schizophrenia: factor structure, clinical and functional correlates. *Journal of Mental Health, 25*(4), 330–337. https://doi.org/10.3109/09638237.2015.1124397.

Buonocore, M., Bosia, M., Baraldi, M. A., Bechi, M., Spangaro, M., Cocchi, F., … Cavallaro, R. (2018). Achieving recovery in patients with schizophrenia through psychosocial interventions: a retrospective study. *Psychiatry and Clinical Neurosciences, 72*(1), 28–34.

Buonocore, M., Bosia, M., Bechi, M., Spangaro, M., Cavedoni, S., Cocchi, F., … Cavallaro, R. (2017). Targeting anxiety to improve quality of life in patients with schizophrenia. *European Psychiatry, 45*, 129–135. https://doi.org/10.1016/j.eurpsy.2017.06.014.

Buonocore, M., Bosia, M., Riccaboni, R., Bechi, M., Spangaro, M., Piantanida, M., … Smeraldi, E. (2015). Combined neurocognitive and metacognitive rehabilitation in schizophrenia: effects on bias against disconfirmatory evidence. *European Psychiatry, 30*(5), 615–621.

Buonocore, M., Spangaro, M., Bechi, M., Baraldi, M. A., Cocchi, F., Guglielmino, C., … Cavallaro, R. (2018). Integrated cognitive remediation and standard rehabilitation therapy in patients of schizophrenia: persistence after 5 years. *Schizophrenia Research, 192*, 335–339.

Cavallaro, R., Anselmetti, S., Poletti, S., Bechi, M., Ermoli, E., Cocchi, F., … Smeraldi, E. (2009). Computer-aided neurocognitive remediation as an enhancing strategy for schizophrenia rehabilitation. *Psychiatry Research, 169*(3), 191–196. https://doi.org/10.1016/j.psychres.2008.06.027. S0165-1781(08)00197-2 [pii].

Combs, D. R., Adams, S. D., Penn, D. L., Roberts, D., Tiegreen, J., & Stem, P. (2007). Social Cognition and Interaction Training (SCIT) for inpatients with schizophrenia spectrum disorders: preliminary findings. *Schizophrenia Research, 91*(1–3), 112–116. S0920-9964(06)00517-2 [pii] https://doi.org/10.1016/j.schres.2006.12.010.

Coryell, W., Scheftner, W., Keller, M., Endicott, J., Maser, J., & Klerman, G. L. (1993). The enduring psychosocial consequences of mania and depression. *The American Journal of Psychiatry, 150*(5), 720–727. https://doi.org/10.1176/ajp.150.5.720.

Cotter, J., Bartholomeusz, C., Papas, A., Allott, K., Nelson, B., Yung, A. R., & Thompson, A. (2017). Examining the association between social cognition and functioning in individuals at ultra-high risk for psychosis. *The Australian and New Zealand Journal of Psychiatry, 51*(1), 83–92. https://doi.org/10.1177/0004867415622691. 0004867415622691 [pii].

Cotter, J., Drake, R. J., Bucci, S., Firth, J., Edge, D., & Yung, A. R. (2014). What drives poor functioning in the at-risk mental state? A systematic review. *Schizophrenia Research, 159*(2–3), 267–277. https://doi.org/10.1016/j.schres.2014.09.012. S0920-9964(14)00478-2 [pii].

Couture, S. M., Penn, D. L., & Roberts, D. L. (2006). The functional significance of social cognition in schizophrenia: a review. *Schizophrenia Bulletin, 32*(Suppl. 1), S44–S63. sbl029 [pii] https://doi.org/10.1093/schbul/sbl029.

de Achaval, D., Costanzo, E. Y., Villarreal, M., Jauregui, I. O., Chiodi, A., Castro, M. N., … Guinjoan, S. M. (2010). Emotion processing and theory of mind in schizophrenia patients and their unaffected first-degree relatives. *Neuropsychologia, 48*(5), 1209–1215. https://doi.org/10.1016/j.neuropsychologia.2009.12.019. S0028-3932(09)00495-3 [pii].

DelBello, M. P., Hanseman, D., Adler, C. M., Fleck, D. E., & Strakowski, S. M. (2007). Twelve-month outcome of adolescents with bipolar disorder following first hospitalization for a manic or mixed episode. *The American Journal of Psychiatry, 164*(4), 582–590. 164/4/582 [pii] https://doi.org/10.1176/ajp.2007.164.4.582.

Dickerson, F. B., Boronow, J. J., Stallings, C. R., Origoni, A. E., Cole, S., & Yolken, R. H. (2004). Association between cognitive functioning and employment status of persons with bipolar disorder. *Psychiatric Services, 55*(1), 54–58. https://doi.org/10.1176/appi.ps.55.1.54.

Dodell-Feder, D., Tully, L. M., & Hooker, C. I. (2015). Social impairment in schizophrenia: new approaches for treating a persistent problem. *Current Opinion in Psychiatry, 28*(3), 236–242. https://doi.org/10.1097/YCO.0000000000000154.

Fett, A. K., & Maat, A. (2013). Social cognitive impairments and psychotic symptoms: what is the nature of their association? *Schizophrenia Bulletin, 39*(1), 77–85. https://doi.org/10.1093/schbul/sbr058. sbr058 [pii].

Fett, A. K., Viechtbauer, W., Dominguez, M. D., Penn, D. L., van Os, J., & Krabbendam, L. (2011). The relationship between neurocognition and social cognition with functional outcomes in schizophrenia: a meta-analysis. *Neuroscience and Biobehavioral Reviews, 35*(3), 573–588. https://doi.org/10.1016/j.neubiorev.2010.07.001. S0149-7634(10)00114-4 [pii].

Foussias, G., & Remington, G. (2010). Negative symptoms in schizophrenia: avolition and Occam's razor. *Schizophrenia Bulletin, 36*(2), 359–369. https://doi.org/10.1093/schbul/sbn094. sbn094 [pii].

Fusar-Poli, P., Bonoldi, I., Yung, A. R., Borgwardt, S., Kempton, M. J., Valmaggia, L., … McGuire, P. (2012). Predicting psychosis: meta-analysis of transition outcomes in individuals at high clinical risk. *Archives of General Psychiatry, 69*(3), 220–229. https://doi.org/10.1001/archgenpsychiatry.2011.1472. 69/3/220 [pii].

Galderisi, S., Bucci, P., Mucci, A., Kirkpatrick, B., Pini, S., Rossi, A., … Maj, M. (2013). Categorical and dimensional approaches to negative symptoms of schizophrenia: focus on long-term stability and functional outcome. *Schizophrenia Research, 147*(1), 157–162. S0920-9964(13)00166-7 [pii] https://doi.org/10.1016/j.schres.2013.03.020.

Galderisi, S., Vignapiano, A., Mucci, A., & Boutros, N. N. (2014). Physiological correlates of positive symptoms in schizophrenia. *Current Topics in Behavioral Neurosciences, 21*, 103–128. https://doi.org/10.1007/7854_2014_322.

Gard, D. E., Kring, A. M., Gard, M. G., Horan, W. P., & Green, M. F. (2007). Anhedonia in schizophrenia: distinctions between anticipatory and consummatory pleasure. *Schizophrenia Research, 93*(1–3), 253–260. S0920-9964(07)00125-9 [pii] https://doi.org/10.1016/j.schres.2007.03.008.

Getz, G. E., Shear, P. K., & Strakowski, S. M. (2003). Facial affect recognition deficits in bipolar disorder. *Journal of the International Neuropsychological Society, 9*(4), 623–632. https://doi.org/10.1017/S1355617703940021.

Gillberg, I. C., Hellgren, L., & Gillberg, C. (1993). Psychotic disorders diagnosed in adolescence. Outcome at age 30 years. *Journal of Child Psychology and Psychiatry, 34*(7), 1173–1185.

Gitlin, M. J., & Miklowitz, D. J. (2017). The difficult lives of individuals with bipolar disorder: a review of functional outcomes and their implications for treatment. *Journal of Affective Disorders, 209*, 147–154.

Glenthoj, L. B., Fagerlund, B., Hjorthoj, C., Jepsen, J. R. M., Bak, N., Kristensen, T. D., … Nordentoft, M. (2016). Social cognition in patients at ultra-high risk for psychosis: what is the relation to social skills and functioning? *Schizophrenia Research: Cognition, 5*, 21–27. https://doi.org/10.1016/j.scog.2016.06.004. S2215-0013(16)30010-5 [pii].

Grant, C., Addington, J., Addington, D., & Konnert, C. (2001). Social functioning in first- and multiepisode schizophrenia. *Canadian Journal of Psychiatry*, *46*(8), 746–749. https://doi.org/10.1177/070674370104600808.
Green, M. F. (1996). What are the functional consequences of neurocognitive deficits in schizophrenia? *The American Journal of Psychiatry*, *153*(3), 321–330. https://doi.org/10.1176/ajp.153.3.321.
Green, M. F. (2016). Impact of cognitive and social cognitive impairment on functional outcomes in patients with schizophrenia. *The Journal of Clinical Psychiatry*, 77(Suppl. 2), 8–11. https://doi.org/10.4088/JCP.14074su1c.02.
Green, M. F., Bearden, C. E., Cannon, T. D., Fiske, A. P., Hellemann, G. S., Horan, W. P. , … Nuechterlein, K. H. (2012). Social cognition in schizophrenia, Part 1: performance across phase of illness. *Schizophrenia Bulletin*, *38*(4), 854–864. https://doi.org/10.1093/schbul/sbq171. sbq171 [pii].
Green, M. F., Horan, W. P., & Lee, J. (2015). Social cognition in schizophrenia. *Nature Reviews. Neuroscience*, *16*(10), 620–631. https://doi.org/10.1038/nrn4005. nrn4005 [pii].
Green, M. F., Kern, R. S., Braff, D. L., & Mintz, J. (2000). Neurocognitive deficits and functional outcome in schizophrenia: are we measuring the "right stuff"? *Schizophrenia Bulletin*, *26*(1), 119–136.
Green, M. F., Kern, R. S., & Heaton, R. K. (2004). Longitudinal studies of cognition and functional outcome in schizophrenia: implications for MATRICS. *Schizophrenia Research*, *72*(1), 41–51. S0920-9964(04)00344-5 [pii] https://doi.org/10.1016/j.schres.2004.09.009.
Gupta, M., Holshausen, K., Mausbach, B., Patterson, T. L., & Bowie, C. R. (2012). Predictors of change in functional competence and functional behavior after functional adaptation skills training for schizophrenia. *The Journal of Nervous and Mental Disease*, *200*(8), 705–711. https://doi.org/10.1097/NMD.0b013e3182613f79. 00005053-201208000-00009 [pii].
Harrington, L., Siegert, R. J., & McClure, J. (2005). Theory of mind in schizophrenia: a critical review. *Cognitive Neuropsychiatry*, *10*(4), 249–286. HKR6122W7M2819N8 [pii] https://doi.org/10.1080/13546800444000056.
Harvey, P. D. (2009). When does cognitive decline occur in the period prior to the first episode of schizophrenia? *Psychiatry (Edgmont)*, *6*(7), 12–14.
Harvey, P. D., Bowie, C. R., & Friedman, J. I. (2001). Cognition in schizophrenia. *Current Psychiatry Reports*, *3*(5), 423–428.
Harvey, P. D., & Penn, D. (2010). Social cognition: the key factor predicting social outcome in people with schizophrenia? *Psychiatry (Edgmont)*, 7(2), 41–44.
Harvey, P. D., Velligan, D., & Bellack, A. S. (2007). Performance-based measures of functional skills: usefulness in clinical treatment studies. *Schizophrenia Bulletin*, *33*(5), 1138–1148. https://doi.org/10.1093/schbul/sbm040.
Healey, K. M., Bartholomeusz, C. F., & Penn, D. L. (2016). Deficits in social cognition in first episode psychosis: a review of the literature. *Clinical Psychology Review*, *50*, 108–137. S0272-7358(16)30254-9 [pii] https://doi.org/10.1016/j.cpr.2016.10.001.
Heinrichs, D. W., Hanlon, T. E., & Carpenter, W. T., Jr. (1984). The Quality of Life Scale: an instrument for rating the schizophrenic deficit syndrome. *Schizophrenia Bulletin*, *10*(3), 388–398.
Henry, J. D., Bailey, P. E., & Rendell, P. G. (2008). Empathy, social functioning and schizotypy. *Psychiatry Research*, *160*(1), 15–22. https://doi.org/10.1016/j.psychres.2007.04.014. S0165-1781(07)00126-6 [pii].
Hoe, M., Nakagami, E., Green, M. F., & Brekke, J. S. (2012). The causal relationships between neurocognition, social cognition and functional outcome over time in schizophrenia: a latent difference score approach. *Psychological Medicine*, *42*(11), 2287–2299. https://doi.org/10.1017/S0033291712000578. S0033291712000578 [pii].

Hoertnagl, C. M., Muehlbacher, M., Biedermann, F., Yalcin, N., Baumgartner, S., Schwitzer, G., … Hofer, A. (2011). Facial emotion recognition and its relationship to subjective and functional outcomes in remitted patients with bipolar I disorder. *Bipolar Disorders, 13*(5–6), 537–544. https://doi.org/10.1111/j.1399-5618.2011.00947.x.

Hofer, A., Benecke, C., Edlinger, M., Huber, R., Kemmler, G., Rettenbacher, M. A., … Wolfgang Fleischhacker, W. (2009). Facial emotion recognition and its relationship to symptomatic, subjective, and functional outcomes in outpatients with chronic schizophrenia. *European Psychiatry, 24*(1), 27–32. https://doi.org/10.1016/j.eurpsy.2008.06.008. S0924-9338(08)01574-5 [pii].

Hooker, C., & Park, S. (2002). Emotion processing and its relationship to social functioning in schizophrenia patients. *Psychiatry Research, 112*(1), 41–50.

Horan, W. P., Green, M. F., DeGroot, M., Fiske, A., Hellemann, G., Kee, K., … Nuechterlein, K. H. (2012). Social cognition in schizophrenia, Part 2: 12-month stability and prediction of functional outcome in first-episode patients. *Schizophrenia Bulletin, 38*(4), 865–872. https://doi.org/10.1093/schbul/sbr001. sbr001 [pii].

Horan, W. P., Reise, S. P., Kern, R. S., Lee, J., Penn, D. L., & Green, M. F. (2015). Structure and correlates of self-reported empathy in schizophrenia. *Journal of Psychiatric Research, 66–67*, 60–66. https://doi.org/10.1016/j.jpsychires.2015.04.016. S0022-3956(15)00119-3 [pii].

Horan, W. P., Subotnik, K. L., Snyder, K. S., & Nuechterlein, K. H. (2006). Do recent-onset schizophrenia patients experience a "social network crisis". *Psychiatry, 69*(2), 115–129. https://doi.org/10.1521/psyc.2006.69.2.115.

Irani, F., Seligman, S., Kamath, V., Kohler, C., & Gur, R. C. (2012). A meta-analysis of emotion perception and functional outcomes in schizophrenia. *Schizophrenia Research, 137*(1–3), 203–211. https://doi.org/10.1016/j.schres.2012.01.023. S0920-9964(12)00048-5 [pii].

Izard, C. E. (2001). Emotional intelligence or adaptive emotions? *Emotion, 1*(3), 249–257.12934684.

Jaeger, J., & Vieta, E. (2007). Functional outcome and disability in bipolar disorders: ongoing research and future directions. *Bipolar Disorders, 9*(1–2), 1–2.

Jiménez-López, E., Sánchez-Morla, E. M., Aparicio, A. I., López-Villarreal, A., Martínez-Vizcaíno, V., Rodriguez-Jimenez, R., … Santos, J. L. (2018). Psychosocial functioning in patients with psychotic and non-psychotic bipolar I disorder. A comparative study with individuals with schizophrenia. *Journal of Affective Disorders, 15*(229), 177–185. https://doi.org/10.1016/j.jad.2017.12.094.

Judd, L. L., Schettler, P. J., Solomon, D. A., Maser, J. D., Coryell, W., Endicott, J., & Akiskal, H. S. (2008). Psychosocial disability and work role function compared across the long-term course of bipolar I, bipolar II and unipolar major depressive disorders. *Journal of Affective Disorders, 108*(1), 49–58.

Karagianis, J., Novick, D., Pecenak, J., Haro, J. M., Dossenbach, M., Treuer, T., … Lowry, A. J. (2009). Worldwide-Schizophrenia Outpatient Health Outcomes (W-SOHO): baseline characteristics of pan-regional observational data from more than 17,000 patients. *International Journal of Clinical Practice, 63*(11), 1578–1588. https://doi.org/10.1111/j.1742-1241.2009.02191.x. IJCP2191 [pii].

Karpouzian, T. M., Alden, E. C., Reilly, J. L., & Smith, M. J. (2016). High functioning individuals with schizophrenia have preserved social perception but not mentalizing abilities. *Schizophrenia Research, 171*(1–3), 137–139. https://doi.org/10.1016/j.schres.2016.01.029. S0920-9964(16)30031-7 [pii].

Keck, P. E., Jr. (2006). Long-term management strategies to achieve optimal function in patients with bipolar disorder. *The Journal of Clinical Psychiatry, 67*(Suppl. 9), 19–24 [discussion 36-42].

Kee, K. S., Green, M. F., Mintz, J., & Brekke, J. S. (2003). Is emotion processing a predictor of functional outcome in schizophrenia? *Schizophrenia Bulletin, 29*(3), 487–497.

Kee, K. S., Horan, W. P., Salovey, P., Kern, R. S., Sergi, M. J., Fiske, A. P., ... Green, M. F. (2009). Emotional intelligence in schizophrenia. *Schizophrenia Research, 107*(1), 61–68. https://doi.org/10.1016/j.schres.2008.08.016. S0920-9964(08)00348-4 [pii].

Kee, K. S., Kern, R. S., & Green, M. F. (1998). Perception of emotion and neurocognitive functioning in schizophrenia: what's the link? *Psychiatry Research, 81*(1), 57–65. S0165-1781(98)00083-3 [pii].

Keefe, R. S., Perkins, D. O., Gu, H., Zipursky, R. B., Christensen, B. K., & Lieberman, J. A. (2006). A longitudinal study of neurocognitive function in individuals at-risk for psychosis. *Schizophrenia Research, 88*(1–3), 26–35. S0920-9964(06)00309-4 [pii] https://doi.org/10.1016/j.schres.2006.06.041.

Kim, H. S., Shin, N. Y., Jang, J. H., Kim, E., Shim, G., Park, H. Y., ... Kwon, J. S. (2011). Social cognition and neurocognition as predictors of conversion to psychosis in individuals at ultra-high risk. *Schizophrenia Research, 130*(1–3), 170–175. https://doi.org/10.1016/j.schres.2011.04.023. S0920-9964(11)00230-1 [pii].

Kirkpatrick, B., Buchanan, R. W., Ross, D. E., & Carpenter, W. T., Jr. (2001). A separate disease within the syndrome of schizophrenia. *Archives of General Psychiatry, 58*(2), 165–171. yps8375 [pii].

Kohler, C. G., Turner, T. H., Bilker, W. B., Brensinger, C. M., Siegel, S. J., Kanes, S. J., ... Gur, R. C. (2003). Facial emotion recognition in schizophrenia: intensity effects and error pattern. *The American Journal of Psychiatry, 160*(10), 1768–1774. https://doi.org/10.1176/appi.ajp.160.10.1768.

Kohler, C. G., Walker, J. B., Martin, E. A., Healey, K. M., & Moberg, P. J. (2010). Facial emotion perception in schizophrenia: a meta-analytic review. *Schizophrenia Bulletin, 36*(5), 1009–1019. https://doi.org/10.1093/schbul/sbn192. sbn192 [pii].

Konstantakopoulos, G., Ioannidi, N., Typaldou, M., Sakkas, D., & Oulis, P. (2016). Clinical and cognitive factors affecting psychosocial functioning in remitted patients with bipolar disorder. *Psychiatriki, 27*(3), 182–191.

Kurtz, M. M., Gagen, E., Rocha, N. B., Machado, S., & Penn, D. L. (2016). Comprehensive treatments for social cognitive deficits in schizophrenia: a critical review and effect-size analysis of controlled studies. *Clinical Psychology Review, 43*, 80–89. https://doi.org/10.1016/j.cpr.2015.09.003. S0272-7358(15)00121-X [pii].

Lahera, G., Herrera, S., Reinares, M., Benito, A., Rullas, M., Gonzalez-Cases, J., & Vieta, E. (2015). Hostile attributions in bipolar disorder and schizophrenia contribute to poor social functioning. *Acta Psychiatrica Scandinavica, 131*(6), 472–482. https://doi.org/10.1111/acps.12399.

Lavelle, M., Healey, P. G., & McCabe, R. (2014). Nonverbal behavior during face-to-face social interaction in schizophrenia: a review. *The Journal of Nervous and Mental Disease, 202*(1), 47–54. https://doi.org/10.1097/NMD.0000000000000031. 00005053-201401000-00010 [pii].

Lay, B., Blanz, B., Hartmann, M., & Schmidt, M. H. (2000). The psychosocial outcome of adolescent-onset schizophrenia: a 12-year followup. *Schizophrenia Bulletin, 26*(4), 801–816.

Lee, J., Altshuler, L., Glahn, D. C., Miklowitz, D. J., Ochsner, K., & Green, M. F. (2013). Social and nonsocial cognition in bipolar disorder and schizophrenia: relative levels of impairment. *American Journal of Psychiatry. 170*(3) https://doi.org/10.1176/appi.ajp.2012.12040490.

Lee, J., Zaki, J., Harvey, P. O., Ochsner, K., & Green, M. F. (2011). Schizophrenia patients are impaired in empathic accuracy. *Psychological Medicine, 41*(11), 2297–2304. https://doi.org/10.1017/S0033291711000614. S0033291711000614 [pii].

Lee, T. Y., Hong, S. B., Shin, N. Y., & Kwon, J. S. (2015). Social cognitive functioning in prodromal psychosis: a meta-analysis. *Schizophrenia Research*, *164*(1–3), 28–34. https://doi.org/10.1016/j.schres.2015.02.008. S0920-9964(15)00083-3 [pii].

Leeson, V. C., Barnes, T. R., Hutton, S. B., Ron, M. A., & Joyce, E. M. (2009). IQ as a predictor of functional outcome in schizophrenia: a longitudinal, four-year study of first-episode psychosis. *Schizophrenia Research*, *107*(1), 55–60. https://doi.org/10.1016/j.schres.2008.08.014. S0920-9964(08)00381-2 [pii].

Leifker, F. R., Bowie, C. R., & Harvey, P. D. (2009). Determinants of everyday outcomes in schizophrenia: the influences of cognitive impairment, functional capacity, and symptoms. *Schizophrenia Research*, *115*(1), 82–87. https://doi.org/10.1016/j.schres.2009.09.004. S0920-9964(09)00422-8 [pii].

Lembke, A., & Ketter, T. A. (2002). Impaired recognition of facial emotion in mania. *American Journal of Psychiatry*, *159*(2), 302–304. https://doi.org/10.1176/appi.ajp.159.2.302.

Lincoln, T. M., Wilhelm, K., & Nestoriuc, Y. (2007). Effectiveness of psychoeducation for relapse, symptoms, knowledge, adherence and functioning in psychotic disorders: a meta-analysis. *Schizophrenia Research*, *96*(1), 232–245.

Lo, P., & Siu, A. M. (2015). Social cognition and work performance of persons with schizophrenia in a Chinese population. *Work*, *50*(4), 629–636. https://doi.org/10.3233/WOR-141828. 6350367562821202 [pii].

Lysaker, P. H., Bryson, G. J., Marks, K., Greig, T. C., & Bell, M. D. (2004). Coping style in schizophrenia: associations with neurocognitive deficits and personality. *Schizophrenia Bulletin*, *30*(1), 113–121.

Lysaker, P. H., Gumley, A., Luedtke, B., Buck, K. D., Ringer, J. M., Olesek, K., … Dimaggio, G. (2013). Social cognition and metacognition in schizophrenia: evidence of their independence and linkage with outcomes. *Acta Psychiatrica Scandinavica*, *127*(3), 239–247. https://doi.org/10.1111/acps.12012.

Maat, A., Fett, A. K., & Derks, E. (2012). Social cognition and quality of life in schizophrenia. *Schizophrenia Research*, *137*(1–3), 212–218. https://doi.org/10.1016/j.schres.2012.02.017. S0920-9964(12)00113-2 [pii].

MacBeth, A., & Gumley, A. (2008). Premorbid adjustment, symptom development and quality of life in first episode psychosis: a systematic review and critical reappraisal. *Acta Psychiatrica Scandinavica*, *117*(2), 85–99. ACP1134 [pii] https://doi.org/10.1111/j.1600-0447.2007.01134.x.

MacQueen, G. M., Young, L. T., & Joffe, R. T. (2001). A review of psychosocial outcome in patients with bipolar disorder. *Acta Psychiatrica Scandinavica*, *103*(3), 163–170. acp059 [pii].

Mancuso, F., Horan, W. P., Kern, R. S., & Green, M. F. (2011). Social cognition in psychosis: multidimensional structure, clinical correlates, and relationship with functional outcome. *Schizophrenia Research*, *125*(2–3), 143–151. https://doi.org/10.1016/j.schres.2010.11.007. S0920-9964(10)01636-1 [pii].

Martinez-Aran, A., Vieta, E., Torrent, C., Sanchez-Moreno, J., Goikolea, J. M., Salamero, M., … Ayuso-Mateos, J. L. (2007). Functional outcome in bipolar disorder: the role of clinical and cognitive factors. *Bipolar Disorders*, *9*(1–2), 103–113. BDI327 [pii] https://doi.org/10.1111/j.1399-5618.2007.00327.x.

Martino, D. J., Strejilevich, S. A., Fassi, G., Marengo, E., & Igoa, A. (2011). Theory of mind and facial emotion recognition in euthymic bipolar I and bipolar II disorders. *Psychiatry Research*, *189*(3), 379–384.

Marwaha, S., & Johnson, S. (2005). Views and experiences of employment among people with psychosis: a qualitative descriptive study. *International Journal of Social Psychiatry*, *51*(4), 302–316. https://doi.org/10.1177/0020764005057386.

Mausbach, B. T., Harvey, P. D., Pulver, A. E., Depp, C. A., Wolyniec, P. S., Thornquist, M. H., ... Patterson, T. L. (2010). Relationship of the Brief UCSD Performance-based Skills Assessment (UPSA-B) to multiple indicators of functioning in people with schizophrenia and bipolar disorder. *Bipolar Disorders, 12*(1), 45–55. https://doi.org/10.1111/j.1399-5618.2009.00787.x. BDI787 [pii].

McCabe, R., Leudar, I., & Antaki, C. (2004). Do people with schizophrenia display theory of mind deficits in clinical interactions? *Psychological Medicine, 34*(3), 401–412.

McGorry, P., Johanessen, J. O., Lewis, S., Birchwood, M., Malla, A., Nordentoft, M., ... Yung, A. (2010). Early intervention in psychosis: keeping faith with evidence-based health care. *Psychological Medicine, 40*(3), 399–404. https://doi.org/10.1017/S0033291709991346. S0033291709991346 [pii].

McLaughlin, D., Carrion, R. E., Auther, A. M., Olvet, D. M., Addington, J., Bearden, C. E., ... Cornblatt, B. A. (2016). Functional capacity assessed by the map task in individuals at clinical high-risk for psychosis. *Schizophrenia Bulletin, 42*(5), 1234–1242. https://doi.org/10.1093/schbul/sbw039. sbw039 [pii].

Medalia, A., & Saperstein, A. M. (2013). Does cognitive remediation for schizophrenia improve functional outcomes? *Current Opinion in Psychiatry, 26*(2), 151–157. https://doi.org/10.1097/YCO.0b013e32835dcbd4.

Menendez-Miranda, I., Garcia-Portilla, M. P., Garcia-Alvarez, L., Arrojo, M., Sanchez, P., Sarramea, F., ... Bobes, J. (2015). Predictive factors of functional capacity and real-world functioning in patients with schizophrenia. *European Psychiatry, 30*(5), 622–627. https://doi.org/10.1016/j.eurpsy.2014.12.011. S0924-9338(15)00026-7 [pii].

Menezes, N. M., Arenovich, T., & Zipursky, R. B. (2006). A systematic review of longitudinal outcome studies of first-episode psychosis. *Psychological Medicine, 36*(10), 1349–1362. S0033291706007951 [pii] https://doi.org/10.1017/S0033291706007951.

Michaels, T. M., Horan, W. P., Ginger, E. J., Martinovich, Z., Pinkham, A. E., & Smith, M. J. (2014). Cognitive empathy contributes to poor social functioning in schizophrenia: evidence from a new self-report measure of cognitive and affective empathy. *Psychiatry Research*. https://doi.org/10.1016/j.psychres.2014.08.054. S0165-1781(14)00751-3 [pii].

Milev, P., Ho, B. C., Arndt, S., & Andreasen, N. C. (2005). Predictive values of neurocognition and negative symptoms on functional outcome in schizophrenia: a longitudinal first-episode study with 7-year follow-up. *The American Journal of Psychiatry, 162*(3), 495–506. 162/3/495 [pii] https://doi.org/10.1176/appi.ajp.162.3.495.

Miller, P., Byrne, M., Hodges, A., Lawrie, S. M., Owens, D. G., & Johnstone, E. C. (2002). Schizotypal components in people at high risk of developing schizophrenia: early findings from the Edinburgh High-Risk Study. *The British Journal of Psychiatry, 180*, 179–184.

Mueser, K. T., & McGurk, S. R. (2004). Schizophrenia. *Lancet, 363*(9426), 2063–2072. https://doi.org/10.1016/S0140-6736(04)16458-1. S0140-6736(04)16458-1 [pii].

Nuechterlein, K. H., Subotnik, K. L., Green, M. F., Ventura, J., Asarnow, R. F., Gitlin, M. J., ... Mintz, J. (2011). Neurocognitive predictors of work outcome in recent-onset schizophrenia. *Schizophrenia Bulletin, 37*(Suppl. 2), S33–S40. https://doi.org/10.1093/schbul/sbr084 sbr084 [pii].

Ohmuro, N., Katsura, M., Obara, C., Kikuchi, T., Sakuma, A., Iizuka, K., ... Matsumoto, K. (2016). Deficits of cognitive theory of mind and its relationship with functioning in individuals with an at-risk mental state and first-episode psychosis. *Psychiatry Research, 243*, 318–325. https://doi.org/10.1016/j.psychres.2016.06.051. S0165-1781(16)30080-4 [pii].

Patterson, T. L., Moscona, S., McKibbin, C. L., Davidson, K., & Jeste, D. V. (2001). Social skills performance assessment among older patients with schizophrenia. *Schizophrenia Research, 48*(2–3), 351–360. https://doi.org/S0920996400001092 [pii].

Patterson, T. L., Semple, S. J., Shaw, W. S., Halpain, M., Moscona, S., Grant, I., & Jeste, D. V. (1997). Self-reported social functioning among older patients with schizophrenia. *Schizophrenia Research, 27*(2–3), 199–210. S0920-9964(97)00078-9 [pii] https://doi.org/10.1016/S0920-9964(97)00078-9.

Penn, D. L., Combs, D. R., Ritchie, M., Francis, J., Cassisi, J., Morris, S., & Townsend, M. (2000). Emotion recognition in schizophrenia: further investigation of generalized versus specific deficit models. *Journal of Abnormal Psychology, 109*(3), 512–516.

Peuskens, J., & Gorwood, P. (2012). How are we assessing functioning in schizophrenia? A need for a consensus approach. *European Psychiatry, 27*(6), 391–395. https://doi.org/10.1016/j.eurpsy.2011.02.013. S0924-9338(11)00040-X [pii].

Poole, J. H., Tobias, F. C., & Vinogradov, S. (2000). The functional relevance of affect recognition errors in schizophrenia. *Journal of the International Neuropsychological Society, 6*(6), 649–658.

Rabinowitz, J., Levine, S. Z., Garibaldi, G., Bugarski-Kirola, D., Berardo, C. G., & Kapur, S. (2012). Negative symptoms have greater impact on functioning than positive symptoms in schizophrenia: analysis of CATIE data. *Schizophrenia Research, 137*(1–3), 147–150. https://doi.org/10.1016/j.schres.2012.01.015. S0920-9964(12)00040-0 [pii].

Reichenberg, A., Harvey, P. D., Bowie, C. R., Mojtabai, R., Rabinowitz, J., Heaton, R. K., & Bromet, E. (2009). Neuropsychological function and dysfunction in schizophrenia and psychotic affective disorders. *Schizophrenia Bulletin, 35*(5), 1022–1029. https://doi.org/10.1093/schbul/sbn044. sbn044 [pii].

Rocca, P., Galderisi, S., Rossi, A., Bertolino, A., Rucci, P., Gibertoni, D., … Maj, M. (2016). Social cognition in people with schizophrenia: a cluster-analytic approach. *Psychological Medicine, 46*(13), 2717–2729. https://doi.org/10.1017/S0033291716001100. S0033291716001100 [pii].

Saavedra, J., Lopez, M., Gonzales, S., & Cubero, R. (2016). Does employment promote recovery? Meanings from work experience in people diagnosed with serious mental illness. *Culture, Medicine and Psychiatry, 40*(3), 507–532. https://doi.org/10.1007/s11013-015-9481-4, 10.1007/s11013-015-9481-4 [pii].

Sabbag, S., Twamley, E. M., Vella, L., Heaton, R. K., Patterson, T. L., & Harvey, P. D. (2011). Assessing everyday functioning in schizophrenia: not all informants seem equally informative. *Schizophrenia Research, 131*(1–3), 250–255. https://doi.org/10.1016/j.schres.2011.05.003. S0920-9964(11)00258-1 [pii].

Sanchez-Moreno, J., Martinez-Aran, A., Tabares-Seisdedos, R., Torrent, C., Vieta, E., & Ayuso-Mateos, J. L. (2009). Functioning and disability in bipolar disorder: an extensive review. *Psychotherapy and Psychosomatics, 78*(5), 285–297. https://doi.org/10.1159/000228249. 000228249 [pii].

Savla, G. N., Vella, L., Armstrong, C. C., Penn, D. L., & Twamley, E. W. (2013). Deficits in domains of social cognition in schizophrenia: a meta-analysis of the empirical evidence. *Schizophrenia Bulletin, 39*(5), 979–992. https://doi.org/10.1093/schbul/sbs080. sbs080 [pii].

Schmidt, S. J., Mueller, D. R., & Roder, V. (2011). Social cognition as a mediator variable between neurocognition and functional outcome in schizophrenia: empirical review and new results by structural equation modeling. *Schizophrenia Bulletin, 37*(Suppl. 2), S41–S54. https://doi.org/10.1093/schbul/sbr079. sbr079 [pii].

Schneider, F., Gur, R. C., Koch, K., Backes, V., Amunts, K., Shah, N. J., … Habel, U. (2006). Impairment in the specificity of emotion processing in schizophrenia. *The American Journal of Psychiatry, 163*(3), 442–447. 163/3/442 [pii] https://doi.org/10.1176/appi.ajp.163.3.442.

Schultz, D., Izard, C. E., Ackerman, B. P., & Youngstrom, E. A. (2001). Emotion knowledge in economically disadvantaged children: self-regulatory antecedents and relations to social difficulties and withdrawal. *Development and Psychopathology, 13*(1), 53–67.

Schultze-Lutter, F., Ruhrmann, S., Berning, J., Maier, W., & Klosterkotter, J. (2010). Basic symptoms and ultrahigh risk criteria: symptom development in the initial prodromal state. *Schizophrenia Bulletin*, *36*(1), 182–191. https://doi.org/10.1093/schbul/sbn072. sbn072 [pii].

Shamay-Tsoory, S. G., Shur, S., Barcai-Goodman, L., Medlovich, S., Harari, H., & Levkovitz, Y. (2007). Dissociation of cognitive from affective components of theory of mind in schizophrenia. *Psychiatry Research*, *149*(1–3), 11–23. S0165-1781(06) 00193-4 [pii] https://doi.org/10.1016/j.psychres.2005.10.018.

Smith, M. J., Horan, W. P., Cobia, D. J., Karpouzian, T. M., Fox, J. M., Reilly, J. L., & Breiter, H. C. (2014). Performance-based empathy mediates the influence of working memory on social competence in schizophrenia. *Schizophrenia Bulletin*, *40*(4), 824–834. https://doi.org/10.1093/schbul/sbt084. sbt084 [pii].

Smith, M. J., Horan, W. P., Karpouzian, T. M., Abram, S. V., Cobia, D. J., & Csernansky, J. G. (2012). Self-reported empathy deficits are uniquely associated with poor functioning in schizophrenia. *Schizophrenia Research*, *137*(1–3), 196–202. https://doi.org/10.1016/j.schres.2012.01.012. S0920-9964(12)00037-0 [pii].

Smith, T. E., Hull, J. W., Huppert, J. D., & Silverstein, S. M. (2002). Recovery from psychosis in schizophrenia and schizoaffective disorder: symptoms and neurocognitive rate-limiters for the development of social behavior skills. *Schizophrenia Research*, *55*(3), 229–237. S0920996401002316 [pii].

Sparks, A., McDonald, S., Lino, B., O'Donnell, M., & Green, M. J. (2010). Social cognition, empathy and functional outcome in schizophrenia. *Schizophrenia Research*, *122*(1–3), 172–178. https://doi.org/10.1016/j.schres.2010.06.011. S0920-9964(10)01366-6 [pii].

Sprong, M., Schothorst, P., Vos, E., Hox, J., & van Engeland, H. (2007). Theory of mind in schizophrenia: meta-analysis. *The British Journal of Psychiatry*, *191*, 5–13. 191/1/5 [pii] https://doi.org/10.1192/bjp.bp.107.035899.

Stefanopoulou, E., Lafuente, A. R., Fonseca, A. S., Keegan, S., Vishnick, C., & Huxley, A. (2011). Global assessment of psychosocial functioning and predictors of outcome in schizophrenia. *International Journal of Psychiatry in Clinical Practice*, *15*(1), 62–68. https://doi.org/10.3109/13651501.2010.519035.

Stouten, L. H., Veling, W., Laan, W., Van der Helm, M., & Van der Gaag, M. (2014). Psychotic symptoms, cognition and affect as predictors of psychosocial problems and functional change in first-episode psychosis. *Schizophrenia Research*, *158*(1–3), 113–119. https://doi.org/10.1016/j.schres.2014.06.023. S0920-9964(14)00315-6 [pii].

Tabarés-Seisdedos, R., Balanzá-Martínez, V., Sánchez-Moreno, J., Martinez-Aran, A., Salazar-Fraile, J., Selva-Vera, G., … Vieta, E. (2008). Neurocognitive and clinical predictors of functional outcome in patients with schizophrenia and bipolar I disorder at one-year follow-up. *Journal of Affective Disorders*, *109*(3), 286–299.

Tas, C., Brown, E., Cubukcuoglu, Z., Aydemir, O., Danaci, A. E., & Brune, M. (2013). Towards an integrative approach to understanding quality of life in schizophrenia: the role of neurocognition, social cognition, and psychopathology. *Comprehensive Psychiatry*, *54*(3), 262–268. https://doi.org/10.1016/j.comppsych.2012.08.001. S0010-440X(12) 00202-7 [pii].

Thompson, A. D., Bartholomeusz, C., & Yung, A. R. (2011). Social cognition deficits and the 'ultra high risk' for psychosis population: a review of literature. *Early Intervention in Psychiatry*, *5*(3), 192–202. https://doi.org/10.1111/j.1751-7893.2011.00275.x.

Tohen, M., Greenfield, S. F., Weiss, R. D., Zarate, C. A., Jr., & Vagge, L. M. (1998). The effect of comorbid substance use disorders on the course of bipolar disorder: a review. *Harvard Review of Psychiatry*, *6*(3), 133–141.

Tohen, M., Jacobs, T. G., & Feldman, P. D. (2000). Onset of action of antipsychotics in the treatment of mania. *Bipolar Disorders*, *2*(3 Pt 2), 261–268.

van Donkersgoed, R. J., Wunderink, L., Nieboer, R., Aleman, A., & Pijnenborg, G. H. (2015). Social cognition in individuals at ultra-high risk for psychosis: a meta-analysis. *PLoS One*, *10*(10), e0141075. https://doi.org/10.1371/journal.pone.0141075. PONE-D-15-10569 [pii].

Van Winkel, R., Myin-Germeys, I., De Hert, M., Delespaul, P., Peuskens, J., & Van Os, J. (2007). The association between cognition and functional outcome in first-episode patients with schizophrenia: mystery resolved? *Acta Psychiatrica Scandinavica*, *116*(2), 119–124.

Vauth, R., Rusch, N., Wirtz, M., & Corrigan, P. W. (2004). Does social cognition influence the relation between neurocognitive deficits and vocational functioning in schizophrenia? *Psychiatry Research*, *128*(2), 155–165. S0165178104001349 [pii] https://doi.org/10.1016/j.psychres.2004.05.018.

Velthorst, E., Koeter, M., van der Gaag, M., Nieman, D. H., Fett, A. K., Smit, F., ... de Haan, L. (2015). Adapted cognitive-behavioural therapy required for targeting negative symptoms in schizophrenia: meta-analysis and meta-regression. *Psychological Medicine*, *45*(3), 453–465. https://doi.org/10.1017/S0033291714001147. S0033291714001147 [pii].

Ventura, J., Ered, A., Gretchen-Doorly, D., Subotnik, K. L., Horan, W. P., Hellemann, G. S., & Nuechterlein, K. H. (2015). Theory of mind in the early course of schizophrenia: stability, symptom and neurocognitive correlates, and relationship with functioning. *Psychological Medicine*, *45*(10), 2031–2043. https://doi.org/10.1017/S0033291714003171. S0033291714003171 [pii].

Ventura, J., Hellemann, G. S., Thames, A. D., Koellner, V., & Nuechterlein, K. H. (2009). Symptoms as mediators of the relationship between neurocognition and functional outcome in schizophrenia: a meta-analysis. *Schizophrenia Research*, *113*(2–3), 189–199. https://doi.org/10.1016/j.schres.2009.03.035. S0920-9964(09)00121-2 [pii].

Ventura, J., Wood, R. C., & Hellemann, G. S. (2013). Symptom domains and neurocognitive functioning can help differentiate social cognitive processes in schizophrenia: a meta-analysis. *Schizophrenia Bulletin*, *39*(1), 102–111. https://doi.org/10.1093/schbul/sbr067. sbr067 [pii].

Vyas, N. S., Hadjulis, M., Vourdas, A., Byrne, P., & Frangou, S. (2007). The Maudsley early onset schizophrenia study. Predictors of psychosocial outcome at 4-year follow-up. *European Child & Adolescent Psychiatry*, *16*(7), 465–470. https://doi.org/10.1007/s00787-007-0621-4.

Wingo, A. P., Harvey, P. D., & Baldessarini, R. J. (2009). Neurocognitive impairment in bipolar disorder patients: functional implications. *Bipolar Disorders*, *11*(2), 113–125. https://doi.org/10.1111/j.1399-5618.2009.00665.x. BDI665 [pii].

Woolverton, C. B., Bell, E. K., Moe, A. M., Harrison-Monroe, P., & Breitborde, N. J. (2017). Social cognition and the course of social functioning in first-episode psychosis. *Early Intervention in Psychiatry*, *12*(6), 1151–1156. https://doi.org/10.1111/eip.12432.

Yung, A. R., & McGorry, P. D. (2007). Prediction of psychosis: setting the stage. *The British Journal of Psychiatry. Supplement*, *51*, s1–s8. https://doi.org/10.1192/bjp.191.51.s1. 191/51/s1 [pii].

Zarate, C. A., Jr., Tohen, M., Land, M., & Cavanagh, S. (2000). Functional impairment and cognition in bipolar disorder. *The Psychiatric Quarterly*, *71*(4), 309–329.

Zhang, T., Cui, H., Wei, Y., Tang, Y., Xu, L., Tang, X., ... Wang, J. (2017). Progressive decline of cognition during the conversion from prodrome to psychosis with a characteristic pattern of the theory of mind compensated by neurocognition. *Schizophrenia Research*. https://doi.org/10.1016/j.schres.2017.08.020. S0920-9964(17)30496-6 [pii].

CHAPTER FIVE

# Shared neural substrates of deficits in social cognition and negative symptoms in schizophrenia

**Andrea Pelletier-Baldelli*, Daphne J. Holt†,‡,§**
*University of North Carolina at Chapel Hill, Department of Psychiatry, Chapel Hill, NC, United States
†Massachusetts General Hospital, Department of Psychiatry, Boston, MA, United States
‡Harvard Medical School, Boston, MA, United States
§The Athinoula A. Martinos Center for Biomedical Imaging, Charlestown, MA, United States

The negative symptoms of schizophrenia broadly consist of deficits in the initiation and maintenance of activities, emotional experience, and verbal and nonverbal communication. Specifically, evidence has accumulated (Blanchard & Cohen, 2006; Kirkpatrick, Fenton, Carpenter Jr., & Marder, 2006; Kirkpatrick & Fischer, 2006) to suggest that there are five primary negative symptoms: anhedonia (diminished intensity/frequency of pleasure), avolition (diminished initiation and persistence in goal-directed activity), asociality (reduced desire/frequency for social interaction), blunted affect (diminished facial, vocal, and body expression), and alogia (reduced quantity of speech). Negative symptoms represent a central feature of schizophrenia and related psychotic disorders and predict poorer recovery (i.e., functional outcomes), subjective well-being, and quality of life in these conditions (Fervaha, Foussias, Agid, & Remington, 2014; Foussias, Agid, Fervaha, & Remington, 2014; Millan, Fone, Steckler, & Horan, 2014; Strauss, Harrow, Grossman, & Rosen, 2010). Some studies have demonstrated that negative symptoms fall roughly into two dimensions, based on factor analyses: the expressive (blunted affect, alogia) and the experiential (anhedonia, avolition, asociality) domains (Messinger et al., 2011; Strauss et al., 2013). Initial work using the two-factor framework suggests that it is the experiential (i.e., anhedonia, asociality, and avolition), rather than the expressive, domain that is more closely tied to functional outcome in schizophrenia (Rassovsky, Horan, Lee, Sergi, & Green, 2011).

Some evidence suggests that negative symptoms may arise from fundamental deficits in social cognition. For example, facial affect recognition deficits have been linked to all five negative symptom domains (Andrzejewska,

https://doi.org/10.1016/B978-0-12-815315-4.00005-7

Wójciak, Domowicz, & Rybakowski, 2017). Other types of social cognition, such as social and emotional perception, show a moderate correlation with overall negative symptom burden ($r=0.36$) (Sergi et al., 2007). Mentalization, which is the ability to interpret thoughts, emotions, and intentions of others, has also been associated with negative symptoms (Ventura et al., 2015). Together, these results suggest that negative symptoms and social cognition overlap in some fundamental way, but are also separable domains (Green et al., 2008).

However, it is difficult to measure the degree of overlap between negative symptoms and social cognition. One challenge is the heterogeneity of clinical presentations and cognitive profiles within cohorts of individuals with schizophrenia. For example, the severity of negative symptoms and cognitive deficits varies greatly, and variation in IQ or other cognitive abilities can impact performance on self-report and laboratories' measures of social cognition (Pousa et al., 2008). Also, the delineation between negative symptoms and social cognition is complicated by the fact that the primary method of assessing each domain differs. Negative symptoms are typically measured using interview-based rating scales, which focus on outward behavior or collateral reports from clinicians or family members. In contrast, domains of social cognition are commonly measured using laboratory-based experimental paradigms. These methodological issues raise the question—are these two domains separable enough to be considered independent? In other words, when we assess asociality or social anhedonia in schizophrenia, are we merely measuring subjective reporting of impairments in social cognition?

Studies of brain function may help to clarify this question by determining whether common or distinct neural substrates are linked to negative symptoms and social cognitive impairment. To date, neuroimaging data suggests that both negative symptoms and deficits in social cognitive processes (such as mentalization and social cue perception) are associated with changes in the functioning of fronto-temporal circuitry (Adolphs, 2009; Green, Horan, & Lee, 2015; Lieberman, 2007; Millan et al., 2014). This shared neural basis supports the idea that these two domains are closely linked. However, there are limited data regarding *how* these circuits generate cognitive and behavioral processes that manifest as negative symptoms and social cognitive impairment. Thus, a comprehensive model for understanding the neurobiological and behavioral basis of negative symptoms and social cognition deficits in psychotic illness does not currently exist.

However, there are several different theories about how disruptions of specific neural processes could give rise to negative symptoms and social cognition. One perspective emphasizes higher-order cognitive impairments as the primary source of these deficits. Another model proposes a "cascade" process that highlights early sensory processing impairments as the source of downstream changes in these domains. In the following section we present each model in turn, as well as an "integrated" model, followed by suggestions for testing these alternatives.

## A bottom-up model

Schizophrenia is associated with abnormalities in lower-level visual and auditory processes (Butler & Javitt, 2005; Butler, Silverstein, & Dakin, 2008; Doniger, Silipo, Rabinowicz, Snodgrass, & Javitt, 2001; Jahshan, Wynn, & Green, 2013; Shin et al., 2012), including an altered ability to filter incoming sensory information (Bramon, Rabe-Hesketh, Sham, Murray, & Frangou, 2004; Schubring, Popov, Miller, & Rockstroh, 2018). For example, a wide range of abnormalities in basic, early visual processes, including deficits in contrast detection, gain control, motion and object recognition, and detection of facial features (Bortolon, Capdevielle, & Raffard, 2015; Butler et al., 2007, 2008; Butler & Javitt, 2005; Javitt, 2009; Matsumoto, Takahashi, Murai, & Takahashi, 2015) have been reported in schizophrenia.

Recent functional magnetic resonance imaging (fMRI) research has also examined regions involved in associative, multisensory processing (Bremmer et al., 2001; Silver & Kastner, 2009), including frontoparietal circuitry known to monitor the space near the body or "personal space" (Holt et al., 2014). Personal space size, which corresponds to the distance that a person prefers to stand from another person, is enlarged in schizophrenia; this enlargement is significantly correlated with negative symptom levels, as well as with the magnitude of responses of the dorsal parietal cortex to personal space intrusions in individuals with schizophrenia (Holt et al., 2015). Thus, abnormalities in sensory processes necessary for responding to social cues and abiding by social norms during routine face-to-face social interactions may contribute to negative symptoms.

Deficits in early auditory processes are also evident and have been associated with poorer cognition (including social cognition) and negative symptoms in schizophrenia (Matsumoto et al., 2015; Thomas et al., 2017). For example, people with schizophrenia show impairments in pitch and rhythm perception (Green et al., 2015; Kantrowitz et al., 2013; Schnakenberg Martin et al., 2017). Additionally, a large number of studies have assessed the auditory mismatch negativity (MMN) response in schizophrenia, which is measured using EEG. Broadly, the MMN response represents an alteration in neuronal firing that occurs when one's expectation for incoming sensory information does not match the actual sensory input. For example, when hearing repetitive sounds interrupted by a novel sound, the brain responds accordingly (i.e., by generating the MMN response). Overall, studies suggest that there is a significant reduction in the amplitude of the MMN signal in schizophrenia (Umbricht & Krljes, 2005). There are also investigations examining how people with schizophrenia filter incoming auditory information (Bramon et al., 2004; Brockhaus-Dumke et al., 2008). Generally, the prediction of this work is that, following an auditory stimulus, the second exposure to that same sound should result in a reduced neural response (i.e., successful sensory gating and suppression of the P50 EEG component). Individuals with schizophrenia consistently show deficits in this early sensory processing domain relative to unaffected healthy controls (as evidenced by an abnormally larger P50 ratio, reflecting difficulty suppressing the P50 response to the second/similar stimulus) (Bramon et al., 2004).

Studies have sought to identify the changes in brain circuitry underlying these auditory impairments. The blunted MMN observed in schizophrenia is related to impairments in automatic processing mediated by temporal-prefrontal circuitry, whereby the brain compares current input to stored memories of previous stimuli (Garrido, Kilner, Stephan, & Friston, 2009). Deficits in auditory sensory gating in schizophrenia have been linked to functional changes in the auditory cortex, prefrontal cortex, and the hippocampus (Grunwald et al., 2003; Korzyukov et al., 2007; Mayer et al., 2009; Williams, Nuechterlein, Subotnik, & Yee, 2011).

The specific impact of these early sensory deficits on social cognition, negative symptoms, and daily functioning is not yet fully understood. One hypothesis is that these deficits cause downstream effects on the perception and interpretation of socially relevant sensory information and the production of social behavior. In support of this model, there is emerging evidence of associations between impaired early sensory processing and impaired social cognition and negative symptoms. For example, in one study, investigators examined eye movements during a biological motion perception task, and found a reduction in accuracy and fixation frequency in individuals with schizophrenia compared with healthy control subjects; both impairments in visual perception were significantly associated with lower scores on a test of affective empathy (Matsumoto et al., 2015). Furthermore, structural equation modeling (SEM, a statistical method used to test for the direction of associations among measurable and latent variables) has shown evidence that deficits in both early visual and auditory perception lead to impairments in cognition and negative symptoms in schizophrenia (Green, Hellemann, Horan, Lee, & Wynn, 2012; Thomas et al., 2017).

In light of this evidence, we can briefly consider the implications of altered visual and auditory processing during an interaction with another person, where the goal is to try to understand and interpret the intention of that other person. During such an interaction, individuals with schizophrenia could have difficulty accurately perceiving facial expressions, identifying changes in pitch or rhythm in someone's voice, or filtering auditory information coming from speech or the environment. Difficulties in gauging how far to stand from another person and how much eye contact to maintain could also arise. The transmission of inaccurate information from sensory cortical areas to areas involved in social cognitive judgments, including the prefrontal cortex, insula, and temporal parietal junction, may produce altered social perceptions. For example, faulty perceptual input may influence and potentially compromise the ability to accurately infer the other person's intentions, beliefs, and/or emotions, which could, in turn, lead to an experience of social interactions as less pleasurable (social anhedonia), or inhibit the desire to engage with others (asociality), impacting relationships with family, friends, co-workers, and/or romantic partners. In sum, this bottom-up model provides a plausible set of mechanisms by which early sensory impairment could cause impairments in social cognition, that then lead to negative symptoms and poorer quality of life in schizophrenia (Fig. 1).

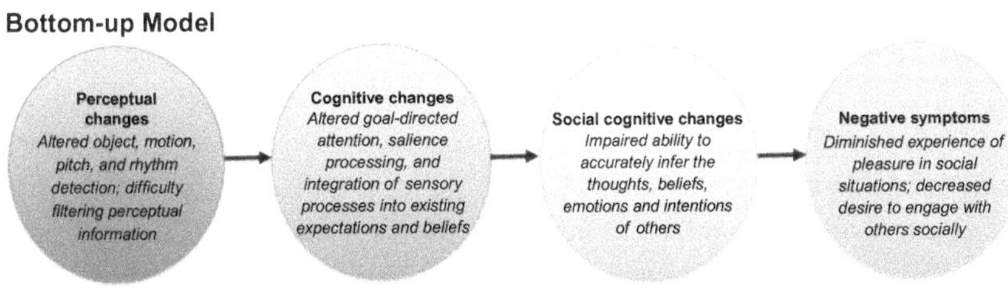


**Fig. 1** Schematic illustration of associations identified in prior studies that support a bottom-up model of social cognitive deficits and negative symptoms in schizophrenia. Studies using statistical modeling approaches (e.g., structural equation modeling) have shown evidence for causal associations among these variables (i.e., that deficits in sensory processing give rise to cognitive and social cognitive impairments and negative symptoms) (Green et al., 2012; Thomas et al., 2017).

## Top-down model

An alternate model for understanding the influence of social cognition and negative symptoms on functioning emphasizes the role of higher-order cognition as opposed to lower-level sensory information. This distinction stems from the dual process theory of cognition, which states that there are two types of cognition—low-level, autonomous processes such as visual and auditory perception, and high-level, effortful processes such as working memory, executive function, goal-directed attention, judgment, and reasoning (Evans & Stanovich, 2013). Thus a general "top-down" model of impairments in social cognition and associated negative symptoms hypothesizes that these abnormalities originate from altered higher-order cognitive functions.

Higher-order cognition is primarily orchestrated by regions of the prefrontal cortex and their targets (Badre, 2008). Cognitive control and related executive functions are mediated by a frontal (e.g., dorsolateral prefrontal cortex) and parietal (e.g., lateral parietal cortex) cortical network—often referred to as the frontoparietal (Giesbrecht, Woldorff, Song, & Mangun, 2003) or central executive (Seeley et al., 2007) network. Another higher-order process, goal-directed attention, is thought to be driven by the dorsal attention network, which is comprised of a separate set of fronto-parietal cortical pathways (involving superior frontal and intraparietal cortices) (Corbetta & Shulman, 2002). A top-down model of brain functioning generally proposes that fronto-parietal cortices represent expectations and learned knowledge of the world, and use that information to influence lower-level sensory processing.

There is a great deal of literature establishing the presence of deficits in higher-order cognition in schizophrenia, both in medicated and unmedicated patients (Fatouros-Bergman, Cervenka, Flyckt, Edman, & Farde, 2014; Schaefer, Giangrande, Weinberger, & Dickinson, 2013). These include impairments in working memory, task switching, error monitoring, and other types of cognitive control (Barch & Ceaser, 2012; Corigliano et al., 2014; Fatouros-Bergman et al., 2014; Kalkstein, Hurford, & Gur, 2010; Lesh, Niendam, Minzenberg, & Carter, 2011). Also, individuals with schizophrenia show abnormalities in goal-directed attention (Kreither et al., 2017; Sawaki et al., 2017). For example, in one study, subjects were instructed to attend only to red circles in the center of a screen; individuals with schizophrenia focused on the "red circle" portion of the instruction, attending to all red circles, regardless of their location (Sawaki et al., 2017). This finding is notable in light of the evidence that automatic, "bottom-up," stimulus-driven attention to visual stimuli is modulated by goal-directed (top-down) attention (e.g., information from the prefrontal cortex regarding the location in space to focus attention) (Kuhn, Teszka, Tenaw, & Kingstone, 2016). The implication is that if the ability of the prefrontal cortex to modulate attention is compromised, there may be reduced allocation of attention to incoming sensory information.

Within this framework, the central hypothesis of this model is that altered function of fronto-parietal regions underlies social cognitive deficits and negative symptoms in psychosis. In support of this hypothesis, studies have shown evidence that higher-order cognition has a modulatory influence on sensory processes and social perception (Gilbert & Sigman, 2007; Hoe, Nakagami, Green, & Brekke, 2012; Mano & Brown, 2013). For example, one study evaluated mentalization ability with the "Reading the Mind in the Eyes Task" (Baron-Cohen, Wheelwright, Hill, Raste, & Plumb, 2001), and measured top-down inhibitory control using an auditory task that required the participant to report on differing vowel-consonant sounds (e.g., ba) heard simultaneously in both ears. In participants with schizophrenia, reduced inhibition of auditory processes was associated with lower mentalization ability (Rominger et al., 2016).

Thus, because lower-level perception is being continuously modulated by higher-order cognitive processes, such as goal-directed attention, disruption of this modulation may then generate changes in perception in schizophrenia. For example, impaired functioning of the dorsal attention network may affect the perception of facial expressions (the mouth may receive more attention than the eyes, gestures, etc.). Such aberrant top-down modulation of visual perception may distort and limit incoming information about another person, disrupting the ability to accurately perceive the other person's thoughts, beliefs, and emotions. Similar to the downstream consequences of impaired "bottom up" processing, such altered mentalizing could then give rise to a desire to withdraw from and avoid others, leading, over time, to a gradually worsening impairment in the ability to initiate or maintain social relationships, interfering ultimately with engagement and success at work, school, or other role obligations (Fig. 2).

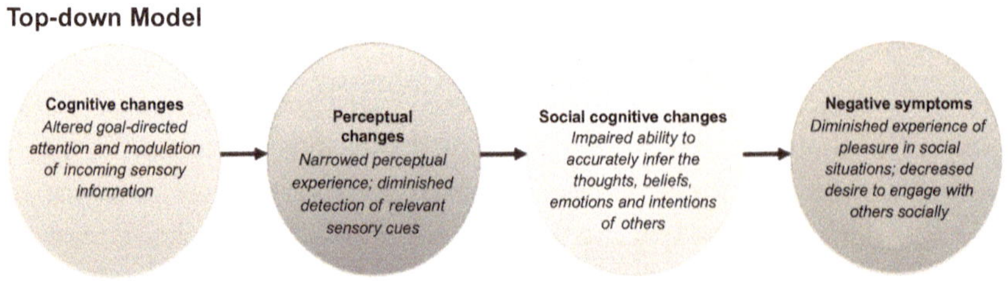


**Fig. 2** Schematic illustration of associations identified in prior studies that support a top-down model of social cognitive deficits and negative symptoms in schizophrenia. The influence of higher-order cognitive processes on lower-level sensory perception has been demonstrated by numerous studies (Gilbert & Sigman, 2007; Hoe et al., 2012; Rominger et al., 2016).

## Integrated model

A third, integrated model is based on the evidence that deficits exist in both higher-order and lower-level domains in schizophrenia (Elshaikh, Sponheim, Chafee, & MacDonald, 2015), and these wide-ranging impairments may interact to bring about impaired social cognition, negative symptoms, and diminished quality of life in schizophrenia.

This integrated model is consistent with evidence for a "switching" process employed by the human brain to engage either top-down or bottom-up processes, based on contextual factors (Corbetta, Patel, & Shulman, 2008; Corbetta & Shulman, 2002). To illustrate this switching process, consider the following scenario: when reading a recipe, top-down, goal-directed processes are recruited while searching for the correct quantity of an ingredient. Then if a hot dish is accidentally touched, bottom-up sensory processes take over and override the goal-direction attention to the recipe, in order to focus on the immediately salient, aversive stimulus in the environment. This switching between top-down and bottom-up processes is referred to as the reorienting response, which shifts and allocates attention to where it is needed (Corbetta et al., 2008). Thus, it is possible that either deficient top-down or bottom-up processes in individuals with schizophrenia could interfere with social cognition, depending on the situational context and type of stimuli present in the immediate environment.

An integrated framework would also better reflect the fluidly dynamic interactions between these two processes that are known to occur (Corbetta & Shulman, 2002; Mechelli, Price, Friston, & Ishai, 2004).

The interaction between bottom-up and top-down processing has been studied extensively using predictive coding models (Clark, 2013). These models represent prior expectations (top-down signaling), along with errors (bottom-up signaling), which reflect discrepancies between expectations and incoming sensory information, and the interaction, synthesis, and co-construction of representations using both perceptual and higher-order cognitive information (Corlett, Honey, & Fletcher, 2016; Friston, Stephan, Montague, & Dolan, 2014; Powers III, Kelley, & Corlett, 2016). Specifically, prior expectations are communicated to lower-level sensory areas, which in turn signal to higher-order areas the presence of any mismatch; this mismatch then influences the expectations subsequently relayed to lower-level sensory regions (Clark, 2013). Based on this model, negative symptoms are proposed to arise from faulty integration of these expectations and error signals, resulting in a learned experience of uncertainty regarding the prediction of outcomes (Corlett et al., 2016). Uncertainty regarding outcomes then leads to inaction, ultimately resulting in avolition (Corlett & Fletcher, 2015; Griffin & Fletcher, 2017).

A large number of studies have used this predictive coding framework to investigate learning in schizophrenia. One body of literature has focused on disruptions of associative learning following receipt of rewards (i.e., money, pleasant scenery, smiling faces). Impairments in reward learning are associated with greater negative symptom levels (Gold et al., 2012), as well as deficits in social cognition in schizophrenia (Lewandowski et al., 2016). Other work has examined associative learning using Pavlovian fear conditioning paradigms (Holt et al., 2009; Jensen et al., 2008), with poorer associative learning (fear learning in this case) again linked to greater levels of negative symptoms (Holt, Coombs, Zeidan, Goff, & Milad, 2012). In sum, associative learning is a basic mechanism that reflects a dynamic interplay of bottom-up and top-down processes; the disruption of this interplay and the associative learning processes that rely on it may underlie negative symptoms and social cognition deficits.

To illustrate this integrated model, consider a social scenario in which prior expectations about what other people will say have not been met. Similarly, rewarding facial cues such as smiles or laughter have not been as forthcoming as anticipated. Subsequently, due to the inability to account for these mismatches between expectations and experience, uncertainty regarding one's ability to accurately interpret and understand other people arises. This uncertainty leads to disengagement and halting or inappropriate responses during the interaction. Also, awareness of relevant social cues is reduced

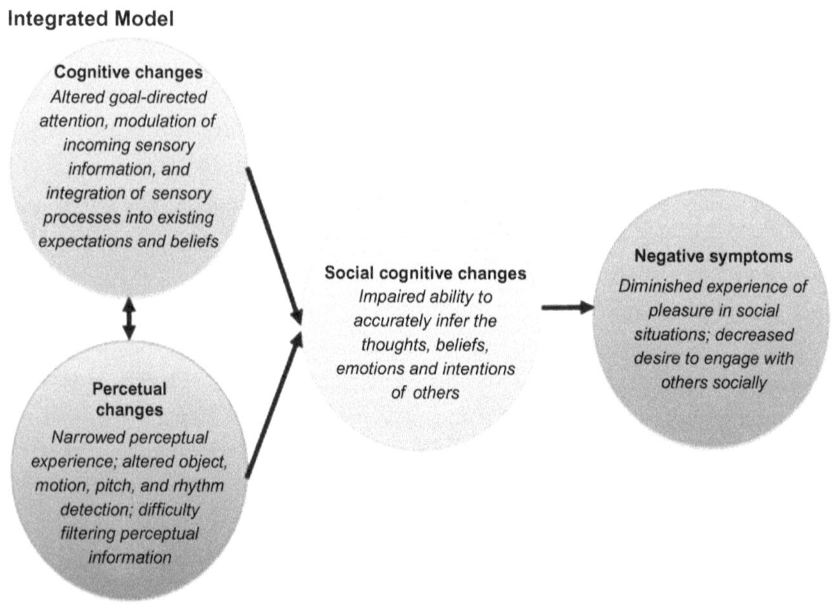


**Fig. 3** Schematic illustration of research that supports an integrated understanding of the origin of social cognitive deficits and negative symptoms in schizophrenia. Studies revealing dynamic relationships between higher-order and lower-level processes suggest that interactions between deficits in these processes may contribute to changes in social cognition and negative symptoms.

as a consequence of disrupted goal-directed attention and impaired perception of salient stimuli. Thus, the interaction becomes one that is dominated by poor learning, uncertainty, and impaired attention resulting from changes in both top-down and bottom-up information streams, and, ultimately leads to inaction. The resulting state or set of states, as they persist and recur over time, can then give rise to difficulties initiating and persisting in activities (avolition) and social interactions (asociality) and ultimately, impaired day-to-day social functioning (Fig. 3).

## The role of negative affect and dysfunctional attitudes

Recent work has also demonstrated the contribution of negative affect, anxiety, and related self-defeating beliefs to social cognitive deficits and negative symptoms in schizophrenia.

First, defeatist beliefs (e.g., believing that one is incapable of particular skills or unable to perform certain goal-directed actions (Campellone,

Sanchez, & Kring, 2016)) and altered, negative expectations (schemas) about oneself have been shown to contribute to the development and/or maintenance of negative symptoms. Specifically, consistently failing to achieve one's goals can lead to a persistent sense of ineffectiveness, which can in turn lead to avolition and impaired functioning (Grant & Beck, 2009). There is evidence that these types of defeatist beliefs mediate the relationship between negative symptoms and poor functional outcomes (Campellone et al., 2016). Further, negative, pessimistic expectations regarding performance and success (e.g., beliefs that one has limited cognitive resources and ability to persist) are associated with anhedonia, avolition, and asociality, but not with blunted affect and alogia (Couture, Blanchard, & Bennett, 2011). These findings shed light on how beliefs may directly impact social behavior and social cognition, and vice-versa. Specifically, impairments in social cognition may play a role in the development of defeatist beliefs, which then may lead to diminished social engagement and subsequent worsening of social skills, leading to further reinforcement of defeatist beliefs (Campellone et al., 2016; Green et al., 2012).

Second, an important role of depression and anxiety in both negative symptoms and social cognition has been demonstrated (Barch, Yodkovik, Sypher-Locke, & Hanewinkel, 2008; Millan et al., 2014; Sutliff, Roy, & Achim, 2015; Wyer Jr. & Srull, 2014). Also, higher levels of depression and anxiety have been linked with defeatist beliefs and negative expectations about the self in patients with psychotic disorders (Couture et al., 2011; Grant & Beck, 2009). Taken together, this body of work suggests that cognitive and affective processes underlying negative beliefs and expectancies about the self, and related symptoms of depression and anxiety, also play an important role in the development, maintenance, and/or progression of social cognitive deficits, negative symptoms, and the associated functional impairment.

## Conclusions and future directions

In this chapter, we have described three alternative models (bottom-up, top-down, and integrated) for understanding how deficits in social cognition and negative symptoms may arise in schizophrenia. In addition, all three models highlight the close links between social cognition deficits and negative symptoms, that is, evidence that alterations in social cognition are manifested over time as negative symptoms. Also, because negative symptoms mediate the relationship between social cognition deficits and functional outcomes (Lin et al., 2013), understanding the neurobiological mechanisms underlying these associations represents a key step in the overall effort to develop new approaches for improving day-to-day functioning and quality of life for people with psychotic illness.

Despite the associations consistently observed between negative symptoms and deficits in social cognition, there is no consensus regarding which model best represents the underlying neurobiology. The integrated model may account for the largest portion of the data, that is, the well-replicated findings of deficits in both basic sensory and higher-order processes in schizophrenia. However, it is possible that one set of abnormalities precedes the other, as the pathophysiology of the disease emerges and progresses during its earliest stages. In support of a primarily top-down model is a recent meta-analysis reporting the presence of cognitive deficits early in the illness trajectory, developing before the initial emergence of symptoms (Bora & Murray, 2014). However, other studies suggest that subtle changes in perception are present at early stages (Keri & Benedek, 2007; Mittal, Gupta, Keane, & Silverstein, 2015; Shin et al., 2012), consistent with a bottom-up model as the primary, etiological framework. To date, sensory and higher-order cognitive processes have not been measured concurrently in longitudinal studies of the prodromal phases of illness; thus, the timing of changes in sensory versus higher-order processes in psychotic illness remains unclear. Studies with such a longitudinal design are currently ongoing, and will likely provide some answers to these questions in the coming years.

In summary, a central gap in our understanding of the models presented in this chapter arises from the limited information available about the neurobiological mechanisms underlying the associations detected in prior work (e.g., the statistical associations illustrated in Figs. 1–3). We have described some illustrative scenarios for the purpose of conveying how these associations might result from certain hypothetical sequences of sensory and cognitive processes. However, to date, the specific neural processes accounting for these statistical relationships are largely unknown. Hypothesis-driven studies examining the effects of interventions that perturb the neural systems implicated by earlier, correlative neuroimaging work may reveal that altering certain circuits, but not others, cause downstream changes in social cognition and negative symptoms. Studies using genetic and/or behavioral manipulations in animals could also complement such intervention work in humans. In addition, studies that identify social cognitive mechanisms linked to distinct negative symptoms, such as asociality versus avolition, and anticipatory versus consummatory aspects of anhedonia, will further advance our understanding of these relationships. These types of convergent, multidisciplinary efforts can then allow novel or existing treatments, such as facial affect recognition training (Frommann, Streit, & Wolwer, 2003; Wolwer et al., 2005), cognitive and social cognitive remediation (Glenthoj, Hjorthoj, Kristensen, Davidson, & Nordentoft, 2017; Lindenmayer et al., 2013; Wykes, Huddy, Cellard, McGurk, & Czobor, 2011), and neuromodulatory interventions (Dokucu, 2015) to be further developed or refined to target the key neural processes underlying these debilitating symptoms.

## References

Adolphs, R. (2009). The social brain: neural basis of social knowledge. *Annual Review of Psychology, 60*, 693–716.

Andrzejewska, M., Wójciak, P., Domowicz, K., & Rybakowski, J. (2017). Emotion recognition and theory of mind in chronic schizophrenia: association with negative symptoms. *Archives of Psychiatry and Psychotherapy, 19*(4), 7–12.

Badre, D. (2008). Cognitive control, hierarchy, and the rostro-caudal organization of the frontal lobes. *Trends in Cognitive Sciences, 12*(5), 193–200.

Barch, D. M., & Ceaser, A. (2012). Cognition in schizophrenia: core psychological and neural mechanisms. *Trends in Cognitive Sciences, 16*(1), 27–34.

Barch, D. M., Yodkovik, N., Sypher-Locke, H., & Hanewinkel, M. (2008). Intrinsic motivation in schizophrenia: relationships to cognitive function, depression, anxiety, and personality. *Journal of Abnormal Psychology, 117*(4), 776–787.

Baron-Cohen, S., Wheelwright, S., Hill, J., Raste, Y., & Plumb, I. (2001). The "Reading the Mind in the Eyes" Test revised version: a study with normal adults, and adults with Asperger syndrome or high-functioning autism. *Journal of Child Psychology and Psychiatry, 42*(2), 241–251.

Blanchard, J. J., & Cohen, A. S. (2006). The structure of negative symptoms within schizophrenia: implications for assessment. *Schizophrenia Bulletin, 32*(2), 238–245.

Bora, E., & Murray, R. M. (2014) Meta-analysis of cognitive deficits in ultra-high risk to psychosis and first-episode psychosis: do the cognitive deficits progress over, or after, the onset of psychosis? *Schizophrenia Bulletin, 40*(4), 744–755.

Bortolon, C., Capdevielle, D., & Raffard, S. (2015). Face recognition in schizophrenia disorder: a comprehensive review of behavioral, neuroimaging and neurophysiological studies. *Neuroscience and Biobehavioral Reviews, 53*, 79–107.

Bramon, E., Rabe-Hesketh, S., Sham, P., Murray, R. M., & Frangou, S. (2004). Meta-analysis of the P300 and P50 waveforms in schizophrenia. *Schizophrenia Research, 70*(2–3), 315–329.

Bremmer, F., Schlack, A., Shah, N. J., Zafiris, O., Kubischik, M., Hoffmann, K., ... Fink, G. R. (2001). Polymodal motion processing in posterior parietal and premotor cortex: a human fMRI study strongly implies equivalencies between humans and monkeys. *Neuron, 29*(1), 287–296.

Brockhaus-Dumke, A., Schultze-Lutter, F., Mueller, R., Tendolkar, I., Bechdolf, A., Pukrop, R., ... Ruhrmann, S. (2008). Sensory gating in schizophrenia: P50 and N100 gating in antipsychotic-free subjects at risk, first-episode, and chronic patients. *Biological Psychiatry, 64*(5), 376–384.

Butler, P. D., & Javitt, D. C. (2005). Early-stage visual processing deficits in schizophrenia. *Current Opinion in Psychiatry, 18*(2), 151–157.

Butler, P. D., Martinez, A., Foxe, J. J., Kim, D., Zemon, V., Silipo, G., ... Javitt, D. C. (2007). Subcortical visual dysfunction in schizophrenia drives secondary cortical impairments. *Brain, 130*(Pt 2), 417–430.

Butler, P. D., Silverstein, S. M., & Dakin, S. C. (2008). Visual perception and its impairment in schizophrenia. *Biological Psychiatry, 64*(1), 40–47.

Campellone, T. R., Sanchez, A. H., & Kring, A. M. (2016). Defeatist performance beliefs, negative symptoms, and functional outcome in schizophrenia: a meta-analytic review. *Schizophrenia Bulletin, 42*(6), 1343–1352.

Clark, A. (2013). Whatever next? Predictive brains, situated agents, and the future of cognitive science. *The Behavioral and Brain Sciences, 36*(3), 181–204.

Corbetta, M., Patel, G., & Shulman, G. L. (2008). The reorienting system of the human brain: from environment to theory of mind. *Neuron, 58*(3), 306–324.

Corbetta, M., & Shulman, G. L. (2002). Control of goal-directed and stimulus-driven attention in the brain. *Nature Reviews. Neuroscience, 3*(3), 201–215.

Corigliano, V., De Carolis, A., Trovini, G., Dehning, J., Di Pietro, S., Curto, M., ... Comparelli, A. (2014). Neurocognition in schizophrenia: from prodrome to multi-episode illness. *Psychiatry Research, 220*(1–2), 129–134.

Corlett, P. R., & Fletcher, P. C. (2015). Delusions and prediction error: clarifying the roles of behavioural and brain responses. *Cognitive Neuropsychiatry, 20*(2), 95–105.

Corlett, P. R., Honey, G. D., & Fletcher, P. C. (2016). Prediction error, ketamine and psychosis: an updated model. *Journal of Psychopharmacology, 30*(11), 1145–1155.

Couture, S. M., Blanchard, J. J., & Bennett, M. E. (2011). Negative expectancy appraisals and defeatist performance beliefs and negative symptoms of schizophrenia. *Psychiatry Research, 189*(1), 43–48.

Dokucu, M. E. (2015). Neuromodulation treatments for schizophrenia. *Current Treatment Options in Psychiatry, 2*(3), 339–348.

Doniger, G. M., Silipo, G., Rabinowicz, E. F., Snodgrass, J. G., & Javitt, D. C. (2001). Impaired sensory processing as a basis for object-recognition deficits in schizophrenia. *American Journal of Psychiatry, 158*(11), 1818–1826.

Elshaikh, A. A., Sponheim, S. R., Chafee, M. V., & MacDonald, A. W. (2015). Spatial attentional control is not impaired in schizophrenia: dissociating specific deficits from generalized impairments. *Journal of Abnormal Psychology, 124*(2), 302–308.

Evans, J. S., & Stanovich, K. E. (2013). Dual-process theories of higher cognition: advancing the debate. *Perspectives on Psychological Science, 8*(3), 223–241.

Fatouros-Bergman, H., Cervenka, S., Flyckt, L., Edman, G., & Farde, L. (2014). Meta-analysis of cognitive performance in drug-naive patients with schizophrenia. *Schizophrenia Research, 158*(1–3), 156–162.

Fervaha, G., Foussias, G., Agid, O., & Remington, G. (2014). Impact of primary negative symptoms on functional outcomes in schizophrenia. *European Psychiatry, 29*(7), 449–455.

Foussias, G., Agid, O., Fervaha, G., & Remington, G. (2014). Negative symptoms of schizophrenia: clinical features, relevance to real world functioning and specificity versus other CNS disorders. *European Neuropsychopharmacology, 24*(5), 693–709.

Friston, K. J., Stephan, K. E., Montague, R., & Dolan, R. J. (2014). Computational psychiatry: the brain as a phantastic organ. *The Lancet Psychiatry, 1*(2), 148–158.

Frommann, N., Streit, M., & Wolwer, W. (2003). Remediation of facial affect recognition impairments in patients with schizophrenia: a new training program. *Psychiatry Research, 117*(3), 281–284.

Garrido, M. I., Kilner, J. M., Stephan, K. E., & Friston, K. J. (2009). The mismatch negativity: a review of underlying mechanisms. *Clinical Neurophysiology, 120*(3), 453–463.

Giesbrecht, B., Woldorff, M. G., Song, A. W., & Mangun, G. R. (2003). Neural mechanisms of top-down control during spatial and feature attention. *NeuroImage, 19*(3), 496–512.

Gilbert, C. D., & Sigman, M. (2007). Brain states: top-down influences in sensory processing. *Neuron, 54*(5), 677–696.
Glenthoj, L. B., Hjorthoj, C., Kristensen, T. D., Davidson, C. A., & Nordentoft, M. (2017). The effect of cognitive remediation in individuals at ultra-high risk for psychosis: a systematic review. *NPJ Schizophrenia, 3*, 20.
Gold, J. M., Waltz, J. A., Matveeva, T. M., Kasanova, Z., Strauss, G. P., Herbener, E. S., ... Frank, M. J. (2012). Negative symptoms and the failure to represent the expected reward value of actions: behavioral and computational modeling evidence. *Archives of General Psychiatry, 69*(2), 129–138.
Grant, P. M., & Beck, A. T. (2009). Defeatist beliefs as a mediator of cognitive impairment, negative symptoms, and functioning in schizophrenia. *Schizophrenia Bulletin, 35*(4), 798–806.
Green, M. F., Hellemann, G., Horan, W. P., Lee, J., & Wynn, J. K. (2012). From perception to functional outcome in schizophrenia: modeling the role of ability and motivation. *Archives of General Psychiatry, 69*(12), 1216–1224.
Green, M. F., Horan, W. P., & Lee, J. (2015). Social cognition in schizophrenia. *Nature Reviews. Neuroscience, 16*(10), 620–631.
Green, M. F., Penn, D. L., Bentall, R., Carpenter, W. T., Gaebel, W., Gur, R. C., ... Heinssen, R. (2008). Social cognition in schizophrenia: an NIMH workshop on definitions, assessment, and research opportunities. *Schizophrenia Bulletin, 34*(6), 1211–1220.
Griffin, J. D., & Fletcher, P. C. (2017). Predictive processing, source monitoring, and psychosis. *Annual Review of Clinical Psychology, 13*, 265–289.
Grunwald, T., Boutros, N. N., Pezer, N., von Oertzen, J., Fernandez, G., Schaller, C., & Elger, C. E. (2003). Neuronal substrates of sensory gating within the human brain. *Biological Psychiatry, 53*(6), 511–519.
Hoe, M., Nakagami, E., Green, M. F., & Brekke, J. S. (2012). The causal relationships between neurocognition, social cognition and functional outcome over time in schizophrenia: a latent difference score approach. *Psychological Medicine, 42*(11), 2287–2299.
Holt, D. J., Boeke, E. A., Coombs, G., 3rd., DeCross, S. N., Cassidy, B. S., Stufflebeam, S., ... Tootell, R. B. (2015). Abnormalities in personal space and parietal-frontal function in schizophrenia. *Neuroimage Clinical, 9*, 233–243. https://doi.org/10.1016/j.nicl.2015.07.008.
Holt, D. J., Cassidy, B. S., Yue, X., Rauch, S. L., Boeke, E. A., Nasr, S., ... Coombs, G., 3rd. (2014). Neural correlates of personal space intrusion. *The Journal of Neuroscience, 34*(12), 4123–4134.
Holt, D. J., Coombs, G., Zeidan, M. A., Goff, D. C., & Milad, M. R. (2012). Failure of neural responses to safety cues in schizophrenia. *Archives of General Psychiatry, 69*(9), 893–903.
Holt, D. J., Lebron-Milad, K., Milad, M. R., Rauch, S. L., Pitman, R. K., Orr, S. P., ... Goff, D. C. (2009). Extinction memory is impaired in schizophrenia. *Biological Psychiatry, 65*(6), 455–463.
Jahshan, C., Wynn, J. K., & Green, M. F. (2013). Relationship between auditory processing and affective prosody in schizophrenia. *Schizophrenia Research, 143*(2–3), 348–353.
Javitt, D. C. (2009). Sensory processing in schizophrenia: neither simple nor intact. *Schizophrenia Bulletin, 35*(6), 1059–1064.
Jensen, J., Willeit, M., Zipursky, R. B., Savina, I., Smith, A. J., Menon, M., ... Kapur, S. (2008). The formation of abnormal associations in schizophrenia: neural and behavioral evidence. *Neuropsychopharmacology, 33*(3), 473–479.
Kalkstein, S., Hurford, I., & Gur, R. C. (2010). Neurocognition in schizophrenia. *Current Topics in Behavioral Neurosciences, 4*, 373–390.
Kantrowitz, J. T., Leitman, D. I., Lehrfeld, J. M., Laukka, P., Juslin, P. N., Butler, P. D., ... Javitt, D. C. (2013). Reduction in tonal discriminations predicts receptive emotion

processing deficits in schizophrenia and schizoaffective disorder. *Schizophrenia Bulletin, 39*(1), 86–93.

Keri, S., & Benedek, G. (2007). Visual contrast sensitivity alterations in inferred magnocellular pathways and anomalous perceptual experiences in people at high-risk for psychosis. *Visual Neuroscience, 24*(2), 183–189.

Kirkpatrick, B., Fenton, W. S., Carpenter, W. T., Jr., & Marder, S. R. (2006). The NIMH-MATRICS consensus statement on negative symptoms. *Schizophrenia Bulletin, 32*(2), 214–219.

Kirkpatrick, B., & Fischer, B. (2006). Subdomains within the negative symptoms of schizophrenia: commentary. *Schizophrenia Bulletin, 32*(2), 246–249.

Korzyukov, O., Pflieger, M. E., Wagner, M., Bowyer, S. M., Rosburg, T., Sundaresan, K., … Boutros, N. N. (2007). Generators of the intracranial P50 response in auditory sensory gating. *NeuroImage, 35*(2), 814–826.

Kreither, J., Lopez-Calderon, J., Leonard, C. J., Robinson, B. M., Ruffle, A., Hahn, B., … Luck, S. J. (2017). Electrophysiological evidence for hyperfocusing of spatial attention in schizophrenia. *The Journal of Neuroscience, 37*(14), 3813–3823.

Kuhn, G., Teszka, R., Tenaw, N., & Kingstone, A. (2016). Don't be fooled! Attentional responses to social cues in a face-to-face and video magic trick reveals greater top-down control for overt than covert attention. *Cognition, 146*, 136–142.

Lesh, T. A., Niendam, T. A., Minzenberg, M. J., & Carter, C. S. (2011). Cognitive control deficits in schizophrenia: mechanisms and meaning. *Neuropsychopharmacology, 36*(1), 316–338.

Lewandowski, K. E., Whitton, A. E., Pizzagalli, D. A., Norris, L. A., Ongur, D., & Hall, M. H. (2016). Reward learning, neurocognition, social cognition, and symptomatology in psychosis. *Frontiers in Psychiatry*, 7, 100.

Lieberman, M. D. (2007). Social cognitive neuroscience: a review of core processes. *Annual Review of Psychology, 58*, 259–289.

Lin, C. -H., Huang, C. -L., Chang, Y. -C., Chen, P. -W., Lin, C. -Y., Tsai, G. E., & Lane, H. -Y. (2013). Clinical symptoms, mainly negative symptoms, mediate the influence of neurocognition and social cognition on functional outcome of schizophrenia. *Schizophrenia Research, 146*(1), 231–237.

Lindenmayer, J. P., McGurk, S. R., Khan, A., Kaushik, S., Thanju, A., Hoffman, L., … Herrmann, E. (2013). Improving social cognition in schizophrenia: a pilot intervention combining computerized social cognition training with cognitive remediation. *Schizophrenia Bulletin, 39*(3), 507–517.

Mano, Q. R., & Brown, G. G. (2013). Cognition-emotion interactions in schizophrenia: emerging evidence on working memory load and implicit facial-affective processing. *Cognition & Emotion, 27*(5), 875–899.

Matsumoto, Y., Takahashi, H., Murai, T., & Takahashi, H. (2015). Visual processing and social cognition in schizophrenia: relationships among eye movements, biological motion perception, and empathy. *Neuroscience Research, 90*, 95–100.

Mayer, A. R., Hanlon, F. M., Franco, A. R., Teshiba, T. M., Thoma, R. J., Clark, V. P., & Canive, J. M. (2009). The neural networks underlying auditory sensory gating. *NeuroImage, 44*(1), 182–189.

Mechelli, A., Price, C. J., Friston, K. J., & Ishai, A. (2004). Where bottom-up meets top-down: neuronal interactions during perception and imagery. *Cerebral Cortex, 14*(11), 1256–1265.

Messinger, J. W., Tremeau, F., Antonius, D., Mendelsohn, E., Prudent, V., Stanford, A. D., & Malaspina, D. (2011). Avolition and expressive deficits capture negative symptom phenomenology: implications for DSM-5 and schizophrenia research. *Clinical Psychology Review, 31*(1), 161–168.

Millan, M. J., Fone, K., Steckler, T., & Horan, W. P. (2014). Negative symptoms of schizophrenia: clinical characteristics, pathophysiological substrates, experimental models and prospects for improved treatment. *European Neuropsychopharmacology, 24*(5), 645–692.

Mittal, V. A., Gupta, T., Keane, B. P., & Silverstein, S. M. (2015). Visual context processing dysfunctions in youth at high risk for psychosis: resistance to the Ebbinghaus illusion and its symptom and social and role functioning correlates. *Journal of Abnormal Psychology, 124*(4), 953–960.

Pousa, E., Duno, R., Blas Navarro, J., Ruiz, A. I., Obiols, J. E., & David, A. S. (2008). Exploratory study of the association between insight and Theory of Mind (ToM) in stable schizophrenia patients. *Cognitive Neuropsychiatry, 13*(3), 210–232.

Powers, A. R., III, Kelley, M., & Corlett, P. R. (2016). Hallucinations as top-down effects on perception. *Biological Psychiatry: Cognitive Neuroscience and Neuroimaging, 1*(5), 393–400.

Rassovsky, Y., Horan, W. P., Lee, J., Sergi, M. J., & Green, M. F. (2011). Pathways between early visual processing and functional outcome in schizophrenia. *Psychological Medicine, 41*(3), 487–497.

Rominger, C., Bleier, A., Fitz, W., Marksteiner, J., Fink, A., Papousek, I., & Weiss, E. M. (2016). Auditory top-down control and affective theory of mind in schizophrenia with and without hallucinations. *Schizophrenia Research, 174*(1–3), 192–196.

Sawaki, R., Kreither, J., Leonard, C. J., Kaiser, S. T., Hahn, B., Gold, J. M., & Luck, S. J. (2017). Hyperfocusing of attention on goal-related information in schizophrenia: evidence from electrophysiology. *Journal of Abnormal Psychology, 126*(1), 106–116.

Schaefer, J., Giangrande, E., Weinberger, D. R., & Dickinson, D. (2013). The global cognitive impairment in schizophrenia: consistent over decades and around the world. *Schizophrenia Research, 150*(1), 42–50.

Schnakenberg Martin, A. M., Bartolomeo, L., Howell, J., Hetrick, W. P., Bolbecker, A. R., Breier, A., … O'Donnell, B. F. (2017). Auditory feature perception and auditory hallucinatory experiences in schizophrenia spectrum disorder. *European Archives of Psychiatry and Clinical Neuroscience, 268*(7), 653–661.

Schubring, D., Popov, T., Miller, G. A., & Rockstroh, B. (2018). Consistency of abnormal sensory gating in first-admission and chronic schizophrenia across quantification methods. *Psychophysiology, 55*(4).

Seeley, W. W., Menon, V., Schatzberg, A. F., Keller, J., Glover, G. H., Kenna, H., … Greicius, M. D. (2007). Dissociable intrinsic connectivity networks for salience processing and executive control. *The Journal of Neuroscience, 27*(9), 2349–2356.

Sergi, M. J., Rassovsky, Y., Widmark, C., Reist, C., Erhart, S., Braff, D. L., … Green, M. F. (2007). Social cognition in schizophrenia: relationships with neurocognition and negative symptoms. *Schizophrenia Research, 90*(1–3), 316–324.

Shin, K. S., Kim, J. S., Kim, S. N., Koh, Y., Jang, J. H., An, S. K., … Kwon, J. S. (2012). Aberrant auditory processing in schizophrenia and in subjects at ultra-high-risk for psychosis. *Schizophrenia Bulletin, 38*(6), 1258–1267.

Silver, M. A., & Kastner, S. (2009). Topographic maps in human frontal and parietal cortex. *Trends in Cognitive Sciences, 13*(11), 488–495.

Strauss, G. P., Harrow, M., Grossman, L. S., & Rosen, C. (2010). Periods of recovery in deficit syndrome schizophrenia: a 20-year multi-follow-up longitudinal study. *Schizophrenia Bulletin, 36*(4), 788–799.

Strauss, G. P., Horan, W. P., Kirkpatrick, B., Fischer, B. A., Keller, W. R., Miski, P., … Carpenter, W. T., Jr. (2013). Deconstructing negative symptoms of schizophrenia: avolition-apathy and diminished expression clusters predict clinical presentation and functional outcome. *Journal of Psychiatric Research, 47*(6), 783–790.

Sutliff, S., Roy, M. A., & Achim, A. M. (2015). Social anxiety disorder in recent onset schizophrenia spectrum disorders: the relation with symptomatology, anxiety, and social rank. *Psychiatry Research, 227*(1), 39–45.

Thomas, M. L., Green, M. F., Hellemann, G., Sugar, C. A., Tarasenko, M., Calkins, M. E., … Light, G. A. (2017). Modeling deficits from early auditory information processing to psychosocial functioning in schizophrenia. *JAMA Psychiatry, 74*(1), 37–46.

Umbricht, D., & Krljes, S. (2005). Mismatch negativity in schizophrenia: a meta-analysis. *Schizophrenia Research, 76*(1), 1–23.

Ventura, J., Subotnik, K. L., Gitlin, M. J., Gretchen-Doorly, D., Ered, A., Villa, K. F., … Nuechterlein, K. H. (2015). Negative symptoms and functioning during the first year after a recent onset of schizophrenia and 8 years later. *Schizophrenia Research, 161*(2–3), 407–413.

Williams, T. J., Nuechterlein, K. H., Subotnik, K. L., & Yee, C. M. (2011). Distinct neural generators of sensory gating in schizophrenia. *Psychophysiology, 48*(4), 470–478.

Wolwer, W., Frommann, N., Halfmann, S., Piaszek, A., Streit, M., & Gaebel, W. (2005). Remediation of impairments in facial affect recognition in schizophrenia: efficacy and specificity of a new training program. *Schizophrenia Research, 80*(2–3), 295–303.

Wyer, R. S., Jr., & Srull, T. K. (2014). Social cognition and clinical psychology: anxiety, depression, and the processing of social information. In *Handbook of social cognition* (2nd ed., pp. 305–354): Psychology Press.

Wykes, T., Huddy, V., Cellard, C., McGurk, S. R., & Czobor, P. (2011). A meta-analysis of cognitive remediation for schizophrenia: methodology and effect sizes. *The American Journal of Psychiatry, 168*(5), 472–485.

CHAPTER SIX

# Role of oxytocin in social cognition in psychosis spectrum disorders


Caridad Benavides*, Sharely Fred-Torres*, Sarah Barbara Rutter*, Emmett Larsen*, Jarrett Fastman*, Andrea Bulbena-Cabre*, Lauren Lepow*, Katie Mahon*, Manuela Russo*, Allison K. Ungar*, Katherine Elizabeth Burdick†, Maria Mercedes Perez-Rodriguez*

*Department of Psychiatry, Icahn School of Medicine at Mount Sinai, New York, NY, United States
†Brigham and Women's Hospital, Department of Psychiatry, Harvard Medical School, Boston, MA, United States


## Oxytocin and social cognition

### Biology and mechanism

Oxytocin is a nonapeptide hormone synthesized by magnocellular and parvocellular neurons in the paraventricular and supraoptic nuclei of the hypothalamus. These neurons project to the posterior pituitary, where newly synthesized oxytocin is packaged and stored into secretory vesicles before being released into the periphery. Oxytocin-producing neurons also project to subcortical brain regions, including the amygdala, hippocampus, striatum, suprachiasmatic nucleus, bed nucleus of stria terminalis, and brainstem, and have recently been demonstrated to have cortical projections as well (Rogers et al., 2018). Additionally, oxytocin released from dendrites diffuses through the extracellular space to act on local and distant brain targets (Meyer-Lindenberg, Domes, Kirsch, & Heinrichs, 2011). The functional anatomy of the oxytocin system reflects the complexity of its actions as a neuromodulator and neurotransmitter, and much of the research into oxytocin's effects on the brain has focused on its role in social cognition and behavior. It is believed that the activities of oxytocin and its closely related neuropeptide, vasopressin, are balanced in dynamic fashion within a number of key brain regions that regulate social functioning (Neumann, Maloumby, Beiderbeck, Lukas, & Landgraf, 2013). There are four different receptors for oxytocin and vasopressin in humans: the OTR, and three vasopressin receptors 1A (AVPR1A), 1B (AVPR1B), and 2 (AVPR2). Though


*Social Cognition in Psychosis*
https://doi.org/10.1016/B978-0-12-815315-4.00006-9

there are a number of regions in the human brain that are known to receive oxytocinergic and vasopressinergic input, the distribution of these receptors has not yet been fully characterized (Meyer-Lindenberg et al., 2011). Current evidence suggests that oxytocin receptors are localized in multiple brain regions, including areas involved in reward and social attachment (e.g., Nucleus accumbens, NAcc), social cognition (amygdala, PFC—prefrontal cortex) and visual processing regions in humans (Bethlehem et al., 2017; Boccia, Petrusz, Suzuki, Marson, & Pedersen, 2013; Lee et al., 2017; Uhrig et al., 2016).

Oxytocin has been shown to play an important role in regulating several aspects of social cognition, including trust and cooperation, social memory, stress response, and the ability to recognize and interpret emotions and mental states in others (Bartz, Zaki, Bolger, & Ochsner, 2011; Bos, Panksepp, Bluthe, & van Honk, 2012; Gumley, Braehler, & Macbeth, 2014; Meyer-Lindenberg et al., 2011). Higher plasma oxytocin levels have recently been associated with better performance on difficult items in a social interaction task (Deuse et al., 2018). Studies on the effects of exogenous oxytocin administration in humans have generally shown "prosocial" effects such as increased trust, reduced social anxiety, and promotion of approach behaviors (Bartz, Zaki, et al., 2011; Bos et al., 2012; Gumley et al., 2014; Meyer-Lindenberg et al., 2011). However, two recent studies found that exogenous oxytocin administered intranasally in healthy subjects increased aggressive and antagonistic behaviors in response to provocation (Ne'eman, Perach-Barzilay, Fischer-Shofty, Atias, & Shamay-Tsoory, 2016; Pfundmair, Reinelt, DeWall, & Feldmann, 2018). Taken together, these findings suggest that oxytocin has a broad, complex role in human behavior centered on modulating the salience of social stimuli, and may not exert exclusively prosocial effects.

Initial studies on the effects of intravenously administered exogenous oxytocin failed to produce significant results (Bakermans-Kranenburg & van I. Jzendoorn, 2013). Given the route of administration, a likely explanation for the absence of behavioral effects is that oxytocin is not able to cross the blood-brain barrier (BBB). Contrastingly, subsequent studies employing intranasal administration of oxytocin have found reproducible effects on social behaviors and cognition (Bakermans-Kranenburg & van I. Jzendoorn, 2013), as well as alterations in brain activity with functional imaging (Ide et al., 2018; Perry et al., 2010; Riem et al., 2011; Rilling, Chen, Chen, & Haroon, 2018; Rutherford et al., 2018). Administering oxytocin intranasally theoretically provides a route for the drug to enter the central nervous system directly via the nasal mucosa(Quintana, Alvares, Hickie, & Guastella, 2015), though there is still debate about whether this actually occurs (Zink & Meyer-Lindenberg, 2012). Of note, intranasal administration of oxytocin has been shown to result in increased oxytocin levels in cerebrospinal fluid (Chang, Barter, Ebitz, Watson, & Platt, 2012; Dal Monte, Noble, Turchi, Cummins, & Averbeck, 2014; Modi, Connor-Stroud, Landgraf, Young, & Parr, 2014; Striepens et al., 2013), and the amygdala and hippocampus (Neumann et al., 2013) in humans (Striepens et al., 2013) and animals (Dal Monte et al., 2014; Modi et al., 2014; Neumann et al., 2013).

Nonhuman studies have provided some insight into the mechanism of action of intranasal oxytocin, painting a complex picture. Behavioral effects have been seen following injection of oxytocin directly into certain brain regions (Bielsky & Young, 2004), and intranasal administration has been found to increase oxytocin concentrations in the brain (Neumann et al., 2013) and cerebrospinal fluid (CSF) (Modi et al., 2014). Complicating these results is a recent study in macaques that found comparable levels of exogenous oxytocin in CSF after both intranasal and intravenous administration (Lee et al., 2017). Importantly, there was no increase seen in endogenous oxytocin levels following the administration by either route, providing evidence against the notion that exogenous oxytocin exerts its effects on the brain by causing central release of endogenous oxytocin (Lee et al., 2017). Studies in rodents have found divergent effects on social behaviors and oxytocin receptor (OTR) expression, with varying dose and treatment schedules, suggesting that chronic treatment and high dosages may produce effects on the brain and behavior that directly oppose those seen with acute treatment at carefully titrated dosages (Guoynes et al., 2018; Huang et al., 2013). Taken together, these results seem to indicate that intranasal oxytocin is capable of acting on the central nervous system directly, though the mechanism remains poorly understood, and much more work will need to be done to determine the relevance of these animal studies to humans (Guastella et al., 2013).

The available evidence suggests that intranasally administered oxytocin modulates a social cognitive network within the brain by acting on central targets (Love, 2014). This social cognitive network, which is important for social functioning (Pinkham, Hopfinger, Ruparel, & Penn, 2008), is thought to be regulated largely by nuclei in the amygdala (Li, Chan, McAlonan, & Gong, 2010). Oxytocin has been shown to play an important role in social cognition and emotion regulation via its effects on frontolimbic network dynamics, particularly on communication between the amygdala and frontal emotion-processing regions such as the medial prefrontal and anterior cingulate cortices (Eckstein et al., 2017; Meyer-Lindenberg et al., 2011). Intranasal oxytocin is capable of causing complex shifts in network dynamics in healthy controls (HCs), increasing functional connectivity between the frontal cortex and amygdala, and it has been found to reduce coupling of the ventral attention network to regions involved in processing internal stimuli, and increase its coupling with regions involved in processing external stimuli (Brodmann, Gruber, & Goya-Maldonado, 2017; Eckstein et al., 2017; Sripada et al., 2013). The extent to which intranasal oxytocin increases functional connectivity between the amygdala and mediofrontal cortex has been negatively correlated with depressive symptom severity, suggesting that oxytocin is less effective at modulating amygdala function in patients with affective disorders and weakened functional connectivity, which may be a useful biomarker for this population (Eckstein et al., 2017; Sripada et al., 2013).

Oxytocin also has an important role in the stress response, especially in social contexts, as it interacts with serotonergic and corticotropin-releasing factor systems to produce anxiolytic and antidepressant effects, in addition to influencing the stress-coping style (Lefevre, Mottolese, et al., 2017; Neumann & Landgraf, 2012). In response to social stressors, oxytocin has been shown to dampen the activity of the hypothalamic-pituitary-adrenal (HPA) axis and limbic regions, including the amygdala (Ditzen et al., 2009; Quirin, Kuhl, & Dusing, 2011; Smith et al., 2016; Zink & Meyer-Lindenberg, 2012). This reduction in activity is thought to lessen social anxiety and promote approach behavior (Radke et al., 2017). Positive social interactions have been shown to have stress-protective anxiolytic effects, and data suggest that oxytocin mediates this phenomenon (Meyer-Lindenberg et al., 2011).

The "prosocial" effects seen after oxytocin administration are likely to be a manifestation of enhancements in cognitive processes that underlie the ability to recognize and understand the emotions and mental states of others. These cognitive changes may promote prosocial behavior by increasing the salience of social stimuli and enhancing the effects of social interaction on the brain's reward system (Groppe et al., 2013; Ne'eman et al., 2016; Stavropoulos & Carver, 2013). Exogenous oxytocin has been shown to modulate dopamine release in the mesocorticolimbic reward pathway (Gordon et al., 2016; Love, 2014; Shamay-Tsoory & Abu-Akel, 2016). Oxytocinergic neurons project from the paraventricular nucleus (PVN) to key limbic sites on the dopaminergic mesocorticolimbic pathway (e.g., VTA—ventral tegmental area, NAcc, and amygdala), allowing it to modulate dopaminergic activity. Through modulating the PFC, motivational factors can influence attention to and perception of social cues (Gordon et al., 2016). Consistent with this model (Gordon et al., 2016), intranasal oxytocin changes fMRI activity (Wang, Yan, Li, & Ma, 2017) and connectivity, both within the mesocorticolimbic reward pathway, and also between mesolimbic sites and cortical regions that contribute to social perception (e.g., ventro medial PFC and orbitofrontal cortex, OFC) (Bethlehem et al., 2017; Gordon et al., 2016). An increase in NAcc-vmPFC/OFC connectivity under oxytocin may indicate increased valuation of the social stimulus under oxytocin versus a placebo (Gordon et al., 2016).

There is evidence that social cognitive enhancement is correlated with pupil dilation and an increased firing rate in the locus coeruleus following oxytocin administration (Leknes et al., 2013; Prehn et al., 2013). Acute oxytocin treatment causes an attentional shift to the eye region of facial stimuli (Bertsch et al., 2013; Brune et al., 2013; Domes et al., 2013; Tollenaar, Chatzimanoli, van der Wee, & Putman, 2013), and has recently been demonstrated to increase attention to faces to the level of HCs in patients with autism spectrum disorders. Evidence suggests that oxytocin produces its effects on attentional components of social cognition by modulating the activity of the amygdala and causing shifts in the coupling of subcortical attention networks that favor orientation to external stimuli (Eckstein et al., 2017; Li et al., 2010; Pinkham et al., 2011).

The effects of oxytocin on social cognition have been described using two opposing theoretical models (Bartz, Zaki, et al., 2011; Meyer-Lindenberg et al., 2011; Perez-Rodriguez, Mahon, Russo, Ungar, & Burdick, 2014). The "interactionist" model (Bartz, Zaki, et al., 2011) posits that an individual's baseline level of social cognitive functioning (and presumably the baseline oxytocinergic "tone") predicts the effects of oxytocin on social cognition for that individual (Bartz et al., 2010; Fischer-Shofty et al., 2013; Leknes et al., 2013). This model posits that there is an inverted-U shaped dose-response relationship between brain oxytocin levels and neural activity/social cognitive performance (Wolf, Brune, & Assion, 2010). It assumes the existence of an oxytocinergic "sweet spot" where social cognition is optimized, implying that impairments in social cognition may improve with exogenous oxytocin, and excessive or distorted social cognition may be exacerbated by exogenous oxytocin (Bartz, Simeon, et al., 2011).

The "optimizing" model (Meyer-Lindenberg et al., 2011; Simeon et al., 2011) diverges from the interactionist model in theorizing that baseline social cognition is irrelevant, as oxytocin acts to optimize social cognition regardless of whether it is deficient or excessive/distorted at the baseline (Meyer-Lindenberg et al., 2011).

The interactionist model partitions individuals into three categories according to baseline levels of social cognitive functioning (Wolf et al., 2010). Those in the social cognitive "sweet spot" (e.g., HCs) demonstrate optimal social cognitive function, as assessed by processing of and attention to salient social stimuli, mentalizing accuracy, emotion recognition, and activity in brain networks important for social cognition. The second category includes individuals with baseline social cognitive impairment (e.g., schizophrenia with negative symptoms (Montag et al., 2011) and autism spectra), as defined by deficiencies in paying attention to and processing social stimuli, reduced activity in social cognitive networks, poor mentalizing accuracy and emotion recognition, and hypomentalizing errors (Montag et al., 2011). Some, but not all, studies suggest that oxytocin administration can alleviate these social cognitive deficits (Feifel et al., 2010; Feifel, Macdonald, Cobb, & Minassian, 2012; Goldman, Gomes, Carter, & Lee, 2011; Leknes et al., 2013; Macdonald & Feifel, 2012; Pedersen et al., 2011). The third category includes individuals with distorted social cognitive functioning at baseline (e.g., borderline personality disorder, schizophrenia (SZ) with prominent positive symptoms) (Montag et al., 2011). These individuals are excessively attentive to social stimuli (Bartz, Simeon, et al., 2011; Bartz, Zaki, et al., 2011), and their social functioning is impaired by distorted and inaccurate mentalizing (hypermentalizing) (Sharp et al., 2011). Studies on the effects of oxytocin in this third category have shown mixed results. In some cases, oxytocin produced positive effects (reducing excessive attention and tempering emotional reactions to social stimuli), while in others it produced negative effects (decreasing trust and cooperativity) (Bartz, Simeon, et al., 2011; Bertsch et al., 2013; Brune et al., 2013; Ebert et al., 2013; Simeon et al., 2011). Though there are data to suggest that patients with bipolar disorder (BD) (Stange et al., 2013) and actively psychotic paranoid SZ exhibit distorted social cognition and attentional biases toward emotional stimuli (Garety & Freeman, 1999; Kaney & Bentall, 1989), it remains unclear whether individuals with these conditions are part of this third category.

Additionally, there are a number of factors capable of interacting with the oxytocinergic system and augmenting the effects of exogenous oxytocin. These factors include an individual's gender, social context, attachment style, and history of childhood trauma and adversity (Bakermans-Kranenburg & van I. Jzendoorn, 2013; Bartz, Zaki, et al., 2011; Domes et al., 2010). For example, a study in healthy individuals found that effects of intranasal oxytocin on functional connectivity were limited to subjects who had supportive family backgrounds (Riem et al., 2011). Psychiatric or neurological illness can also interact in a complex fashion with oxytocin administration, as differential effects of intranasal oxytocin have been correlated with the presence and severity of social anxiety and subclinical depressive symptoms (Eckstein et al., 2017; Kanat et al., 2017).

Concomitant medications are likely a key confounder in clinical trials of exogenous oxytocin. For example, dopaminergic D2 blockers (such as antipsychotic medications commonly used in psychotic disorders) have been shown to block the effects of intranasal oxytocin on social behavior in animal models (Liu & Wang, 2003). In fact, the few studies that have examined the impact of antipsychotic dosage on the effects of oxytocin on social cognition have found a negative correlation between antipsychotic dosage and oxytocin efficacy (Bradley & Woolley, 2017).

## Oxytocinergic abnormalities in schizophrenia spectrum and bipolar disorders

Given the critical role of oxytocin in many aspects of social cognition, it is plausible that abnormalities in the oxytocin system contribute to psychopathology in SZ and BD, which often involve impairments in social cognition and social functioning, as described elsewhere in this book. Studies on endogenous oxytocin in SZ patients have shown varied results, suggesting a complex dysregulation (Aydın, Lysaker, Balıkçı, Ünal-Aydın, & Esen-Danacı, 2018; Rubin et al., 2018; Yang et al., 2017). While the relationship between endogenous oxytocin levels and the categorical diagnosis of SZ itself is unclear, there is some evidence of a link between oxytocin levels and symptom severity. Of note, the methods for assaying oxytocin levels differ significantly across studies. The validity of some of the assays and some of the sample processing procedures (i.e., forgoing extraction) have recently been criticized (Leng & Sabatier, 2016). It should also be noted that there is some controversy about whether there is a correlation between peripheral (e.g., serum or saliva) and central (e.g., CSF) oxytocin levels (Carson et al., 2015; Kagerbauer et al., 2013; Lefevre, Richard, et al., 2017; Martin et al.,

2018; Rutigliano et al., 2016; Valstad, Alvares, Andreassen, Westlye, & Quintana, 2016; Ziegler, 2018).

Of the studies that have assessed peripheral oxytocin levels in SZ patients compared with HCs, one reported an association between low oxytocin levels and severity of positive and negative symptoms, while higher oxytocin levels were associated with worse scores on cognitive measures (Rubin et al., 2018). These findings may be related to social cognition, because both negative and positive symptoms of schizophrenia have been shown to be correlated with social cognitive impairments (Montag et al., 2011; Perez-Rodriguez et al., 2017). Another study found lower oxytocin levels among SZ patients compared with HC, and that low oxytocin correlated with impairments in metacognitive function, but not social cognitive function in the SZ group (Aydın et al., 2018).

Notably, of the three published studies that have compared CSF oxytocin levels in SZ patients and HCs, only one found a significant difference between oxytocin levels in patients and controls (Beckmann, Lang, & Gattaz, 1985). Two of these studies did not explore the relationship between CSF oxytocin and negative SZ symptoms, cognition, or social functioning (Beckmann et al., 1985; Glovinsky, Kalogeras, Kirch, Suddath, & Wyatt, 1994). The third study found an inverse relationship between CSF oxytocin and negative SZ symptoms, although oxytocin levels were not different in SZ compared with HCs (Sasayama et al., 2012).

These results should be interpreted in the context of a broader emerging literature supporting the role of the neuropeptides vasopressin and oxytocin as biomarkers of social behavior and social functioning in human and non-human primates (Clark et al., 2013; Madrid et al., 2017; Parker et al., 2018). For example, several studies have shown a correlation between CSF oxytocin and vasopressin levels and social behavior and functioning in healthy individuals and across psychiatric disorders (Clark et al., 2013; Davis et al., 2013; Parker et al., 2014; Sasayama et al., 2012; Yuen et al., 2014).

Genetic and genomic studies also suggest a dysregulation of the oxytocinergic system in schizophrenia. One study found that oxytocin receptor expression is downregulated in SZ in the left posterior medial temporal gyrus, which is involved in social cognition (Uhrig et al., 2016). Another recent study found that both oxytocin and oxytocin receptor mRNA were elevated in the periphery of first-episode SZ patients compared with HCs, indicating that oxytocinergic abnormalities in SZ are complicated, and may change over the course of the illness (Yang et al., 2017). In an attempt to elucidate oxytocin's role in SZ pathogenesis, one recent study assessed polymorphisms in the oxytocin receptor gene (Veras et al., 2018). All five subjects with rare single nucleotide variants showed reduced negative symptomology, but similar positive symptomology compared with SZ patients in the study without this single-nucleotide polymorphism (SNP), and all had experienced early childhood trauma (Veras et al., 2018). Though very little research has been done in this area, these early results suggest that polymorphisms in components of the oxytocin system are relevant to SZ symptoms, and may predispose individuals to developing SZ in combination with other genetic risk factors and environmental risk factors such as childhood trauma or substance use.

There is very limited data available on oxytocinergic abnormalities in nonpsychotic patients within the schizophrenia spectrum (e.g., schizotypal personality disorder (SPD), psychometric schizotypy). Only one study (Tseng et al., 2014) has examined associations between plasma oxytocin levels and schizotypal traits in healthy individuals. They found that oxytocin levels were significantly positively correlated with the total and interpersonal domain scores on the Schizotypal Personality Questionnaire (SPQ) only in females.

Research on schizotypy in nonpsychotic forms provides insight into genetic and environmental vulnerability and protective factors for the development of schizophrenia-spectrum disorders. Moreover, studies of schizotypy avoid the confounds associated with schizophrenia research, such as antipsychotic medication use, chronic psychosis, and long-term institutionalization. Therefore, future studies examining putative oxytocinergic abnormalities in schizotypy are sorely needed.

If the role of oxytocin in SZ psychopathology remains unclear, its role in BD remains even more so. Only three studies have examined the relationship between endogenous serum oxytocin and BD (Lien et al., 2017; Ozsoy, Esel, & Kula, 2009; Turan, Uysal, Asdemir, & Kilic, 2013). One of these studies (Turan et al., 2013) found that compared with HCs, BD patients have elevated serum oxytocin levels, found to be positively associated with fearful emotion recognition accuracy in remitted BD I patients. Furthermore, this study (Turan et al., 2013) reported that patients experiencing a manic episode demonstrated significantly higher levels of oxytocin compared with remitted and depressed patients with BD I. Oxytocin in this study was measured at baseline, and after patients achieved a 50% reduction of symptoms after treatment with appropriate medications. No differences in pre- and posttreatment oxytocin levels were found, so that even after patients had positively responded to treatment, levels of oxytocin in BD I patients continued to be higher than baseline oxytocin levels in controls. Second, Lien and colleagues (Lien et al., 2017) extended these findings by reporting significantly higher baseline plasma oxytocin levels in drug-naive depressed patients with bipolar II disorder, compared with drug-naive patients with major depression and HCs. After improvement of depressive symptoms following treatment, oxytocin levels decreased only in patients with major depressive disorder (MDD). Thus, posttreatment oxytocin levels of BD II patients remained significantly higher than MDD and HCs. Contrasting both of these findings, the first of these studies that was conducted (Ozsoy et al., 2009) found that compared with HCs, serum oxytocin levels were decreased in patients with MDD and BD, pre- and posttreatment. Oxytocin was measured at baseline, and after patients received appropriate treatment with antidepressants and electroconvulsive therapy (ECT). There were no significant differences in oxytocin serum levels between MDD and BD.

Furthermore, in a population-based birth cohort study, administration of perinatal oxytocin intended to induce labor was associated with an increased risk of offspring with BD (Freedman, Brown, Shen, & Schaefer, 2015). Despite these preliminary findings pointing to oxytocinergic abnormalities in BD, more work will be needed before any conclusions can be drawn about the mechanisms by which oxytocin dysregulation interacts with symptomology. Furthermore, given the presence of social cognitive deficits in BD, it may be worthwhile to explore intranasal oxytocin as a potential treatment for these deficits.

An additional area of investigation with regard to oxytocin in SZ and BD involves its role in metabolic syndrome. Metabolic syndrome is the constellation of hypertension, high cholesterol, high blood sugar, and obesity. It is diagnosed in approximately one third of patients with psychotic disorders, and evidence suggests that disruptions in the normal functioning of the oxytocin system may represent a shared pathogenic mechanism between these conditions (Quintana, Dieset, Elvsåshagen, Westlye, & Andreassen, 2017). Given findings of reduced peripheral oxytocin in SZ patients, a recent study tested whether intranasal oxytocin increases satiety and reduces meal size in SZ patients with metabolic syndrome, but found no effect (Warren et al., 2018). Additional data will be needed to determine whether oxytocin may be effective at treating metabolic syndrome. Given the significant contribution of metabolic syndrome to morbidity and mortality in patients with psychotic disorders, a better understanding of oxytocin's role may eventually allow for metabolic and psychotic symptoms to be treated simultaneously by targeting the oxytocin system.

## Clinical trials of exogenous oxytocin in SZ

Clinical trials of intranasal oxytocin in SZ are relatively recent, and to date about 20 clinical trials (summarized in Table 1; see Bradley & Woolley, 2017; Bürkner, Williams, Simmons, & Woolley, 2017 for a review) have been conducted, specifically looking at social cognitive deficits and clinical symptoms.

A range of doses have been used, with the most common being 20–24 IU (International Units) and 40 IU. Most studies have demonstrated an effect of intranasal oxytocin improving facial affect recognition in SZ compared with placebo groups (Averbeck, Bobin, Evans, & Shergill, 2012; Fischer-Shofty, Shamay-Tsoory, & Levkovitz, 2013; Goldman et al., 2011), with a particular improvement in recognition of fearful faces in one study (Fischer-Shofty, Shamay-Tsoory, & Levkovitz, 2013). Only one study (Davis et al., 2013) failed to demonstrate significant improvements in facial affect recognition under oxytocin (OXT). However, this same study showed a large effect size ($d = 1.0$) of oxytocin on high-level social cognitive processes assessed by a Theory of Mind (ToM) task (the sarcasm subscale of The Awareness of Social Inference Test (TASIT-III)) (McDonald et al., 2006) and an empathy task; the Emotional Perspective Taking Task (EPTT) (Derntl et al., 2009). Pedersen and colleagues (Pedersen et al., 2011) also found a positive effect of OXT on ToM in SZ, but only for the more complex tasks of second-order beliefs. Others found an improved ability to determine the nature of relationships (e.g., platonic versus romantic) after OXT administration (Fischer-Shofty, Brune, et al., 2013). There are promising results for the effects of oxytocin on mentalizing (a high-level social cognitive domain) in SZ patients, and an overall trend for a significant impact of even single doses of OXT on higher-level social cognitive domains (Brambilla et al., 2016; Davis et al., 2013; Fischer-Shofty, Brune, et al., 2013; Gibson et al., 2014; Guastella et al., 2013; Pedersen et al., 2011). A recent metaanalysis confirmed that, across 12 randomized control trials (RCTs) of oxytocin in SZ, there is more of an effect on high- (e.g., ToM) versus low-level social cognitive domains (e.g., emotion recognition); and the effect of OXT on high-level domains is significant across trials (Bürkner et al., 2017). In terms of the effects of OXT on non-social cognition, some have found that OXT can have a positive effect on

Table 1 Overview of clinical trials on the effect of OXYTOCIN (OT) on social cognition and psychopathology in schizophrenia.

| Study | Sample[a] | Oxytocin intranasal dose | Design | Outcome measure | Results |
|---|---|---|---|---|---|
| Goldman et al. (2011) | 13 SZ<br>-. 5 with PD; age = 53 ± 3; males = 3<br>-. 8 without PD; age = 44 ± 9; males = 4<br>11 HCs<br>-. Age = 38 ± 13; males = 4 | 10 IU or 20 IU once a week | Double-blind, placebo-controlled crossover 3-week trial | -. Facial affect recognition | Compared with HCs, low OT dose (10 IU) increased facial affect intensity in PD group, but not non-PD group.<br>The 20 IU dose decreased intensity in the PD group relative to the non-PD group, but there were no differences compared with HCs.<br>OT reduced "fear" recognition in the PD group compared with non-PD.<br>Overall recognition worsened in both SZ groups with 10 IU dose, but improved in the SCZ PD group with 20 IU dose.<br>Overall, emotion recognition improved with the 20 IU dose in PD patients compared with non-PD patients. |

| | | | | | |
|---|---|---|---|---|---|
| Feifel et al. (2012) | 15 SZ<br>-. 6 in OT group<br>-. 9 in placebo group<br>Sample age = 48 ± 8.9; males = 12 | 20 IU twice a day (first week); 40 IU twice a day for subsequent 2 weeks | Double-blind, placebo-controlled crossover 3-week trial | -. Positive And Negative Symptoms Scale (PANSS)<br>-. Clinical Global Impression (CGI) | In OT group, compared with placebo group, a reduction of PANSS symptoms and CGI scores was seen at 3 week endpoint, suggesting that OT has antipsychotic properties with effects on both negative and positive symptoms, with more robust effects on positive symptoms. |
| Pedersen et al. (2011) | 20 SZ<br>-. 11 in OT group; age = 39 ± 11.2; males = 9<br>-. 9 in placebo group age = 35.8 ± 9.5; males = 8 | 24 IU twice daily | Randomized, double-blind, placebo-controlled 2-week trial | -. Brune Theory of Mind Picture Stories Task<br>-. Trustworthiness Task<br>-. Positive And Negative Symptoms Scale (PANSS)<br>-. Paranoia Scale | The OT group improved in identification of second-order false beliefs (Brune Task).<br>The OT group also showed reduction in PANSS (positive, total, general subscales, and suspiciousness/persecutory items) and Paranoia Scale scores. |

*Continued*

**Table 1** Overview of clinical trials on the effect of OXYTOCIN (OT) on social cognition and psychopathology in schizophrenia.—cont'd

| Study | Sample | Oxytocin intranasal dose | Design | Outcome measure | Results |
|---|---|---|---|---|---|
| Averbeck et al. (2012) | 21 SZ<br>-. 10 in OT group<br>-. 11 in placebo group<br>Sample age = 38.2 ± 1.8; males = 21 | 24 IU single dose | Double-blind, placebo-controlled crossover 2-week trial | -. Facial Affect Recognition: facial expression of emotion in still images from the standardized items developed by Ekman (Wolkind, 1973) | Compared with placebo, the OT group showed improved performance on overall facial emotion recognition, with no differences recognizing individual emotions. |
| Feifel et al. (2012) | 15 SZ<br>-. 6 in OT group<br>-. 9 placebo in placebo group<br>Sample age = 48 ± 8.9; males = 12 | 20 IU twice a day (first week); 40 IU twice a day for the other 2 weeks | Double-blind, placebo-controlled crossover study | -. Verbal memory: California Verbal Learning Test (CVLT)<br>-. Working Memory: LNS (letter number sequence) (from WAIS-III) | Compared with the placebo group, the OT group showed improvement on CVLT, particularly on short-term recall, but not on LNS, suggesting that OT modulates neural processes involved in short-term verbal memory, specifically. |
| Modabbernia et al. (2013) | 40 SZ<br>-. 20 in OT group | 20 IU twice daily for the first week; 40 IU twice daily for the | Randomized, double-blind, placebo-controlled | -. Positive And Negative Symptoms Scale (PANSS)<br>-. Extrapyramidal Symptom Rating Scale (ESRS) | OT group showed 11.2% total PANSS reduction by week 4, with a 20%, 7%, and 8% reduction in positive and negative |

| | | | | | |
|---|---|---|---|---|---|
| | age = 32.3 ± 7.4; males = 17<br>-. 20 in placebo group age = 33.2 ± 6.9; males = 16 | following 7 weeks | 8-week trial | | symptoms, and general psychopathology, respectively, by week 6. No changes were detected in ESRS score between the OT and the placebo group. |
| Davis et al. (2013) | 23 SZ<br>-. 11 in OT group age = 48.6 ± 6.6; males = 11<br>-. 12 in placebo group age = 48.9 ± 9.1; males = 12 | Single dose of 40 IU | Randomized, double-blind, placebo-controlled study | -. Theory of Mind: The Awareness of Social Inference Test (TASIT-III)<br>-. Empathy: Emotional Perspective Taking Task (EPTT)<br>-. Social Perception: Half Profile of Nonverbal Sensitivity (Half-PONS)<br>-. Facial affect recognition: assessed asking participants to identify facial expression of emotion in still images from the standardized items developed by Wolkind (1973) | OT did not improve performance in social composite measure or basic-level social cognitive tasks (TASIT-Part III detection of lies, Half-PONS and Facial Affect Recognition Task). However, OT significantly improved performance on high-level social cognition (TASIT-Part III detection of sarcasm, EPTT), with a large effect size ($d = 1.0$). OT did not improve positive, negative, or general symptoms. |

*Continued*

**Table 1** Overview of clinical trials on the effect of OXYTOCIN (OT) on social cognition and psychopathology in schizophrenia.—cont'd

| Study | Sample | Oxytocin intranasal dose | Design | Outcome measure | Results |
|---|---|---|---|---|---|
| Fischer-Shofty, Brune, et al. (2013) | 35 SZ<br>-. Age = 32.4 ± 6.4; Males = 31<br>48 HCs<br>-. Age = 30.4 ± 5.8; Males = 39 | 24 IU single dose | Double-blind, within-subjects crossover 2-week trial | -. Social Perception: the Interpersonal Perception Task (IPT). Participants needed to discriminate between two categories of relationship: kinship (familial relationship) and intimacy (romantic relationship)<br>-. Depression: Depression Adjective Check List (DACL) | There was no general effect of OT on mood.<br>Improved social perception performance was seen in the OT group, compared with placebo.<br>The patient OT and placebo groups differed significantly in identification of kinship relationships, but not intimacy. This effect was not detected in the control group. |
| Fischer-Shofty, Shamay-Tsoory, and Levkovitz (2013) | 30 SZ<br>-. Age = 31.8 ± 6.5; Males = 27<br>35 HCs<br>-. Age = 29.5 ± 5.6; Males = 32 | 24 IU single dose | Double-blind, within-subjects crossover 2-week trial | -. Facial Emotion Recognition: the Facemorphing Task<br>-. Depression: Depression Adjective Check List (DACL) | There was no general effect of OT on mood.<br>In all participants, OT administration resulted in more accurate recognition of "fearful" but not "happy" faces, even in participants with baseline fear recognition below the median. |

| | | | | | |
|---|---|---|---|---|---|
| Lee et al. (2013) | 28 SZ or SAD<br>-. 13 (6 inpatients and 7 outpatients) in OT group; age = 44.7 ± 11.8; males = 9<br>-. 15 (6 inpatients and 9 outpatients) in placebo groups; age = 35.1 ± 8.2; males = 11 | 20 IU twice daily | Randomized, double-blind, placebo-controlled 3-week trial | -. Olfactory identification ability (neutral, pleasant and unpleasant odors): the University of Pennsylvania Smell Identification Test (UPSIT)<br>-. Symptomatology: Brief Psychiatric Rating Scale (BPRS)<br>-. Scale for the Assessment of Negative Symptoms (SANS) | Improvement in total UPSIT score and in detection of pleasant smells was seen the OT group.<br>The inpatient group that received OT had a better total SANS ($d = 0.85$) at week 3, while no differences were detected for the outpatient group. The OT inpatient group ($d = 0.74$) reported higher scores in the motivation/pleasure subscale of the SANS. |
| Gibson et al. (2014) | 14 SZ<br>-. 8 in OT group; age = 38.9 ± 7.22; males = 6<br>-. 6 in placebo group; age = 35.6 ± 9.0; males = 5 | 24 IU twice daily | Between-subjects 6-week trial | -. PANSS | OT improved negative symptoms. |

*Continued*

**Table 1** Overview of clinical trials on the effect of OXYTOCIN (OT) on social cognition and psychopathology in schizophrenia.—cont'd

| Study | Sample | Oxytocin intranasal dose | Design | Outcome measure | Results |
|---|---|---|---|---|---|
| Davis et al. (2013) | 27 SZ<br>-. 13 in OT group; age = 37 ± 10.8; males = 13<br>-. 14 in placebo group; age = 37 ± 10.8; males = 14 | 40 IU daily | Between-subjects 6-week trial | -. BPRS<br>-. CAINS | OT did not have any effect on positive or negative symptoms. |
| Horta de Macedo, Zuardi, Machado-de-Sousa, Chagas, and Hallak (2014) | 20 SZ 20 HC<br>-. Males = 40<br>-. Age = SZ: 29.6 ± 6.83; controls: 29.7 ± 9.29 | 48 IU | Single dose, within-subjects design | -. Facial emotion Matching | OT resulted in no improvement on emotion matching in either group. |
| Woolley et al. (2014) | 29 SZ 31 HC<br>-. Males = 60 | 40 IU | Single dose, within-subjects design | -. RMET<br>-. TASIT | Under OT, social cognition was improved. |
| Shin et al. (2015) | 16 SZ 15 HC<br>-. Males = 31<br>-. | | | Age = 32.0 ± 7.8 years | 40 IU |

| | | | | | |
|---|---|---|---|---|---|
| Single dose, within-subjects design | -. Emotion Recognition Test (C-KFEE stimuli) | Under OT, amygdala activity while viewing emotional faces was attenuated. | | | |
| Guastella et al. (2015) | 21 SZ<br>-. Males = 21<br>-. | | | Age = 37.42 ± 11.14 | 24 IU |
| Single dose, within-subjects trial | -. DANVA FEEST-. RMET FBPSTL-. Hinting task<br>-. Faux Pas Recognition | OT improved | | performance on higher-order social cognition tasks and paralinguistic aspects of DANVA. | |
| Cacciotti-Saija et al. (2015) | 52 SZ<br>-. 27 in OT group; age = 21.5 ± 4.2; males = 18<br>-. 26 in placebo group; age = 22.3 ± 4.4; males = 18 | 24 IU twice daily | Between-subjects 6-week trial | -. SAPS<br>-. SANS | OT did not improve positive symptoms, but increased OT was correlated with less severe negative symptoms. |

*Continued*

**Table 1** Overview of clinical trials on the effect of OXYTOCIN (OT) on social cognition and psychopathology in schizophrenia.—cont'd

| Study | Sample | Oxytocin intranasal dose | Design | Outcome measure | Results |
|---|---|---|---|---|---|
| Dagani et al. (2016) | 32 SZ<br>-. Males = 26<br>-. Age = 30.4 ± 6.7 | 40 IU daily | Within-subjects 16-week trial | -. PANSS | OT did not improve positive, negative, or general symptoms. |
| Brambilla et al. (2016) | | | 4-month trial | -. MSCEIT | OT improved emotional processing. |
| Williams and Burkner (2017) | 238 SCZ<br>-. Males = 188<br>-. Age = 34.1 | 24–80 IU daily | Metaanalysis of eight RCT's | -. Five with PANSS<br>-. Two with SANS<br>-. One with CAINS<br>-. One with SAPS | No effect on symptoms, with evidence in favor of the null (no effect) on negative symptoms. |

[a]All SZ subjects across the clinical trials reported in the table maintained their prestudy antipsychotic medication.
*HCs*, Healthy Controls; *IU*, International Unit; *OT* = oxytocin; *PD*, polydipsia; *SAD*, Schizoaffective; *SZ*, Schizophrenia.

short-term verbal memory, but not working memory (Feifel et al., 2012). Of note, the mechanism by which oxytocin enhances social cognition in these studies remains unknown, and there are no established biomarkers to predict or monitor treatment response. Emerging evidence supports the use of mu rhythm suppression (a reduction in activity over central EEG electrodes in response to social versus nonsocial stimuli) during a biological motion Mu-suppression task (i.e., identifying the gender, emotion, or direction of walking of point-light animations of human movement) as an index of central social processing resources that is modulated by intranasal oxytocin (Wynn, Green, Hellemann, Reavis, & Marder, 2018). Specifically, the authors found that single-dose intranasal oxytocin at doses of 36 and 48 IU significantly enhanced mu suppression in response to social stimuli compared with placebo during a biological motion task.

In contrast with the data supporting an effect of oxytocin on social cognitive impairments, described herein, the effects of oxytocin on clinical symptoms in psychotic patients are inconsistent. Using higher doses and longer treatment duration, some researchers have seen improvement in overall psychopathology (Feifel et al., 2010; Lee et al., 2013; Modabbernia et al., 2013; Pedersen et al., 2011). Pedersen and colleagues (Pedersen et al., 2011) found a significant impact in a 12-week trial for negative symptoms. Studies using only a single dose did not report any significant effect on symptoms (Davis et al., 2013; Fischer-Shofty, Brune, et al., 2013). A metaanalysis of the effects of oxytocin on the symptoms of schizophrenia was conducted (8 RCTs with 238 patients) and found no effect on positive, negative, or general symptoms (Bürkner et al., 2017). A prior metaanalysis also found no effect on positive, negative, or general symptoms (Oya, Matsuda, Matsunaga, Kishi, & Iwata, 2016). Many of these studies are limited by antipsychotic use, variable doses and dosing schemes, inconsistency in symptom measures, and the gender and sociodemographic makeup of the patient populations study.

## Clinical trials of exogenous oxytocin in SPD/schizotypy/genetic risk/clinical high risk

To date, there is no literature regarding oxytocin trials for social cognition in Schizoid, SPD, or Clinical High Risk SZ spectrum disorders. Our unpublished clinical trial pilot data suggested that administration of a single dose of intranasal oxytocin (24 or 40 IU) to SPD tended to reverse baseline mentalizing abnormalities (Perez-Rodriguez et al., 2017). Furthermore, a naturalistic fMRI task of our same study showed that 40 IU oxytocin

(OT) improves deficits in visual attention to social cues seen in SPD patients, measured with in-MRI eye tracking (Kundu et al., 2017).

## Clinical trials combining intranasal oxytocin and psychotherapy/cognitive behavioral interventions

Given the growing body of evidence that acute treatment with oxytocin can attenuate social cognitive deficits in clinical populations, it seems plausible that the beneficial effects of psychotherapy could be enhanced by augmentation with oxytocin. A recent study found that higher plasma oxytocin levels predicted greater symptom improvement following a course of cognitive behavioral therapy in patients with major depressive disorder, suggesting that oxytocin administration may increase the efficacy of psychotherapy in patients with low baseline oxytocin levels (Jobst et al., 2018). Two studies to date have directly assessed the impact of oxytocin on psychotherapy, with results indicating that psychotherapy combined with oxytocin improves social cognition and self-perceptions during the induction of social anxiety as compared with placebo (Guastella, Howard, Dadds, Mitchell, & Carson, 2009; MacDonald et al., 2013). Another study examined the effects of oxytocin on hypnosis and found enhancements in hypnotizability, suggesting a potential use for oxytocin in the treatment of chronic pain and other disorders that may be improved by hypnosis, such as anxiety and substance use disorders (Bryant & Hung, 2013). Additionally, it has been found that oxytocin administration combined with social support is more effective at improving cortisol responses to social stress than either intervention alone (Heinrichs, Baumgartner, Kirschbaum, & Ehlert, 2003). Despite these promising initial data, more work is needed to determine the specific effects of oxytocin on various forms of psychotherapy and their underlying neural mechanisms. Oxytocin has not yet been combined with evidence-based psychotherapy in treating depression and BD, and it remains unknown how chronic daily oxytocin treatment impacts outcomes. Comprehensive studies combining oxytocin with psychotherapy that use standardized validated assessments of social cognition before and after treatment are also needed.

In conclusion, the effect of endogenous and exogenous oxytocin on social cognition is probably more complex than initially thought in prior studies, and many questions about its absorption and mechanism of action remain unanswered. Moreover, individual and contextual factors, as well as concomitant medications, likely play a key role as potential confounders/moderators of the effect of oxytocin on social cognition. An important limitation of clinical trials of exogenous oxytocin is that it has never before been used as a monotherapy in unmedicated SZ patients, only as an *add-on* to antipsychotics (Bradley & Woolley, 2017). However, growing evidence suggests that concomitant antipsychotic treatment is a critical confounder, particularly in studies targeting cognitive or negative symptoms of schizophrenia (Bradley & Woolley, 2017; Davidson et al., 2017; Rehse et al., 2016). First, there is some evidence that some antipsychotic medications increase oxytocin secretion in animal models (Uvnas-Moberg, Alster, & Svensson, 1992). In humans, CSF oxytocin levels in schizophrenia patients are significantly negatively correlated with a second-generation antipsychotic dose (Sasayama et al., 2012). An antipsychotic dose may also impact the effect of exogenous oxytocin on social cognition, such that those under higher doses of antipsychotics show the highest improvements in social cognition after oxytocin administration (Bradley & Woolley, 2017). Finally, the effect of antipsychotics on the dopaminergic reward system may block the effects of endogenous oxytocin on social cognition. This is consistent with animal data showing that a dopaminergic D2 blocker, similar to antipsychotics, blocks the effect of exogenous oxytocin on partner preference formation in the prairie vole (Liu & Wang, 2003). Therefore, it is critical to conduct clinical trials of oxytocin in monotherapy in patients with psychosis spectrum disorders. Milder schizophrenia spectrum disorders such as SPD are an ideal "human model" of SZ in which to identify therapeutic targets and test new compounds as monotherapy, avoiding the confounders inherent to schizophrenia (SCZ). For example, SPD populations have allowed researchers to conduct proof-of-concept studies of cognitive enhancing drugs that could be later used in SCZ (e.g., the D1 agonist dihydrexidine) (Rosell et al., 2015).

## References

Averbeck, B. B., Bobin, T., Evans, S., & Shergill, S. S. (2012). Emotion recognition and oxytocin in patients with schizophrenia. *Psychological Medicine*, *42*(2), 259–266. https://doi.org/10.1017/S0033291711001413.

Aydın, O., Lysaker, P. H., Balıkçı, K., Ünal-Aydın, P., & Esen-Danacı, A. (2018). Associations of oxytocin and vasopressin plasma levels with neurocognitive, social cognitive and meta cognitive function in schizophrenia. *Psychiatry Research*. https://doi.org/10.1016/j.psychres.2018.03.048.

Bakermans-Kranenburg, M. J., & van I. Jzendoorn, M. H. (2013). Sniffing around oxytocin: review and meta-analyses of trials in healthy and clinical groups with implications for pharmacotherapy. *Translational Psychiatry*, *3*(5), e258. https://doi.org/10.1038/tp.2013.34.

Bartz, J., Simeon, D., Hamilton, H., Kim, S., Crystal, S., Braun, A., ... Hollander, E. (2011). Oxytocin can hinder trust and cooperation in borderline personality disorder. *Social Cognitive and Affective Neuroscience*, *6*(5), 556–563. https://doi.org/10.1093/scan/nsq085.

Bartz, J. A., Zaki, J., Bolger, N., Hollander, E., Ludwig, N. N., Kolevzon, A., & Ochsner, K. N. (2010). Oxytocin selectively improves empathic accuracy. *Psychological Science*, *21*(10), 1426–1428. https://doi.org/10.1177/0956797610383439.

Bartz, J. A., Zaki, J., Bolger, N., & Ochsner, K. N. (2011). Social effects of oxytocin in humans: context and person matter. *Trends in Cognitive Sciences*, *15*(7), 301–309. https://doi.org/10.1016/j.tics.2011.05.002.

Beckmann, H., Lang, R. E., & Gattaz, W. F. (1985). Vasopressin-oxytocin in cerebrospinal fluid of schizophrenic patients and normal controls. *Psychoneuroendocrinology*, *10*(2), 187–191. https://doi.org/10.1016/0306-4530(85)90056-3.

Bertsch, K., Gamer, M., Schmidt, B., Schmidinger, I., Walther, S., Kastel, T., ... Herpertz, S. C. (2013). Oxytocin and reduction of social threat hypersensitivity in women with borderline personality disorder. *The American Journal of Psychiatry*, *170*(10), 1169–1177. https://doi.org/10.1176/appi.ajp.2013.13020263.

Bethlehem, R. A. I., Lombardo, M. V., Lai, M. C., Auyeung, B., Crockford, S. K., Deakin, J., ... Baron-Cohen, S. (2017). Intranasal oxytocin enhances intrinsic corticostriatal functional connectivity in women. *Translational Psychiatry*, 7(4), e1099. https://doi.org/10.1038/tp.2017.72.

Bielsky, I. F., & Young, L. J. (2004). Oxytocin, vasopressin, and social recognition in mammals. *Peptides*, *25*(9), 1565–1574. https://doi.org/10.1016/j.peptides.2004.05.019.

Boccia, M. L., Petrusz, P., Suzuki, K., Marson, L., & Pedersen, C. A. (2013). Immunohistochemical localization of oxytocin receptors in human brain. *Neuroscience*, *253*, 155–164. https://doi.org/10.1016/j.neuroscience.2013.08.048.

Bos, P. A., Panksepp, J., Bluthe, R. M., & van Honk, J. (2012). Acute effects of steroid hormones and neuropeptides on human social-emotional behavior: a review of single administration studies. *Frontiers in Neuroendocrinology*, *33*(1), 17–35. https://doi.org/10.1016/j.yfrne.2011.01.002.

Bradley, E. R., & Woolley, J. D. (2017). Oxytocin effects in schizophrenia: reconciling mixed findings and moving forward. *Neuroscience and Biobehavioral Reviews*, *80*, 36–56. https://doi.org/10.1016/j.neubiorev.2017.05.007.

Brambilla, M., Cotelli, M., Manenti, R., Dagani, J., Sisti, D., Rocchi, M., ... de Girolamo, G. (2016). Oxytocin to modulate emotional processing in schizophrenia: a randomized, double-blind, cross-over clinical trial. *European Neuropsychopharmacology*, *26*(10), 1619–1628. https://doi.org/10.1016/j.euroneuro.2016.08.001.

Brodmann, K., Gruber, O., & Goya-Maldonado, R. (2017). Intranasal oxytocin selectively modulates large-scale brain networks in humans. *Brain Connectivity*, 7(7), 454–463. https://doi.org/10.1089/brain.2017.0528.

Brune, M., Ebert, A., Kolb, M., Tas, C., Edel, M. A., & Roser, P. (2013). Oxytocin influences avoidant reactions to social threat in adults with borderline personality disorder. *Human Psychopharmacology*, *28*(6), 552–561. https://doi.org/10.1002/hup.2343.

Bryant, R. A., & Hung, L. (2013). Oxytocin enhances social persuasion during hypnosis. *PLoS One*, *8*(4), e60711. https://doi.org/10.1371/journal.pone.0060711.

Bürkner, P. -C., Williams, D. R., Simmons, T. C., & Woolley, J. D. (2017). Intranasal oxytocin may improve high-level social cognition in schizophrenia, but not social cognition or neurocognition in general: a multilevel Bayesian meta-analysis. *Schizophrenia Bulletin*, *43*(6), 1291–1303. https://doi.org/10.1093/schbul/sbx053.

Cacciotti-Saija, C., Langdon, R., Ward, P. B., Hickie, I. B., Scott, E. M., Naismith, S. L., ... Guastella, A. J. (2015). A double-blind randomized controlled trial of oxytocin nasal spray and social cognition training for young people with early psychosis. *Schizophrenia Bulletin*, *41*(2), 483–493. https://doi.org/10.1093/schbul/sbu094.

Carson, D. S., Berquist, S. W., Trujillo, T. H., Garner, J. P., Hannah, S. L., Hyde, S. A., … Parker, K. J. (2015). Cerebrospinal fluid and plasma oxytocin concentrations are positively correlated and negatively predict anxiety in children. *Molecular Psychiatry*, *20*(9), 1085–1090. https://doi.org/10.1038/mp.2014.132.

Chang, S. W. C., Barter, J. W., Ebitz, R. B., Watson, K. K., & Platt, M. L. (2012). Inhaled oxytocin amplifies both vicarious reinforcement and self reinforcement in rhesus macaques (*Macaca mulatta*). *Proceedings of the National Academy of Sciences*, *109*(3), 959.

Clark, C. L., St. John, N., Pasca, A. M., Hyde, S. A., Hornbeak, K., Abramova, M., … Penn, A. A. (2013). Neonatal CSF oxytocin levels are associated with parent report of infant soothability and sociability. *Psychoneuroendocrinology*, *38*(7), 1208–1212. https://doi.org/10.1016/j.psyneuen.2012.10.017.

Dagani, J., Sisti, D., Abelli, M., Di Paolo, L., Pini, S., Raimondi, S., … de Girolamo, G. (2016). Do we need oxytocin to treat schizophrenia? A randomized clinical trial. *Schizophrenia Research*, *172*(1), 158–164. https://doi.org/10.1016/j.schres.2016.02.011.

Dal Monte, O., Noble, P. L., Turchi, J., Cummins, A., & Averbeck, B. B. (2014). CSF and blood oxytocin concentration changes following intranasal delivery in macaque. *PLoS One*, *9*(8), e103677. https://doi.org/10.1371/journal.pone.0103677.

Davidson, M., Saoud, J., Staner, C., Noel, N., Luthringer, E., Werner, S., … Luthringer, R. (2017). Efficacy and safety of MIN-101: a 12-week randomized, double-blind, placebo-controlled trial of a new drug in development for the treatment of negative symptoms in schizophrenia. *The American Journal of Psychiatry*, *174*(12), 1195–1202. https://doi.org/10.1176/appi.ajp.2017.17010122.

Davis, M. C., Lee, J., Horan, W. P., Clarke, A. D., McGee, M. R., Green, M. F., & Marder, S. R. (2013). Effects of single dose intranasal oxytocin on social cognition in schizophrenia. *Schizophrenia Research*, *147*(2–3), 393–397. https://doi.org/10.1016/j.schres.2013.04.023.

Derntl, B., Finkelmeyer, A., Toygar, T. K., Hulsmann, A., Schneider, F., Falkenberg, D. I., & Habel, U. (2009). Generalized deficit in all core components of empathy in schizophrenia. *Schizophrenia Research*, *108*(1–3), 197–206. https://doi.org/10.1016/j.schres.2008.11.009.

Deuse, L., Wudarczyk, O., Rademacher, L., Kaleta, P., Karges, W., Kacheva, S., … Lammertz, S. (2018). Peripheral oxytocin predicts higher-level social cognition in men regardless of empathy quotient. *Pharmacopsychiatry*. https://doi.org/10.1055/a-0590-4850.

Ditzen, B., Schaer, M., Gabriel, B., Bodenmann, G., Ehlert, U., & Heinrichs, M. (2009). Intranasal oxytocin increases positive communication and reduces cortisol levels during couple conflict. *Biological Psychiatry*, *65*(9), 728–731. https://doi.org/10.1016/j.biopsych.2008.10.011.

Domes, G., Lischke, A., Berger, C., Grossmann, A., Hauenstein, K., Heinrichs, M., & Herpertz, S. C. (2010). Effects of intranasal oxytocin on emotional face processing in women. *Psychoneuroendocrinology*, *35*(1), 83–93. https://doi.org/10.1016/j.psyneuen.2009.06.016.

Domes, G., Sibold, M., Schulze, L., Lischke, A., Herpertz, S. C., & Heinrichs, M. (2013). Intranasal oxytocin increases covert attention to positive social cues. *Psychological Medicine*, *43*(8), 1747–1753. https://doi.org/10.1017/s0033291712002565.

Ebert, A., Kolb, M., Heller, J., Edel, M. A., Roser, P., & Brune, M. (2013). Modulation of interpersonal trust in borderline personality disorder by intranasal oxytocin and childhood trauma. *Social Neuroscience*, *8*(4), 305–313. https://doi.org/10.1080/17470919.2013.807301.

Eckstein, M., Markett, S., Kendrick, K. M., Ditzen, B., Liu, F., Hurlemann, R., & Becker, B. (2017). Oxytocin differentially alters resting state functional connectivity between amygdala subregions and emotional control networks: inverse correlation with depressive traits. *NeuroImage*, *149*, 458–467. https://doi.org/10.1016/j.neuroimage.2017.01.078.

Feifel, D., Macdonald, K., Cobb, P., & Minassian, A. (2012). Adjunctive intranasal oxytocin improves verbal memory in people with schizophrenia. *Schizophrenia Research*, *139*(1–3), 207–210. https://doi.org/10.1016/j.schres.2012.05.018.

Feifel, D., Macdonald, K., Nguyen, A., Cobb, P., Warlan, H., Galangue, B., … Hadley, A. (2010). Adjunctive intranasal oxytocin reduces symptoms in schizophrenia patients. *Biological Psychiatry*, *68*(7), 678–680. https://doi.org/10.1016/j.biopsych.2010.04.039.

Fischer-Shofty, M., Brune, M., Ebert, A., Shefet, D., Levkovitz, Y., & Shamay-Tsoory, S. G. (2013). Improving social perception in schizophrenia: the role of oxytocin. *Schizophrenia Research*, *146*(1–3), 357–362. https://doi.org/10.1016/j.schres.2013.01.006.

Fischer-Shofty, M., Shamay-Tsoory, S. G., & Levkovitz, Y. (2013). Characterization of the effects of oxytocin on fear recognition in patients with schizophrenia and in healthy controls. *Frontiers in Neuroscience*, 7, 127. https://doi.org/10.3389/fnins.2013.00127.

Freedman, D., Brown, A. S., Shen, L., & Schaefer, C. A. (2015). Perinatal oxytocin increases the risk of offspring bipolar disorder and childhood cognitive impairment. *Journal of Affective Disorders*, *173*, 65–72. https://doi.org/10.1016/j.jad.2014.10.052.

Garety, P. A., & Freeman, D. (1999). Cognitive approaches to delusions: a critical review of theories and evidence. *The British Journal of Clinical Psychology*, *38*(Pt 2), 113–154. https://doi.org/10.1348/014466599162700.

Gibson, C. M., Penn, D. L., Smedley, K. L., Leserman, J., Elliott, T., & Pedersen, C. A. (2014). A pilot six-week randomized controlled trial of oxytocin on social cognition and social skills in schizophrenia. *Schizophrenia Research*, *156*(2–3), 261–265. https://doi.org/10.1016/j.schres.2014.04.009.

Glovinsky, D., Kalogeras, K. T., Kirch, D. G., Suddath, R., & Wyatt, R. J. (1994). Cerebrospinal fluid oxytocin concentration in schizophrenic patients does not differ from control subjects and is not changed by neuroleptic medication. *Schizophrenia Research*, *11*(3), 273–276. https://doi.org/10.1016/0920-9964(94)90021-3.

Goldman, M. B., Gomes, A. M., Carter, C. S., & Lee, R. (2011). Divergent effects of two different doses of intranasal oxytocin on facial affect discrimination in schizophrenic patients with and without polydipsia. *Psychopharmacology*, *216*(1), 101–110. https://doi.org/10.1007/s00213-011-2193-8.

Gordon, I., Jack, A., Pretzsch, C. M., Vander Wyk, B., Leckman, J. F., Feldman, R., & Pelphrey, K. A. (2016). Intranasal oxytocin enhances connectivity in the neural circuitry supporting social motivation and social perception in children with autism. *Scientific Reports*. *6*(1). https://doi.org/10.1038/srep35054.

Groppe, S. E., Gossen, A., Rademacher, L., Hahn, A., Westphal, L., Grunder, G., & Spreckelmeyer, K. N. (2013). Oxytocin influences processing of socially relevant cues in the ventral tegmental area of the human brain. *Biological Psychiatry*, *74*(3), 172–179. https://doi.org/10.1016/j.biopsych.2012.12.023.

Guastella, A. J., Hickie, I. B., McGuinness, M. M., Otis, M., Woods, E. A., Disinger, H. M., … Banati, R. B. (2013). Recommendations for the standardisation of oxytocin nasal administration and guidelines for its reporting in human research. *Psychoneuroendocrinology*, *38*(5), 612–625. https://doi.org/10.1016/j.psyneuen.2012.11.019.

Guastella, A. J., Howard, A. L., Dadds, M. R., Mitchell, P., & Carson, D. S. (2009). A randomized controlled trial of intranasal oxytocin as an adjunct to exposure therapy for social anxiety disorder. *Psychoneuroendocrinology*, *34*(6), 917–923. https://doi.org/10.1016/j.psyneuen.2009.01.005.

Guastella, A. J., Ward, P. B., Hickie, I. B., Shahrestani, S., Hodge, M. A., Scott, E. M., & Langdon, R. (2015). A single dose of oxytocin nasal spray improves higher-order social cognition in schizophrenia. *Schizophrenia Research, 168*(3), 628–633. https://doi.org/10.1016/j.schres.2015.06.005.

Gumley, A., Braehler, C., & Macbeth, A. (2014). A meta-analysis and theoretical critique of oxytocin and psychosis: prospects for attachment and compassion in promoting recovery. *The British Journal of Clinical Psychology, 53*(1), 42–61. https://doi.org/10.1111/bjc.12041.

Guoynes, C. D., Simmons, T. C., Downing, G. M., Jacob, S., Solomon, M., & Bales, K. L. (2018). Chronic intranasal oxytocin has dose-dependent effects on central oxytocin and vasopressin systems in prairie voles (*Microtus ochrogaster*). *Neuroscience, 369*, 292–302. https://doi.org/10.1016/j.neuroscience.2017.11.037.

Heinrichs, M., Baumgartner, T., Kirschbaum, C., & Ehlert, U. (2003). Social support and oxytocin interact to suppress cortisol and subjective responses to psychosocial stress. *Biological Psychiatry, 54*(12), 1389–1398.

Horta de Macedo, L. R., Zuardi, A. W., Machado-de-Sousa, J. P., Chagas, M. H. N., & Hallak, J. E. C. (2014). Oxytocin does not improve performance of patients with schizophrenia and healthy volunteers in a facial emotion matching task. *Psychiatry Research, 220*(1), 125–128. https://doi.org/10.1016/j.psychres.2014.07.082.

Huang, H., Michetti, C., Busnelli, M., Managò, F., Sannino, S., Scheggia, D., … Papaleo, F. (2013). Chronic and acute intranasal oxytocin produce divergent social effects in mice. *Neuropsychopharmacology, 39*(5), 1102–1114. https://doi.org/10.1038/npp.2013.310.

Ide, J. S., Nedic, S., Wong, K. F., Strey, S. L., Lawson, E. A., Dickerson, B. C., … Mujica-Parodi, L. R. (2018). Oxytocin attenuates trust as a subset of more general reinforcement learning, with altered reward circuit functional connectivity in males. *NeuroImage, 174*, 35–43. https://doi.org/10.1016/j.neuroimage.2018.02.035.

Jobst, A., Sabaß, L., Hall, D., Brücklmeier, B., Buchheim, A., Hall, J., … Padberg, F. (2018). Oxytocin plasma levels predict the outcome of psychotherapy: a pilot study in chronic depression. *Journal of Affective Disorders, 227*, 206–213. https://doi.org/10.1016/j.jad.2017.10.037.

Kagerbauer, S. M., Martin, J., Schuster, T., Blobner, M., Kochs, E. F., & Landgraf, R. (2013). Plasma oxytocin and vasopressin do not predict neuropeptide concentrations in human cerebrospinal fluid. *Journal of Neuroendocrinology, 25*(7), 668–673. https://doi.org/10.1111/jne.12038.

Kanat, M., Spenthof, I., Riedel, A., van Elst, L. T., Heinrichs, M., & Domes, G. (2017). Restoring effects of oxytocin on the attentional preference for faces in autism. *Translational Psychiatry*, 7(4), e1097. https://doi.org/10.1038/tp.2017.67.

Kaney, S., & Bentall, R. P. (1989). Persecutory delusions and attributional style. *The British Journal of Medical Psychology, 62*(Pt 2), 191–198. https://doi.org/10.1111/j.2044-8341.1989.tb02826.x.

Kundu, P., Smith, D. A., Rutter, S., Fisher, A., Lyashenko, M., Benavides, C., … Perez-Rodriguez, M. M. (2017). Neuromodulation by combined oxytocin and a naturalistic social cognition task. *Neuropsychopharmacology, 42*, S476–S652.

Lee, M. R., Scheidweiler, K. B., Diao, X. X., Akhlaghi, F., Cummins, A., Huestis, M. A., … Averbeck, B. B. (2017). Oxytocin by intranasal and intravenous routes reaches the cerebrospinal fluid in rhesus macaques: determination using a novel oxytocin assay. *Molecular Psychiatry, 23*(1), 115–122. https://doi.org/10.1038/mp.2017.27.

Lee, M. R., Wehring, H. J., McMahon, R. P., Linthicum, J., Cascella, N., Liu, F., … Kelly, D. L. (2013). Effects of adjunctive intranasal oxytocin on olfactory identification and clinical symptoms in schizophrenia: results from a randomized double blind placebo controlled pilot study. *Schizophrenia Research, 145*(1–3), 110–115. https://doi.org/10.1016/j.schres.2013.01.001.

Lefevre, A., Mottolese, R., Dirheimer, M., Mottolese, C., Duhamel, J. R., & Sirigu, A. (2017). A comparison of methods to measure central and peripheral oxytocin concentrations in human and non-human primates. *Scientific Reports*, 7(1), 17222. https://doi.org/10.1038/s41598-017-17674-7.

Lefevre, A., Richard, N., Jazayeri, M., Beuriat, P. -A., Fieux, S., Zimmer, L., … Sirigu, A. (2017). Oxytocin and serotonin brain mechanisms in the nonhuman primate. *The Journal of Neuroscience*, *37*(28), 6741–6750. https://doi.org/10.1523/jneurosci.0659-17.2017.

Leknes, S., Wessberg, J., Ellingsen, D. M., Chelnokova, O., Olausson, H., & Laeng, B. (2013). Oxytocin enhances pupil dilation and sensitivity to 'hidden' emotional expressions. *Social Cognitive and Affective Neuroscience*, *8*(7), 741–749. https://doi.org/10.1093/scan/nss062.

Leng, G., & Sabatier, N. (2016). Measuring oxytocin and vasopressin: bioassays, immunoassays and random numbers. *Journal of Neuroendocrinology*. *28*(10)https://doi.org/10.1111/jne.12413.

Li, H., Chan, R. C., McAlonan, G. M., & Gong, Q. Y. (2010). Facial emotion processing in schizophrenia: a meta-analysis of functional neuroimaging data. *Schizophrenia Bulletin*, *36*(5), 1029–1039. https://doi.org/10.1093/schbul/sbn190.

Lien, Y. -J., Chang, H. H., Tsai, H. -C., Kuang Yang, Y., Lu, R. -B., & See Chen, P. (2017). Plasma oxytocin levels in major depressive and bipolar II disorders. *Psychiatry Research*, *258*, 402–406. https://doi.org/10.1016/j.psychres.2017.08.080.

Liu, Y., & Wang, Z. X. (2003). Nucleus accumbens oxytocin and dopamine interact to regulate pair bond formation in female prairie voles. *Neuroscience*, *121*(3), 537–544.

Love, T. M. (2014). Oxytocin, motivation and the role of dopamine. *Pharmacology Biochemistry and Behavior*, *119*, 49–60. https://doi.org/10.1016/j.pbb.2013.06.011.

Macdonald, K., & Feifel, D. (2012). Oxytocin in schizophrenia: a review of evidence for its therapeutic effects. *Acta Neuropsychiatrica*, *24*(3), 130–146. https://doi.org/10.1111/j.1601-5215.2011.00634.x.

MacDonald, K., MacDonald, T. M., Brune, M., Lamb, K., Wilson, M. P., Golshan, S., & Feifel, D. (2013). Oxytocin and psychotherapy: a pilot study of its physiological, behavioral and subjective effects in males with depression. *Psychoneuroendocrinology*, *38*(12), 2831–2843. https://doi.org/10.1016/j.psyneuen.2013.05.014.

Madrid, J. E., Oztan, O., Sclafani, V., Del Rosso, L. A., Calonder, L. A., Chun, K., … Parker, K. J. (2017). Preference for novel faces in male infant monkeys predicts cerebrospinal fluid oxytocin concentrations later in life. *Scientific Reports*, 7(1), 12935. https://doi.org/10.1038/s41598-017-13109-5.

Martin, J., Kagerbauer, S. M., Gempt, J., Podtschaske, A., Hapfelmeier, A., & Schneider, G. (2018). Oxytocin levels in saliva correlate better than plasma levels with concentrations in the cerebrospinal fluid of patients in neurocritical care. *Journal of Neuroendocrinology*, e12596. https://doi.org/10.1111/jne.12596.

McDonald, S., Bornhofen, C., Shum, D., Long, E., Saunders, C., & Neulinger, K. (2006). Reliability and validity of The Awareness of Social Inference Test (TASIT): a clinical test of social perception. *Disability and Rehabilitation*, *28*(24), 1529–1542. https://doi.org/10.1080/09638280600646185.

Meyer-Lindenberg, A., Domes, G., Kirsch, P., & Heinrichs, M. (2011). Oxytocin and vasopressin in the human brain: social neuropeptides for translational medicine. *Nature Reviews Neuroscience*, *12*(9), 524–538. https://doi.org/10.1038/nrn3044.

Modabbernia, A., Rezaei, F., Salehi, B., Jafarinia, M., Ashrafi, M., Tabrizi, M., … Akhondzadeh, S. (2013). Intranasal oxytocin as an adjunct to risperidone in patients with schizophrenia: an 8-week, randomized, double-blind, placebo-controlled study. *CNS Drugs*, *27*(1), 57–65. https://doi.org/10.1007/s40263-012-0022-1.

Modi, M. E., Connor-Stroud, F., Landgraf, R., Young, L. J., & Parr, L. A. (2014). Aerosolized oxytocin increases cerebrospinal fluid oxytocin in rhesus macaques. *Psychoneuroendocrinology, 45*, 49–57. https://doi.org/10.1016/j.psyneuen.2014.02.011.

Montag, C., Dziobek, I., Richter, I. S., Neuhaus, K., Lehmann, A., Sylla, R., … Gallinat, J. (2011). Different aspects of theory of mind in paranoid schizophrenia: evidence from a video-based assessment. *Psychiatry Research, 186*(2–3), 203–209. https://doi.org/10.1016/j.psychres.2010.09.006.

Ne'eman, R., Perach-Barzilay, N., Fischer-Shofty, M., Atias, A., & Shamay-Tsoory, S. G. (2016). Intranasal administration of oxytocin increases human aggressive behavior. *Hormones and Behavior, 80*, 125–131. https://doi.org/10.1016/j.yhbeh.2016.01.015.

Neumann, I. D., & Landgraf, R. (2012). Balance of brain oxytocin and vasopressin: implications for anxiety, depression, and social behaviors. *Trends in Neurosciences, 35*(11), 649–659. https://doi.org/10.1016/j.tins.2012.08.004.

Neumann, I. D., Maloumby, R., Beiderbeck, D. I., Lukas, M., & Landgraf, R. (2013). Increased brain and plasma oxytocin after nasal and peripheral administration in rats and mice. *Psychoneuroendocrinology, 38*(10), 1985–1993. https://doi.org/10.1016/j.psyneuen.2013.03.003.

Oya, K., Matsuda, Y., Matsunaga, S., Kishi, T., & Iwata, N. (2016). Efficacy and safety of oxytocin augmentation therapy for schizophrenia: an updated systematic review and meta-analysis of randomized, placebo-controlled trials. *European Archives of Psychiatry and Clinical Neuroscience, 266*(5), 439–450. https://doi.org/10.1007/s00406-015-0634-9.

Ozsoy, S., Esel, E., & Kula, M. (2009). Serum oxytocin levels in patients with depression and the effects of gender and antidepressant treatment. *Psychiatry Research, 169*(3), 249–252. https://doi.org/10.1016/j.psychres.2008.06.034.

Parker, K. J., Garner, J. P., Libove, R. A., Hyde, S. A., Hornbeak, K. B., Carson, D. S., … Hardan, A. Y. (2014). Plasma oxytocin concentrations and OXTR polymorphisms predict social impairments in children with and without autism spectrum disorder. *Proceedings of the National Academy of Sciences of the United States of America, 111*(33), 12258–12263. https://doi.org/10.1073/pnas.1402236111.

Parker, K. J., Garner, J. P., Oztan, O., Tarara, E. R., Li, J., Sclafani, V., … Capitanio, J. P. (2018). Arginine vasopressin in cerebrospinal fluid is a marker of sociality in nonhuman primates. *Science Translational Medicine, 10*(439). https://doi.org/10.1126/scitranslmed.aam9100.

Pedersen, C. A., Gibson, C. M., Rau, S. W., Salimi, K., Smedley, K. L., Casey, R. L., … Penn, D. L. (2011). Intranasal oxytocin reduces psychotic symptoms and improves Theory of Mind and social perception in schizophrenia. *Schizophrenia Research, 132*(1), 50–53. https://doi.org/10.1016/j.schres.2011.07.027.

Perez-Rodriguez, M. M., Fisher, A., Rutter, S., Smith, D. A., Benavides, C., McClure, M., … Kundu, P. (2017). Intranasal oxytocin modulates social cognitive errors in the psychosis spectrum. *Neuropsychopharmacology, 42*, S476–S652.

Perez-Rodriguez, M. M., Mahon, K., Russo, M., Ungar, A. K., & Burdick, K. E. (2014). Oxytocin and social cognition in affective and psychotic disorders. *European Neuropsychopharmacology*. https://doi.org/10.1016/j.euroneuro.2014.07.012.

Perry, A., Bentin, S., Shalev, I., Israel, S., Uzefovsky, F., Bar-On, D., & Ebstein, R. P. (2010). Intranasal oxytocin modulates EEG mu/alpha and beta rhythms during perception of biological motion. *Psychoneuroendocrinology, 35*(10), 1446–1453. https://doi.org/10.1016/j.psyneuen.2010.04.011.

Pfundmair, M., Reinelt, A., DeWall, C. N., & Feldmann, L. (2018). Oxytocin strengthens the link between provocation and aggression among low anxiety people. *Psychoneuroendocrinology, 93*, 124–132. https://doi.org/10.1016/j.psyneuen.2018.04.025.

Pinkham, A. E., Hopfinger, J. B., Ruparel, K., & Penn, D. L. (2008). An investigation of the relationship between activation of a social cognitive neural network and social functioning. *Schizophrenia Bulletin*, *34*(4), 688–697. https://doi.org/10.1093/schbul/sbn031.
Pinkham, A. E., Loughead, J., Ruparel, K., Overton, E., Gur, R. E., & Gur, R. C. (2011). Abnormal modulation of amygdala activity in schizophrenia in response to direct- and averted-gaze threat-related facial expressions. *The American Journal of Psychiatry*, *168*(3), 293–301. https://doi.org/10.1176/appi.ajp.2010.10060832.
Prehn, K., Kazzer, P., Lischke, A., Heinrichs, M., Herpertz, S. C., & Domes, G. (2013). Effects of intranasal oxytocin on pupil dilation indicate increased salience of socioaffective stimuli. *Psychophysiology*, *50*(6), 528–537. https://doi.org/10.1111/psyp.12042.
Quintana, D. S., Alvares, G. A., Hickie, I. B., & Guastella, A. J. (2015). Do delivery routes of intranasally administered oxytocin account for observed effects on social cognition and behavior? A two-level model. *Neuroscience & Biobehavioral Reviews*, *49*, 182–192. https://doi.org/10.1016/j.neubiorev.2014.12.011.
Quintana, D. S., Dieset, I., Elvsåshagen, T., Westlye, L. T., & Andreassen, O. A. (2017). Oxytocin system dysfunction as a common mechanism underlying metabolic syndrome and psychiatric symptoms in schizophrenia and bipolar disorders. *Frontiers in Neuroendocrinology*, *45*, 1–10. https://doi.org/10.1016/j.yfrne.2016.12.004.
Quirin, M., Kuhl, J., & Dusing, R. (2011). Oxytocin buffers cortisol responses to stress in individuals with impaired emotion regulation abilities. *Psychoneuroendocrinology*, *36*(6), 898–904. https://doi.org/10.1016/j.psyneuen.2010.12.005.
Radke, S., Volman, I., Kokal, I., Roelofs, K., de Bruijn, E. R. A., & Toni, I. (2017). Oxytocin reduces amygdala responses during threat approach. *Psychoneuroendocrinology*, *79*, 160–166. https://doi.org/10.1016/j.psyneuen.2017.02.028.
Rehse, M., Bartolovic, M., Baum, K., Richter, D., Weisbrod, M., & Roesch-Ely, D. (2016). Influence of antipsychotic and anticholinergic loads on cognitive functions in patients with schizophrenia. *Schizophrenia Research and Treatment*, *2016*, 8213165. https://doi.org/10.1155/2016/8213165.
Riem, M. M., Bakermans-Kranenburg, M. J., Pieper, S., Tops, M., Boksem, M. A., Vermeiren, R. R., … Rombouts, S. A. (2011). Oxytocin modulates amygdala, insula, and inferior frontal gyrus responses to infant crying: a randomized controlled trial. *Biological Psychiatry*, *70*(3), 291–297. https://doi.org/10.1016/j.biopsych.2011.02.006.
Rilling, J. K., Chen, X., Chen, X., & Haroon, E. (2018). Intranasal oxytocin modulates neural functional connectivity during human social interaction. *American Journal of Primatology*, e22740. https://doi.org/10.1002/ajp.22740.
Rogers, C. N., Ross, A. P., Sahu, S. P., Siegel, E. R., Dooyema, J. M., Cree, M. A., … Preuss, T. M. (2018). Oxytocin- and arginine vasopressin-containing fibers in the cortex of humans, chimpanzees, and rhesus macaques. *American Journal of Primatology*, e22875. https://doi.org/10.1002/ajp.22875.
Rosell, D. R., Zaluda, L. C., McClure, M. M., Perez-Rodriguez, M. M., Strike, K. S., Barch, D. M., … Siever, L. J. (2015). Effects of the D(1) dopamine receptor agonist dihydrexidine (DAR-0100A) on working memory in schizotypal personality disorder. *Neuropsychopharmacology*, *40*(2), 446–453. https://doi.org/10.1038/npp.2014.192.
Rubin, L. H., Wehring, H. J., Demyanovich, H., Sue Carter, C., Pournajafi-Nazarloo, H., Feldman, S. M., … Kelly, D. L. (2018). Peripheral oxytocin and vasopressin are associated with clinical symptom severity and cognitive functioning in midlife women with chronic schizophrenia. *Schizophrenia Research*, *195*, 409–411. https://doi.org/10.1016/j.schres.2017.09.041.
Rutherford, H. J. V., Guo, X. M., Wu, J., Graber, K. M., Hayes, N. J., Pelphrey, K. A., & Mayes, L. C. (2018). Intranasal oxytocin decreases cross-frequency coupling of neural

oscillations at rest. *International Journal of Psychophysiology, 123*, 143–151. https://doi.org/10.1016/j.ijpsycho.2017.09.017.

Rutigliano, G., Rocchetti, M., Paloyelis, Y., Gilleen, J., Sardella, A., Cappucciati, M., … Fusar-Poli, P. (2016). Peripheral oxytocin and vasopressin: biomarkers of psychiatric disorders? A comprehensive systematic review and preliminary meta-analysis. *Psychiatry Research, 241*, 207–220. https://doi.org/10.1016/j.psychres.2016.04.117.

Sasayama, D., Hattori, K., Teraishi, T., Hori, H., Ota, M., Yoshida, S., … Kunugi, H. (2012). Negative correlation between cerebrospinal fluid oxytocin levels and negative symptoms of male patients with schizophrenia. *Schizophrenia Research, 139*(1–3), 201–206. https://doi.org/10.1016/j.schres.2012.06.016.

Shamay-Tsoory, S. G., & Abu-Akel, A. (2016). The social salience hypothesis of oxytocin. *Biological Psychiatry, 79*(3), 194–202. https://doi.org/10.1016/j.biopsych.2015.07.020.

Sharp, C., Pane, H., Ha, C., Venta, A., Patel, A. B., Sturek, J., & Fonagy, P. (2011). Theory of mind and emotion regulation difficulties in adolescents with borderline traits. *Journal of the American Academy of Child and Adolescent Psychiatry, 50*(6), 563–573 e561. https://doi.org/10.1016/j.jaac.2011.01.017.

Shin, N. Y., Park, H. Y., Jung, W. H., Park, J. W., Yun, J. -Y., Jang, J. H., … Kwon, J. S. (2015). Effects of oxytocin on neural response to facial expressions in patients with schizophrenia. *Neuropsychopharmacology, 40*, 1919. https://doi.org/10.1038/npp.2015.41. https://www.nature.com/articles/npp201541 - supplementary-information.

Simeon, D., Bartz, J., Hamilton, H., Crystal, S., Braun, A., Ketay, S., & Hollander, E. (2011). Oxytocin administration attenuates stress reactivity in borderline personality disorder: a pilot study. *Psychoneuroendocrinology, 36*(9), 1418–1421. https://doi.org/10.1016/j.psyneuen.2011.03.013.

Smith, A. S., Tabbaa, M., Lei, K., Eastham, P., Butler, M. J., Linton, L., … Wang, Z. (2016). Local oxytocin tempers anxiety by activating GABAA receptors in the hypothalamic paraventricular nucleus. *Psychoneuroendocrinology, 63*, 50–58. https://doi.org/10.1016/j.psyneuen.2015.09.017.

Sripada, C. S., Phan, K. L., Labuschagne, I., Welsh, R., Nathan, P. J., & Wood, A. G. (2013). Oxytocin enhances resting-state connectivity between amygdala and medial frontal cortex. *The International Journal of Neuropsychopharmacology, 16*(2), 255–260. https://doi.org/10.1017/S1461145712000533.

Stange, J. P., Sylvia, L. G., Magalhaes, P. V., Frank, E., Otto, M. W., Miklowitz, D. J., … Deckersbach, T. (2013). Extreme attributions predict transition from depression to mania or hypomania in bipolar disorder. *Journal of Psychiatric Research, 47*(10), 1329–1336. https://doi.org/10.1016/j.jpsychires.2013.05.016.

Stavropoulos, K. K., & Carver, L. J. (2013). Research review: social motivation and oxytocin in autism—implications for joint attention development and intervention. *Journal of Child Psychology and Psychiatry, 54*(6), 603–618. https://doi.org/10.1111/jcpp.12061.

Striepens, N., Kendrick, K. M., Hanking, V., Landgraf, R., Wüllner, U., Maier, W., & Hurlemann, R. (2013). Elevated cerebrospinal fluid and blood concentrations of oxytocin following its intranasal administration in humans. *Scientific Reports, 3*(1). https://doi.org/10.1038/srep03440.

Tollenaar, M. S., Chatzimanoli, M., van der Wee, N. J., & Putman, P. (2013). Enhanced orienting of attention in response to emotional gaze cues after oxytocin administration in healthy young men. *Psychoneuroendocrinology, 38*(9), 1797–1802. https://doi.org/10.1016/j.psyneuen.2013.02.018.

Tseng, H. C., Chi, M. H., Lee, L. T., Tsai, H. C., Lee, I. H., Chen, K. C., … Chen, P. S. (2014). Sex-specific associations between plasma oxytocin levels and schizotypal personality features in healthy individuals. *Journal of Psychiatric Research, 51*, 37–41. https://doi.org/10.1016/j.jpsychires.2013.12.011.

Turan, T., Uysal, C., Asdemir, A., & Kilic, E. (2013). May oxytocin be a trait marker for bipolar disorder? *Psychoneuroendocrinology, 38*(12), 2890–2896. https://doi.org/10.1016/j.psyneuen.2013.07.017.

Uhrig, S., Hirth, N., Broccoli, L., von Wilmsdorff, M., Bauer, M., Sommer, C., … Schmitt, A. (2016). Reduced oxytocin receptor gene expression and binding sites in different brain regions in schizophrenia: a post-mortem study. *Schizophrenia Research, 177*(1–3), 59–66. https://doi.org/10.1016/j.schres.2016.04.019.

Uvnas-Moberg, K., Alster, P., & Svensson, T. H. (1992). Amperozide and clozapine but not haloperidol or raclopride increase the secretion of oxytocin in rats. *Psychopharmacology, 109*(4), 473–476.

Valstad, M., Alvares, G. A., Andreassen, O. A., Westlye, L. T., & Quintana, D. S. (2016). The relationship between central and peripheral oxytocin concentrations: a systematic review and meta-analysis protocol. *Systematic Reviews, 5*, 49. https://doi.org/10.1186/s13643-016-0225-5.

Veras, A. B., Getz, M., Froemke, R. C., Nardi, A. E., Alves, G. S., Walsh-Messinger, J., … Malaspina, D. (2018). Rare missense coding variants in oxytocin receptor (OXTR) in schizophrenia cases are associated with early trauma exposure, cognition and emotional processing. *Journal of Psychiatric Research, 97*, 58–64. https://doi.org/10.1016/j.jpsychires.2017.11.011.

Wang, D., Yan, X., Li, M., & Ma, Y. (2017). Neural substrates underlying the effects of oxytocin: a quantitative meta-analysis of pharmaco-imaging studies. *Social Cognitive and Affective Neuroscience, 12*(10), 1565–1573. https://doi.org/10.1093/scan/nsx085.

Warren, K. R., Wehring, H. J., Liu, F., McMahon, R. P., Chen, S., Chester, C., & Kelly, D. L. (2018). Effects of intranasal oxytocin on satiety signaling in people with schizophrenia. *Physiology & Behavior, 189*, 86–91. https://doi.org/10.1016/j.physbeh.2018.03.008.

Williams, D. R., & Burkner, P. C. (2017). Effects of intranasal oxytocin on symptoms of schizophrenia: a multivariate Bayesian meta-analysis. *Psychoneuroendocrinology, 75*, 141–151. https://doi.org/10.1016/j.psyneuen.2016.10.013.

Wolf, F., Brune, M., & Assion, H. J. (2010). Theory of mind and neurocognitive functioning in patients with bipolar disorder. *Bipolar Disorders, 12*(6), 657–666. https://doi.org/10.1111/j.1399-5618.2010.00854.x.

Wolkind, S. (1973). Emotion in the Human Face: Guidelines for Research and an Integration of Findings. By P. Ekman, W. V. Friesen and P. Ellsworth. Pergamon Press. 1971. Pp. 191. Price £3.50. *British Journal of Psychiatry, 122*(566), 108. https://doi.org/10.1192/bjp.122.1.108.

Woolley, J. D., Chuang, B., Lam, O., Lai, W., O'Donovan, A., Rankin, K. P., … Vinogradov, S. (2014). Oxytocin administration enhances controlled social cognition in patients with schizophrenia. *Psychoneuroendocrinology, 47*, 116–125. https://doi.org/10.1016/j.psyneuen.2014.04.024.

Wynn, J. K., Green, M. F., Hellemann, G., Reavis, E. A., & Marder, S. R. (2018). A dose-finding study of oxytocin using neurophysiological measures of social processing. *Neuropsychopharmacology*. https://doi.org/10.1038/s41386-018-0165-y.

Yang, X., Tang, Y., Wei, Q., Lang, B., Tao, H., Zhang, X., … Tang, A. (2017). Up-regulated expression of oxytocin mRNA in peripheral blood lymphocytes from first-episode schizophrenia patients. *Oncotarget, 8*(45). https://doi.org/10.18632/oncotarget.20252.

Yuen, K. W., Garner, J. P., Carson, D. S., Keller, J., Lembke, A., Hyde, S. A., … Parker, K. J. (2014). Plasma oxytocin concentrations are lower in depressed vs. healthy control women and are independent of cortisol. *Journal of Psychiatric Research, 51*, 30–36. https://doi.org/10.1016/j.jpsychires.2013.12.012.

Ziegler, T. E. (2018). Measuring peripheral oxytocin and vasopressin in nonhuman primates. *American Journal of Primatology*, e22871. https://doi.org/10.1002/ajp.22871.

Zink, C. F., & Meyer-Lindenberg, A. (2012). Human neuroimaging of oxytocin and vasopressin in social cognition. *Hormones and Behavior*, *61*(3), 400–409. https://doi.org/10.1016/j.yhbeh.2012.01.016.

CHAPTER SEVEN

# Reward processing and social functioning in psychosis

**Alexis E. Whitton*, Kathryn E. Lewandowski[†,‡]**
*The University of Sydney, Faculty of Medicine and Health, Sydney, NSW, Australia
[†]Harvard Medical School, Department of Psychiatry, Boston, MA, United States
[‡]McLean Hospital, Psychotic Disorders Division, Belmont, MA, United States

Abnormalities in motivation are hallmarks of psychosis (Strauss, Waltz, & Gold, 2013), and have been described as key components of the schizophrenia syndrome since the early observations of Kraepelin and Bleuler (Bleuler, 1950; Kraepelin, 1919). Deficits in motivation have been linked to disturbances in processing rewarding events, stimuli, and environmental cues, and in using reward feedback to guide future behavior (Treadway & Zald, 2013). Accordingly, there has been a growing interest in studying reward processing in psychosis as a way to understand the pathophysiological processes that give rise to motivational disturbances in patients (for a review, see Whitton, Treadway, & Pizzagalli, 2015). Gaining a better understanding of the mechanisms underlying motivational impairment in psychosis is crucial, as motivational symptoms are among the most disabling and difficult to treat, and are strongly linked to poor community outcomes (Fervaha, Foussias, Agid, & Remington, 2014; Foussias et al., 2011).

Of all the domains that motivational impairments affect, perhaps the most striking is the social domain. Motivational deficits can lead to diminished interest in social interaction, thereby interfering with an individual's ability to engage with and navigate the social world. As we discuss in this chapter, emerging evidence highlights a potential association between basic reward processes and social motivation. This link has important implications for our understanding of social deficits in patients with psychotic disorders, as much of the existing literature has conceptualized these deficits as stemming from abnormalities in social cognition, as opposed to abnormalities in motivation (Green, Horan, & Lee, 2015). Decades of animal research provide a rich understanding of the neurobiological basis for dysfunctional reward processing, highlighting a key role for aberrant mesolimbic dopamine (DA) system functioning (Schultz, 2016). Accordingly, evidence of a link between dysfunctional reward processing and social impairment in


https://doi.org/10.1016/B978-0-12-815315-4.00007-0

psychosis may point to novel neurobiological treatment targets for social impairments in individuals with psychotic disorders.

Reward processing is a multifaceted construct comprised of several subprocesses found to have overlapping, yet partially distinct, neurobiological correlates (Cuthbert & Insel, 2013). These subprocesses have been parsed in several ways across a number of different theoretical frameworks; however, two frameworks have gained considerable traction in recent years. The first framework (which has been reviewed extensively by Berridge and Robinson, e.g., Berridge & Robinson, 1998; Berridge & Robinson, 2003; Berridge, Robinson, & Aldridge, 2009) parses reward processing into subprocesses that support reward anticipation (termed "wanting"), reward consumption (termed "liking"), and the ability to modify behavior based on prior reward history (termed "learning"). The second framework is the National Institute of Health's Research Domain Criteria (RDoC; Cuthbert & Insel, 2013). Within this framework, the overarching construct of reward processing is formalized as the Positive Valence System (PVS), and comprises five distinct subprocesses that are in many ways a finer-grained parcellation of the wanting, liking, and learning framework reviewed by Berridge and colleagues. The subprocesses within the RDoC PVS framework are: (1) approach motivation (similar to "wanting"), (2) initial responsiveness to reward attainment, (3) sustained/longer-term responsiveness to reward attainment (both 2 and 3 reflecting "liking" processes), (4) reward learning, and (5) habit-based responding (both 4 and 5 reflecting different forms of "learning"). In this chapter, we will use the RDoC framework, together with the broader concepts of wanting, liking, and learning, to describe the different components of motivation and reward, how they are affected in psychotic disorders, their proposed links with social functioning, and the neurobiological substrates that may underpin co-occurring reward and social deficits in psychotic populations.

## Motivation and reward processing: Wanting, liking, and learning

Approach motivation (a subprocess captured under the umbrella term "wanting") refers to processes involved in the initiation and maintenance of reward-seeking behavior. Individual differences in approach motivation are thought to be determined by four different factors: reward valuation, effort valuation, reward prediction, and action selection (Barch & Dowd, 2010;

Cuthbert & Insel, 2013). Reward valuation is the process by which the probability and benefits of a prospective outcome are computed using external information, social context, and/or prior experience. Effort valuation refers to the process by which the cost or effort required to obtain a reward is computed, and that drives an individual to overcome response costs to obtain something rewarding. Reward prediction refers to the process by which the likelihood of obtaining a reward in a given context is estimated. Reward prediction is most commonly a state that is triggered by cues, contexts, or events that allow for the ability to make a prediction about the likelihood of achieving a certain outcome. Finally, action selection refers to the process by which the potential benefit of choosing one of multiple competing available actions is computed against the cost of choosing that action, as well as the cost of not choosing other actions.

In contrast to approach motivation, initial responsiveness to reward and sustained response to reward (both of which fall under the umbrella of "liking") refer to the processes that govern our capacity to derive pleasure from stimuli or events in the environment. Initial and sustained reward responses differ in their subsequent effects on behavior. Initial responses to reward are often reflected by in-the-moment subjective experiences of enjoyment, as well as behavioral responses indicative of pleasure (e.g., smiling). In the literature, individual differences in initial responses to reward are often referred to as differences in hedonic capacity. In contrast, sustained responses to reward lead to the termination of reward seeking and reward consumption, and reflect satisfaction and satiety.

Finally, reward learning and habit-based responding (both of which fall under the umbrella term of "learning") refer to processes by which external or internal cues, stimuli, or events elicit conditioned behavior. However, reward learning and habit-based responding capture different forms of learning that show different levels of responsiveness to changes in outcome. Reward learning refers to the ability to acquire information about stimuli, actions, and contexts that predict positive outcomes, and by which behavior is altered when outcomes are different than expected. In reward learning, the value of specific stimuli, actions, and contexts is closely tied to the outcomes they elicit, with value being acquired either via explicit or implicit processes, and via instrumental or Pavlovian conditioning (Rescorla & Solomon, 1967). In contrast, habit-based responding refers to sequential, repetitive motor or cognitive behaviors that are triggered by internal or external stimuli. Although habits may be initially acquired through reward learning, they differ in that they can become resistant to changes in outcome value.

Each of these processes interact to drive motivated behavior. Depending on the component affected, abnormalities within one or more of these processes may impair certain aspects of motivation, and of particular interest is understanding which processes impact socially motivated behavior. In the following section we outline evidence for aberrant reward processing in individuals with psychotic disorders, with a particular focus on distinguishing the processes that appear to be deficient from those that appear to be intact in these populations.

## Reward-processing abnormalities in psychosis

Evidence suggests that many, but not all, aspects of reward processing are abnormal in patients with psychosis. Briefly, studies have found that patients with schizophrenia display deficits in anticipatory, but not consummatory pleasure (Cohen & Minor, 2008; Foussias & Remington, 2008; Gard, Kring, Gard, Horan, & Green, 2007; Mote, Minzenberg, Carter, & Kring, 2014); increased threat sensitivity, but typical approach/activation (Scholten, van Honk, Aleman, & Kahn, 2006); impaired reward learning, but intact punishment learning (Waltz, Frank, Robinson, & Gold, 2007); difficulty with rapid, but not gradual reward learning (Gold et al., 2012; Gold, Waltz, Prentice, Morris, & Heerey, 2008); and deficits in the expression, but not experience of pleasure (Kring & Neale, 1996; Mote, Stuart, & Kring, 2014; Norman, Manchanda, Harricharan, & Northcott, 2015). Several of these deficits have been found to extend to patients with affective psychosis, with studies showing evidence of poor ability to integrate reinforcement over time in patients with bipolar disorder (Pizzagalli, Goetz, Ostacher, Iosifescu, & Perlis, 2008). In the following section we outline research into reward processing deficits in psychosis, separated according to the "wanting," "liking," "learning," framework.

### Abnormalities in reward "wanting" in psychosis

Several studies have highlighted deficits in anticipatory pleasure, or "wanting," in individuals with schizophrenia. For example, in a study assessing the ability of individuals with schizophrenia to predict future pleasure in daily life, Gard et al. (2007) found that compared with healthy controls, those with schizophrenia anticipated experiencing less enjoyment from future-goal-directed activities. Studies assessing self-reported trait levels of anticipatory and consummatory pleasure on the Temporal Experience of Pleasure Scale (TEPS; Gard, Gard, Kring, & John, 2006) have also found consistent reductions in trait anticipatory pleasure in individuals with schizophrenia across a range of illness stages, including chronic (Kring, Siegel, & Barrett, 2014) and early stages of schizophrenia (Mote, Minzenberg, et al., 2014), as well as in those at clinical high risk for schizophrenia (Schlosser et al., 2014). Highlighting potential links between aberrant reward anticipation and social motivation, Buck and Lysaker (2013) found that lower scores on the TEPS trait anticipatory pleasure subscale predicted poorer interpersonal functioning at a 6-month follow-up assessment.

Building on these findings, a wealth of studies now highlights evidence of consistent deficits in effort-based decision making (a core subprocess driving approach motivation) in individuals with schizophrenia, which manifests as a reduced willingness to expend effort to obtain rewards. For example, individuals with schizophrenia are less willing to expend physical effort (repeated, rapid button presses with the little finger of the nondominant hand) to obtain high-probability, high-value rewards on the Effort Expenditure for Rewards Task (EEfRT; Treadway, Buckholtz, Schwartzman, Lambert, & Zald, 2009). This also extends to cognitive effort, with schizophrenia patients showing less willingness to choose a hard (versus easy) cognitive task for the possibility of obtaining a larger reward (Barch, Treadway, & Schoen, 2014; Culbreth, Moran, & Barch, 2018; Culbreth, Westbrook, & Barch, 2016; Fervaha et al., 2013; Gold et al., 2013; Reddy et al., 2015; Treadway, Peterman, Zald, & Park, 2015). Importantly, rather than reflecting an overall reduction in willingness to expend effort, individuals with schizophrenia show a reduced willingness to expend effort in contexts in which it is most advantageous to do so; for example, when the probability or value of the reward is high (Gold et al., 2013). This has been found to be especially prominent in individuals displaying severe negative symptomatology (Gold et al., 2013), and has been linked to greater levels of motivational impairment and poorer overall functioning (Barch et al., 2014).

## Intact reward "liking" in psychosis

Although anhedonia is a common symptom in psychotic disorders, and is defined as a loss of pleasure, considerable experimental data suggests that capacity to experience pleasure (or "liking") may be intact in those with psychotic disorders (for reviews, see Cohen & Minor, 2008; Kring & Moran, 2008). For example, a well-replicated finding is that individuals with schizophrenia experience normative levels of positive emotion in response to emotionally evocative stimuli (Cohen & Minor, 2008), and in their daily lives (Gard et al., 2007; Myin-Germeys, Delespaul, & DeVries, 2000). Studies assessing responses on the sweet taste test—a prototypical test of hedonic capacity—have found no differences in hedonic responses to sucrose in individuals with schizophrenia relative to healthy controls, suggesting intact experiences of pleasure in response to sweet tastes (Berlin, Givry-Steiner, Lecrubier, & Puech, 1998). More recently, in a multimodal assessment of anticipatory and consummatory responses to emotionally-evocative images, Moran and Kring (2018) found that in-the-moment ratings of emotion, along with emotion-related blink rate, were normal in individuals with schizophrenia relative to controls, despite evidence of dampened anticipatory responses (as indexed by anticipatory ratings of emotion and blink rate during an anticipation phase). Intact capacity for pleasure has even been found in psychotic patients with blunted affect (Berenbaum & Oltmanns, 1992; Earnst et al., 1996), suggesting that these patients may experience normative liking, even in the absence of overt expressions of pleasure.

However, some studies suggest a more nuanced picture of reward "liking" in psychotic individuals, and suggest that consummatory reward processes may also be impaired in these individuals in specific contexts. Experience-sampling measures, which ascertain people's experiences in the moment (as opposed to self-report measures, which ask people to retrospectively report on perceptions of their experiences), show that patients with psychotic disorders experience liking in daily life, although the frequency and context may differ from controls. For example, patients with schizophrenia show less frequent or intense positive emotions in real-world situations, and the reduced intensity of these emotions tends to be more evident in more demanding situations, such as work or school, compared with less demanding contexts, such as eating (Myin-Germeys et al., 2000). The degree of impairment in hedonic capacity also appears to be associated with anhedonia severity, with studies showing evidence of reduced experience of pleasure in response to positive stimuli in patients with schizophrenia who report high levels of anhedonia (Dowd & Barch, 2012). Overall, while liking appears to be roughly intact in patients with psychosis, liking or pleasure may actually be reduced in a subset of patients with elevated levels of anhedonia, and in contexts involving more complex, goal-oriented behaviors (Tso, Grove, & Taylor, 2014).

## Impairments in some, but not all forms of reward-learning in psychosis

Patients with psychosis also exhibit deficits in some aspects of reward learning, particularly when learning paradigms are challenging. Specifically, impairments tend to be most pronounced when learning involves multiple levels of prediction, reversal of learned contingencies, or when there is a requirement for especially rapid learning (Gold et al., 2008; Morris, Heerey, Gold, & Holroyd, 2008; Waltz & Gold, 2007). Furthermore, deficits in reward learning have been found to vary as a function of whether learning occurs implicitly or explicitly. For example, in a study evaluating explicit and implicit reward learning in a mixed psychosis sample of individuals with schizophrenia, schizoaffective disorder, and bipolar disorder, Barch et al. (2017) found that individuals with psychotic disorders showed intact implicit reward learning. In terms of explicit reward learning, patients with schizophrenia and schizoaffective disorder exhibited deficits relative to controls, whereas patients with bipolar disorder did not differ from controls. Furthermore, worse performance on the explicit, but not the implicit, learning task was related to decreased motivation and pleasure across all diagnostic categories (Barch et al., 2017).

Taken together, these findings show that patients with psychosis exhibit deficits in some, but not all, aspects of reward processing and motivated behavior. Although motivational impairments are evident across the entire spectrum of psychosis, some differences are observed across diagnostic groups and symptom profiles. Specifically, impairments appear to be more pronounced in nonaffective, as opposed to affective, psychotic disorders, and in individuals reporting more severe levels of anhedonia. In the next section, we describe the ways in which deficits in wanting, liking, and learning may lead to abnormalities in social functioning.

## Links between reward processing and social behavior

The drive for social connection has been proposed to be innate in humans (Baumeister & Leary, 1995), and serves several adaptive purposes. In healthy adults, social interactions have been shown to increase positive affect (Fleeson, Malanos, & Achille, 2002), and stronger motivation for social closeness has been linked to greater subjective well-being (Shankar, Rafnsson, & Steptoe, 2015). Individuals with psychotic disorders experience significant difficulties with social relationships, and as a result, experience high levels of social isolation and difficulty forming social bonds (Oliveira, Esteves, & Carvalho, 2015). Although social impairments in psychosis are most commonly attributed to deficits in social cognition (Couture, Granholm, & Fish, 2011), emerging theories posit that social impairment in this population may arise in part from reductions in social motivation. Supporting this notion, evidence has shown that abnormalities in a number of aspects of social reward processing and social motivation are associated with impairments in broader social functioning in patients with psychotic disorders (Campellone & Kring, 2018). Furthermore, social deficits appear to be more closely associated with negative (as opposed to positive) symptoms, which are the symptoms that have been most strongly linked to deficits in reward processing.

This evidence of decreased drive to engage in social contact is sometimes referred to as social anhedonia (Blanchard, Mueser, & Bellack, 1998), and is both common in patients with schizophrenia, and predictive of conversion to schizophrenia in high-risk samples (Kwapil, 1998). Social anhedonia may be associated with reductions in social wanting, social liking, or both. In an experience sampling study of nonpsychotic young adults, higher levels of social anhedonia were associated with reduced positive affect in social situations, and this had implications for social behavior, including more time spent alone, a greater preference for being alone, and higher rates of disengagement during social interactions (Brown, Silvia, Myin-Germeys, & Kwapil, 2007). Similarly, patients with schizophrenia report less positive and more negative affect during social affiliation tasks, with pleasure and motivational deficits being associated with lower feelings of social closeness, as well as greater self-reported social anhedonia (Blanchard, Park, Catalano, & Bennett, 2015; McCarthy et al., 2018). Patients with psychosis also appear to anticipate experiencing less pleasure for social interactions compared with controls, particularly for interactions with social partners who display positive affect (Campellone & Kring, 2018; Engel, Fritzsche, & Lincoln, 2016). Taken together, these findings suggest that patients with schizophrenia, or those at elevated risk for schizophrenia, show reductions in *both* social wanting and social liking, and that both of these deficits have the functional consequence of decreased social interaction. This contrasts somewhat with the nonsocial reward literature, which finds greater reductions in anticipatory pleasure than consummatory pleasure in psychotic patients.

In terms of learning, patients with psychosis have difficulties learning in the context of positive social interactions, but show no differences from controls in negative social learning. For example, on paradigms that require the participant to learn whether or not to trust a social partner, learning appears to be intact when the prior interactions with that social partner were negative, but reduced compared with controls when the earlier interactions were positive (Campellone, Fisher, & Kring, 2016). This is similar to findings showing reduced updating of anticipated social pleasure after interaction with a smiling social partner (Campellone & Kring, 2018). Together, findings from experimental, experience sampling, and self-report data suggest that patients with schizophrenia and bipolar disorder and people with higher levels of social anhedonia exhibit abnormalities in wanting, liking, and learning in social contexts, and that these abnormalities are associated with reduced quantity and quality of social engagement.

Social reward processing across the PVS may be associated with deficits in processing information relevant to the social environment, or may be attributable to deficits in social cognition. Social cognition refers to processes related to the perception of, interpretation of, and response to the social environment, and includes social perception, social knowledge, emotion recognition in the self and others, theory of mind, and attributional bias (Green et al., 2008). Deficits in social cognition are well-described in patients with schizophrenia, particularly in social perception, theory of mind/mental state attribution, and emotion regulation (for a review, see Green et al., 2015), and have been shown to be associated with functional impairment independent from more general neurocognitive deficits (Fett et al., 2011). Although present across the psychosis spectrum, the severity of social cognitive deficits appears to differ across diagnoses, with deficits being less pronounced and more selective in patients with bipolar disorder compared with those with schizophrenia (Lee et al., 2013; Samamé, Martino, & Strejilevich, 2012). Thus, social cognitive deficits may be associated with abnormalities in social reward processing differently across diagnoses. Because the neurobiology of a number of aspects of reward processing has been studied, as described herein, such differences may reveal aspects of distinct neurobiology and/or pathophysiology between disorders.

A recent model hypothesized that wanting, liking, learning, social skills, and social cognition may interact in the social environment to drive reduced social motivation in schizophrenia. In this model, which integrates both social approach and social avoidance, Fulford, Campellone, and Gard (2018) hypothesize that diminished social approach and reward sensitivity, together with heightened social avoidance and punishment sensitivity, are associated with reduced attention to social rewards and increased negative expectances, along with increased attention to negative social cues and interpretation of neutral cues as hostile. Later, patients experience negative bias in the interpretation of the interaction, and diminished positive memory consolidation regarding the social interaction, along with reduced positive and increased negative affect. Thus, abnormalities in expectancies, reward sensitivity, and reward learning interact with deficits in attention, emotion recognition, and theory of mind to reduce motivated social behavior, even in the presence of a desire for social affiliation.

## The neurobiology of reward processing

Different facets of reward processing have been found to draw on overlapping yet partially distinct neural systems (for a review, see Der-Avakian & Markou, 2012). Although a comprehensive review of the neurocircuitry underpinning reward-based behavior is beyond the scope of this chapter, in the following section we briefly outline some of the key neural systems underpinning different facets of reward processing, with a focus on those thought to support aspects of reward wanting, liking, and learning.

The mesolimbic dopamine system, particularly the dopaminergic projections from the ventral tegmental area (VTA) to the ventral striatum (VS), including the nucleus accumbens (NAc), is suggested to play an important role in reward anticipation or wanting. Evidence for this comes from studies using reward-based paradigms that allow for the separation of anticipatory from consummatory phases of reward processing. For example, one paradigm that has been widely used is the Monetary Incentive Delay (MID; Knutson, Westdorp, Kaiser, & Hommer, 2000) task, which requires the participant to react to a target stimulus that is presented after an incentive cue in order to win or avoid losing a reward. This task incorporates both an anticipatory phase and a consummatory phase, allowing the neural correlates of each to be adequately disentangled. A recent meta-analysis of 50 fMRI studies using the MID in 1271 healthy participants found that the ventral and dorsal striatum, particularly the dorsolateral portion of the NAc, was most active during the anticipation phase of the task (Oldham et al., 2018). In addition to the striatum, the anticipation phase elicited consistent activation in the anterior insula, amygdala, and thalamus—regions that are hypothesized to project to the VS (Haber & Knutson, 2010), in order to update predictions and facilitate action selection.

In contrast, studies examining the neural correlates of reward consumption (liking) implicate "hedonic hotspots" within the ventral pallidum and NAc, which are the site of endogenous opioid action (Pecina & Berridge, 2005; Smith, Tindell, Aldridge, & Berridge, 2009). Other regions important for reward consumption include the orbitofrontal cortex (OFC), which has been implicated in encoding reward value (O'Doherty, Deichmann, Critchley, & Dolan, 2002), and the posterior cingulate cortex (PCC), which has been implicated in monitoring the environment, tracking past outcomes, and self-referential processes. By comparing patterns of activation during the outcome relative to the anticipation phase of the MID, Oldham et al. (2018) found that OFC and PCC activity was observed only during the outcome phase, suggesting that these regions may be uniquely implicated in reward consumption.

Finally, phasic dopaminergic signaling in the VS is hypothesized to support reward learning. Specifically, several lines of evidence indicate that phasic firing of striatal and midbrain dopamine neurons signal reward prediction errors that encode the discrepancy between an expected reward and the actual reward received. When a reward is better than expected (a positive prediction error), increased phasic dopaminergic firing is observed, whereas omission of an expected reward (a negative prediction error) leads to transient depression of dopamine firing rates. These changes in dopaminergic signaling serve to either strengthen or weaken action-outcome associations and, subsequently, play an integral role in driving adaptive changes in behavior. Data also implicate a key role for the anterior cingulate cortex (ACC) and medial prefrontal cortex (mPFC) in maximizing adaptive behavior, particularly in terms of using reinforcement history to guide future actions (Kennerley, Walton, Behrens, Buckley, & Rushworth, 2006).

## Evidence for mesolimbic dopamine reward system disruption in psychosis

In addition to supporting a repertoire of different hedonic behaviors, the mesolimbic dopamine reward system is thought to play a critical role in the pathophysiology of psychosis. Several theories on the role of dopamine in psychotic disorders, particularly schizophrenia, posit that increased presynaptic striatal dopamine release and aberrant striatal dopaminergic signaling may form a pathophysiological basis for psychotic symptoms. Evidence for this comes from studies that have used positron emission tomography (PET) to study dopamine release, and have shown that an excess of striatal dopamine, particularly in the associative striatum, is present in individuals with schizophrenia (for a review, see Weinstein et al., 2017). Furthermore, striatal dopaminergic hyperactivity has been found to correlate with the severity of positive psychotic symptoms (Howes et al., 2012), and is present even in medication-naive prodromal patients (Howes et al., 2009), suggesting that it is not merely a by-product of antipsychotic medication use.

In the context of reward learning, one of the key functions subserved by transient changes in striatal dopamine activity is to facilitate the detection of personally salient features of the environment (for a review, see Grace,

2016). As such, an excess of striatal dopamine has been hypothesized to drive psychosis by inducing a state of aberrant salience, wherein irrelevant cues and events are incorrectly ascribed motivational relevance. Support for this theory comes from studies showing evidence of abnormal VS activation—an indirect index of striatal dopamine signaling—to neutral, as opposed to reward-predictive stimuli in individuals with schizophrenia. Specifically, relative to healthy control participants, individuals with schizophrenia have been found to show inappropriately strong VS activation in response to unconditioned stimuli, which has been interpreted to reflect the aberrant assignment of motivational salience to irrelevant environmental cues (Diaconescu et al., 2011). This misattributed salience to neutral cues appears to be coupled with diminished salience processing of reward-relevant cues. For example, individuals with schizophrenia have been found to display reduced VS activation to reward-prediction errors during a reversal learning task (Schlagenhauf et al., 2014), and a meta-analysis has also shown evidence of significant bilateral VS hypoactivation during reward anticipation in patients with schizophrenia relative to healthy controls (Radua et al., 2015).

According to a recent iteration of the aberrant salience hypothesis put forth by Maia and Frank (2017), one of the main consequences of salience misattribution is a disruption in reinforcement learning, and this disruption is proposed to constitute a fundamental mechanism underlying the cognitive and emotional symptoms observed in psychotic patients. Specifically, Maia and Frank suggest that abnormal mesolimbic dopamine function causes a chronic mismatch between events in the environment and phasic dopamine signals, which impedes the discrimination of salient from nonsalient environmental stimuli. This then impairs the ability to utilize relevant cues and environmental feedback to adaptively guide behavior. In the context of social deficits in this population, an important question is whether mesolimbic dopamine reward system dysfunction contributes to deficits in social motivation observed in psychosis. In the following sections, we outline evidence linking mesolimbic dopamine reward system function to social motivation, and current theories on the relationship between reward system dysfunction and social functioning impairment in individuals with psychotic disorders.

## The role of dopamine reward pathways in social functioning

As noted in the preceding section, motivation for, and reward from, social involvement is both innate in humans and associated with subjective well-being. A growing interest in the neuroscience of reward or pleasure in the context of social behavior has produced converging evidence from preclinical studies, as well as from human social and affective neuroscience, suggesting that these appetitive aspects of social behavior are supported by processes occurring within the mesolimbic dopamine system. Furthermore, evidence suggests that this system may mediate the processing of social rewards in a similar manner as it does nonsocial rewards. For example, studies have shown that key regions within the reward circuit, such as the VS, become active when individuals either receive or anticipate receiving positive social feedback (Rademacher et al., 2010). Studies examining the social sequelae of experimental dopaminergic manipulation also converge on an important role of dopamine reward pathways in social behavior. For example, administration of the dopamine precursor L-DOPA in healthy individuals has been found to improve learning about another individual's prosocial preferences in a social interaction task (Eisenegger et al., 2013). Similarly, animal studies have shown that enhancing striatal dopaminergic activity by either administering amphetamine (a dopamine release/reuptake inhibitor) into the NAc (Manduca et al., 2016), or by blocking the dopamine transporter (Achterberg et al., 2016), increases social play behavior in rats. These findings highlight an important role of striatal dopaminergic neurotransmission in adaptive social functioning.

Similar to reinforcement learning in nonsocial contexts, learning about social reinforcers is thought to be driven by striatal dopamine signals that fire in accordance with the reinforcing properties of social stimuli. In rats, exposure to a putatively prosocial 50 kHz ultrasonic vocalization emitted by a conspecific and designed to elicit social approach behavior triggers phasic dopamine release in the NAc of the observer rat (Willuhn et al., 2014). In this context, phasic NAc dopamine signaling was hypothesized to function as a translator of a motivational communicative signal into a prosocial action. Similarly, evidence from human fMRI studies suggests that dopamine-mediated reinforcement learning signals in the VS may facilitate social cohesion. Specifically, using a paradigm in which participants' initial judgments of facial attractiveness were open to influence by group opinion, Klucharev, Hytönen, Rijpkema, Smidts, and Fernández (2009) found that conflict with a group opinion (i.e., being told that their initial judgments of attractiveness differed from a group norm) led to a significant decrease in activity within the NAc, similar to that observed when individuals make a negative reward prediction error. Furthermore, social conflict-related activity in the NAc was found to be linked to subsequent adjustments in task performance wherein the participant altered their initial judgments to conform to the group normative opinion. These findings reflect an important role of the NAc in social learning.

Extending these findings, Jones et al. (2011) conducted a study to specifically test the hypothesis that the process of learning from positive social exchanges mirrored basic reinforcement learning. In this study, they examined the degree to which participants responded more quickly to faces of peers who provided more frequent positive social reinforcement, and examined the brain regions that tracked cues associated with positive social feedback. Using a trial-by-trial modeling approach analogous to that used in studies of nonsocial reinforcement learning, they found that VS and OFC activation tracked prediction errors in social learning (i.e., when positive social feedback was received unexpectedly or when it was unexpectedly not received). These findings indicate that our ability to update our expectations and behavior as a function of social feedback from others likely draws on similar neurobiological mechanisms as those involved in nonsocial reinforcement learning. As noted herein, patients with schizophrenia exhibit greater deficits in reinforcement learning in the context of positive relative to neutral, or negative stimuli in both social and nonsocial settings when compared with controls, supporting the role of this pathway in abnormal social learning and functioning in patients with psychosis.

A large body of research has also focused on the unique role of the neuropeptide oxytocin in social motivation, and mounting evidence indicates that this neuropeptide may modulate the incentive salience of social cues, in part by activating dopaminergic reward pathways, as described in Chapter 6 and also in (Shamay-Tsoory & Abu-Akel, 2016). Supporting this theory, research has shown that intranasally administered oxytocin increases activity in the VTA in response to cues signaling social reward (i.e., a friendly face) or social punishment (i.e., an angry face; Groppe et al., 2013). Similarly, intranasal oxytocin has been found to facilitate learning to use social feedback and also to selectively increase activity in the amygdala, hippocampus, and striatum during the anticipation of social feedback (Hu et al., 2015). Furthermore, in mice, oxytocin release in the VTA was found to enhance the activity of mesolimbic dopaminergic neurons, which influenced social behavior (Hung et al., 2017).

Taken together, these findings suggest that activity within the mesolimbic dopamine reward system may mediate the value of social rewards and may drive social motivation. As such, the abnormalities within this system may also underpin impairments in social motivation observed in individuals with psychotic disorders. In the following section, we outline evidence for a link between reward circuit dysfunction and social impairment in the context of psychosis.

## Links between aberrant reward circuit functioning and social deficits in psychosis

One of the hallmark symptoms of schizophrenia that seems most obviously linked to deficits in reward function given its parallels with anhedonia, is a lack of social motivation, wherein individuals with the disorder appear less willing to seek out social interaction. Individuals with schizophrenia, and nonpsychotic individuals identified as psychometrically at high risk, exhibit a decreased willingness to expend effort for nonsocial rewards (e.g., Treadway et al., 2015); therefore, deficient effort allocation in the context of social rewards may represent a trait-marker of illness that contributes to asociality in patients. In a recent study, Fulford, Treadway, and Woolley (2018) aimed to test this hypothesis in a sample of individuals with schizophrenia using a social vigor test designed to measure effort exertion in the context of live social encouragement. They hypothesized that if asociality in schizophrenia was related to a decreased willingness to expend effort for social rewards, then individuals with schizophrenia would show less effort exertion in response to social encouragement compared to control participants. In contrast to their predictions, however, Fulford, Treadway, et al. (2018) found that patients with schizophrenia did not differ from controls in terms of their increases in vigor in the context of social encouragement. They interpreted these findings as evidence that effort expenditure in the context of social rewards may, in fact, be spared in individuals with schizophrenia, and deficient effort allocation in a social context may not be the central factor driving decreased social motivation in this population.

One way in which abnormalities in reward circuit functioning in psychosis may be linked to deficits in social processing may be via deficits in detecting salient social cues. Similar to Maia and Frank's (2017) account of the role of abnormal phasic dopamine signaling in aberrant salience of nonsocial information in schizophrenia, Rosenfeld, Lieberman, and Jarskog (2010) put forth a similar theory to account for social deficits in schizophrenia, positing that disordered dopaminergic firing renders irrelevant social information salient, and important social cues and social information may be missed. In this sense, deficits in social functioning in psychosis are hypothesized to arise due to impairments in the ability to discriminate meaningful from meaningless social information, which in turn leads to an inability to properly navigate the social world and shape one's own social behavior accordingly. This theory may help to explain several social deficits observed in psychotic individuals. For example, symptoms of paranoia that are observed across many psychotic disorders during states of heightened positive psychotic symptomatology may be driven by the misattribution of irrelevant social cues as being personally or emotionally salient. Furthermore, decreased social motivation and isolation may arise because an inability to attend to and process relevant social information, in particular, positive social feedback, may render social interaction more difficult and less desirable to individuals with psychosis. A lack of ability to distinguish salient from irrelevant social information may also lead to impairments in the ability to use this information to adaptively modulate one's own social behavior.

## Are abnormalities in reward processing a cause of impaired social functioning in psychosis?

Taken together, the findings outlined in the previous section indicate that psychosis is closely linked to abnormalities in reward circuit functioning, and that dopamine reward pathways may play a crucial role in mediating the rewarding nature of social behavior. However, an important question is whether deficits in social functioning in psychosis arise as a consequence of reward system dysfunction, or whether they arise due to aberrant functioning in other domains, such as cognition. Although very few studies have explicitly addressed this question, evidence for the former comes from a study conducted by Gard, Fisher, Garrett, Genevsky, and Vinogradov (2009), who used a path analysis to examine whether deficits in motivation causally influenced social cognitive deficits in schizophrenia, and thereby led to poor functional outcome, or whether motivational deficits mediated the relationship between cognitive deficits and functioning in schizophrenia. They found that the model containing motivational deficits as a causal factor was a poor fit to the data, and instead concluded that motivational impairments fit best as a mediator of the relationship between social cognitive deficits and functioning. A similar argument has been put forth by Nakagami, Xie, Hoe, and Brekke (2008), who used structural equation modeling and found that motivation, again, best fit as a mediator between neurocognition and functioning in schizophrenia, rather than as a causal factor. Each of these studies indicates that links between motivational impairment and deficits in functioning (including social functioning) arise due to the shared causal factor of disturbed neurocognition.

However, in contrast to these two studies that placed motivation as the mediator between social cognitive deficits and social functioning/social reward, a study of reward processing and cognition in bipolar disorder and schizophrenia found no association between reward learning and neurocognition, and an inverse relationship between social cognitive deficits and reward processing (Lewandowski et al., 2016). Additionally, evidence emerging from the field of autism research indicates that reward and motivational impairments may precede social deficits in development, and may have a causal role in disorders characterized by social impairment. Specifically, Chevallier's social motivation theory of autism (Chevallier, Kohls, Troiani, Brodkin, & Schultz, 2012) proposes that motivational deficits, particularly reduced attention to and motivation to engage with social cues, form the basis of social functioning impairment in autism, rather than deficits in social cognition. Chevallier's idea of motivation as a causal factor in social functioning also aligns with the theory of social impairment in schizophrenia put forth by Rosenfeld et al. (2010), which suggests that social cognitive deficits in schizophrenia arise due to an underlying abnormality in processing the salience of social cues—a process that may be rooted within the reward system. This deficit in salience processing (which aligns well with other aberrant salience models that have been applied to nonsocial information processing in schizophrenia), is thought to then result in difficulties perceiving and responding appropriately to social cues, in turn leading to social withdrawal and isolation. As the field is still in its infancy, further research is clearly needed in order to determine whether motivation is a causal or mediating factor contributing to social functioning impairment in individuals with psychosis.

## Future directions

Gaining a better understanding of how deficits in reward processing are linked to social deficits is important for developing better treatments to target these symptoms. For example, although social anhedonia is considered to be a robust risk factor for the development of schizophrenia (Kwapil, 1998), emerging evidence suggests that many people with schizophrenia express a desire for more social connection (Gard et al., 2014), although as noted herein, patients with higher levels of anhedonia may show diminished wanting and liking in social settings. This suggests that at least some aspects of social motivation may be intact in a subset of patients, and may be leveraged in order to increase social functioning in these individuals. For example, a promising implication of Chevallier et al.'s (2012) social motivation theory is the notion that if social cognitive skills are intact, then addressing motivational deficits and abnormalities in attention to appropriate social cues may be sufficient to improve social functioning.

Although deficits in social functioning are present across psychotic disorders, evidence highlights some differences in the nature and severity of social impairment across different diagnostic categories. For example, social impairments may be more severe in individuals with schizophrenia and schizoaffective disorder relative to psychotic individuals with bipolar disorder. Accordingly, characterizing the differences in reward system dysfunction between these disorders may provide some insight into the neurobiological factors that drive differences in social functioning across these different conditions.

Motivational symptoms are often refractory and do not respond robustly to current first-line interventions. Because motivational impairments are strongly associated with poor outcomes, disability, and reduced subjective quality of life, improving treatment for motivational deficits and other negative symptoms is a critical and unmet clinical need. Identification of specific circuits and pathways that are associated with reward and motivation deficits is critical in the identification of actionable targets to ameliorate these symptoms. For example, a small pilot study of a novel intervention aimed at increasing patients' ability to anticipate pleasure through a behavioral skills training paradigm found that training increased anticipatory pleasure and engagement in daily activities in patients with schizophrenia (Favrod, Giuliani, Ernst, & Bonsack, 2010). In a study of cognitive remediation, a short, targeted motivational interviewing intervention aimed at increasing intrinsic motivation for the training improved both task-specific motivation and program session attendance (Fiszdon, Kurtz, Choi, Bell, & Martino, 2015). Identification of specific reward and motivation targets may hasten the development of therapeutics aimed at improving reward and motivation processes, and increase social engagement and functioning in patients.

## References

Achterberg, E. M., Van Kerkhof, L. W., Servadio, M., Van Swieten, M. M., Houwing, D. J., Aalderink, M., ... Vanderschuren, L. J. (2016). Contrasting roles of dopamine and noradrenaline in the motivational properties of social play behavior in rats. *Neuropsychopharmacology*, *41*(3), 858.

Barch, D. M., Carter, C. S., Gold, J. M., Johnson, S. L., Kring, A. M., MacDonald, A. W., III, ... Strauss, M. E. (2017). Explicit and implicit reinforcement learning across the psychosis spectrum. *Journal of Abnormal Psychology, 126*(5), 694.

Barch, D. M., & Dowd, E. C. (2010). Goal representations and motivational drive in schizophrenia: the role of prefrontal–striatal interactions. *Schizophrenia Bulletin, 36*(5), 919–934.

Barch, D. M., Treadway, M. T., & Schoen, N. (2014). Effort, anhedonia, and function in schizophrenia: reduced effort allocation predicts amotivation and functional impairment. *Journal of Abnormal Psychology, 123*(2), 387–397.

Baumeister, R. F., & Leary, M. R. (1995). The need to belong: desire for interpersonal attachments as a fundamental human motivation. *Psychological Bulletin, 117*(3), 497–529.

Berenbaum, H., & Oltmanns, T. F. (1992). Emotional experience and expression in schizophrenia and depression. *Journal of Abnormal Psychology, 101*(1), 37–44.

Berlin, I., Givry-Steiner, L., Lecrubier, Y., & Puech, A. (1998). Measures of anhedonia and hedonic responses to sucrose in depressive and schizophrenic patients in comparison with healthy subjects. *European Psychiatry, 13*(6), 303–309.

Berridge, K. C., & Robinson, T. E. (1998). What is the role of dopamine in reward: hedonic impact, reward learning, or incentive salience? *Brain Research Reviews, 28*(3), 309–369.

Berridge, K. C., & Robinson, T. E. (2003). Parsing reward. *Trends in Neurosciences, 26*(9), 507–513.

Berridge, K. C., Robinson, T. E., & Aldridge, J. W. (2009). Dissecting components of reward:'liking','wanting', and learning. *Current Opinion in Pharmacology, 9*(1), 65–73.

Blanchard, J. J., Mueser, K. T., & Bellack, A. S. (1998). Anhedonia, positive and negative affect, and social functioning in schizophrenia. *Schizophrenia Bulletin, 24*(3), 413–424.

Blanchard, J. J., Park, S. G., Catalano, L. T., & Bennett, M. E. (2015). Social affiliation and negative symptoms in schizophrenia: examining the role of behavioral skills and subjective responding. *Schizophrenia Research, 168*(1–2), 491–497.

Bleuler, E. (1950). In J. Zinkin (Ed.), *Dementia praecox or the group of schizophrenias*. New York, NY: International University Press (Vol. Original work published in 1911).

Brown, L. H., Silvia, P. J., Myin-Germeys, I., & Kwapil, T. R. (2007). When the need to belong goes wrong: the expression of social anhedonia and social anxiety in daily life. *Psychological Science, 18*(9), 778–782.

Buck, B., & Lysaker, P. H. (2013). Consummatory and anticipatory anhedonia in schizophrenia: stability, and associations with emotional distress and social function over six months. *Psychiatry Research, 205*(1–2), 30–35.

Campellone, T. R., Fisher, A. J., & Kring, A. M. (2016). Using social outcomes to inform decision-making in schizophrenia: relationships with symptoms and functioning. *Journal of Abnormal Psychology, 125*(2), 310–321.

Campellone, T. R., & Kring, A. M. (2018). Anticipated pleasure for positive and negative social interaction outcomes in schizophrenia. *Psychiatry Research, 259*, 203–209.

Chevallier, C., Kohls, G., Troiani, V., Brodkin, E. S., & Schultz, R. T. (2012). The social motivation theory of autism. *Trends in Cognitive Sciences, 16*(4), 231–239.

Cohen, A. S., & Minor, K. S. (2008). Emotional experience in patients with schizophrenia revisited: meta-analysis of laboratory studies. *Schizophrenia Bulletin, 36*(1), 143–150.

Couture, S. M., Granholm, E. L., & Fish, S. C. (2011). A path model investigation of neurocognition, theory of mind, social competence, negative symptoms and real-world functioning in schizophrenia. *Schizophrenia Research, 125*(2–3), 152–160.

Culbreth, A., Moran, E., & Barch, D. (2018). Effort-cost decision-making in psychosis and depression: could a similar behavioral deficit arise from disparate psychological and neural mechanisms? *Psychological Medicine, 48*(6), 889–904.

Culbreth, A., Westbrook, A., & Barch, D. (2016). Negative symptoms are associated with an increased subjective cost of cognitive effort. *Journal of Abnormal Psychology, 125*(4), 528–536.

Cuthbert, B. N., & Insel, T. R. (2013). Toward the future of psychiatric diagnosis: the seven pillars of RDoC. *BMC Medicine*, *11*(1), 126–133.

Der-Avakian, A., & Markou, A. (2012). The neurobiology of anhedonia and other reward-related deficits. *Trends in Neurosciences*, *35*(1), 68–77.

Diaconescu, A. O., Jensen, J., Wang, H., Willeit, M., Menon, M., Kapur, S., & Mcintosh, R. (2011). Aberrant effective connectivity in schizophrenia patients during appetitive conditioning. *Frontiers in Human Neuroscience*, *4*, 239.

Dowd, E. C., & Barch, D. M. (2012). Pavlovian reward prediction and receipt in schizophrenia: relationship to anhedonia. *PLoS One*, 7(5), e35622.

Earnst, K. S., Kring, A. M., Kadar, M. A., Salem, J. E., Shepard, D. A., & Loosen, P. T. (1996). Facial expression in schizophrenia. *Biological Psychiatry*, *40*(6), 556–558.

Eisenegger, C., Pedroni, A., Rieskamp, J., Zehnder, C., Ebstein, R., Fehr, E., & Knoch, D. (2013). DAT1 polymorphism determines L-DOPA effects on learning about others' prosociality. *PLoS One*, *8*(7), e67820.

Engel, M., Fritzsche, A., & Lincoln, T. M. (2016). Anticipation and experience of emotions in patients with schizophrenia and negative symptoms. An experimental study in a social context. *Schizophrenia Research*, *170*(1), 191–197.

Favrod, J., Giuliani, F., Ernst, F., & Bonsack, C. (2010). Anticipatory pleasure skills training: a new intervention to reduce anhedonia in schizophrenia. *Perspectives in Psychiatric Care*, *46*(3), 171–181.

Fervaha, G., Foussias, G., Agid, O., & Remington, G. (2014). Motivational and neurocognitive deficits are central to the prediction of longitudinal functional outcome in schizophrenia. *Acta Psychiatrica Scandinavica*, *130*(4), 290–299.

Fervaha, G., Graff-Guerrero, A., Zakzanis, K. K., Foussias, G., Agid, O., & Remington, G. (2013). Incentive motivation deficits in schizophrenia reflect effort computation impairments during cost-benefit decision-making. *Journal of Psychiatric Research*, *47*(11), 1590–1596.

Fett, A. K., Viechtbauer, W., Dominguez, M. D., Penn, D. L., van Os, J., & Krabbendam, L. (2011). The relationship between neurocognition and social cognition with functional outcomes in schizophrenia: a meta-analysis. *Neuroscience & Biobehavioral Reviews*, *35*(3), 573–588. https://doi.org/10.1016/j.neubiorev.2010.07.001.

Fiszdon, J. M., Kurtz, M. M., Choi, J., Bell, M. D., & Martino, S. (2015). Motivational interviewing to increase cognitive rehabilitation adherence in schizophrenia. *Schizophrenia Bulletin*, *42*(2), 327–334.

Fleeson, W., Malanos, A. B., & Achille, N. M. (2002). An intraindividual process approach to the relationship between extraversion and positive affect: is acting extraverted as "good" as being extraverted? *Journal of Personality and Social Psychology*, *83* (6), 1409–1422.

Foussias, G., Mann, S., Zakzanis, K., van Reekum, R., Agid, O., & Remington, G. (2011). Prediction of longitudinal functional outcomes in schizophrenia: the impact of baseline motivational deficits. *Schizophrenia Research*, *132*(1), 24–27.

Foussias, G., & Remington, G. (2008). Negative symptoms in schizophrenia: avolition and Occam's razor. *Schizophrenia Bulletin*, *36*(2), 359–369.

Fulford, D., Campellone, T., & Gard, D. E. (2018). Social motivation in schizophrenia: how research on basic reward processes informs and limits our understanding. *Clinical Psychology Review*, *63*, 12–24.

Fulford, D., Treadway, M., & Woolley, J. (2018). Social motivation in schizophrenia: the impact of oxytocin on vigor in the context of social and nonsocial reinforcement. *Journal of Abnormal Psychology*, *127*(1), 116–128.

Gard, D. E., Fisher, M., Garrett, C., Genevsky, A., & Vinogradov, S. (2009). Motivation and its relationship to neurocognition, social cognition, and functional outcome in schizophrenia. *Schizophrenia Research*, *115*(1), 74–81.

Gard, D. E., Gard, M. G., Kring, A. M., & John, O. P. (2006). Anticipatory and consummatory components of the experience of pleasure: a scale development study. *Journal of Research in Personality*, *40*(6), 1086–1102.

Gard, D. E., Kring, A. M., Gard, M. G., Horan, W. P., & Green, M. F. (2007). Anhedonia in schizophrenia: distinctions between anticipatory and consummatory pleasure. *Schizophrenia Research*, *93*(1–3), 253–260.

Gard, D. E., Sanchez, A. H., Starr, J., Cooper, S., Fisher, M., Rowlands, A., & Vinogradov, S. (2014). Using self-determination theory to understand motivation deficits in schizophrenia: the 'why'of motivated behavior. *Schizophrenia Research*, *156*(2–3), 217–222.

Gold, J. M., Strauss, G. P., Waltz, J. A., Robinson, B. M., Brown, J. K., & Frank, M. J. (2013). Negative symptoms of schizophrenia are associated with abnormal effort-cost computations. *Biological Psychiatry*, *74*(2), 130–136.

Gold, J. M., Waltz, J. A., Matveeva, T. M., Kasanova, Z., Strauss, G. P., Herbener, E. S., … Frank, M. J. (2012). Negative symptoms and the failure to represent the expected reward value of actions: behavioral and computational modeling evidence. *Archives of General Psychiatry*, *69*(2), 129–138.

Gold, J. M., Waltz, J. A., Prentice, K. J., Morris, S. E., & Heerey, E. A. (2008). Reward processing in schizophrenia: a deficit in the representation of value. *Schizophrenia Bulletin*, *34*(5), 835–847.

Grace, A. A. (2016). Dysregulation of the dopamine system in the pathophysiology of schizophrenia and depression. *Nature Reviews Neuroscience*, *17*(8), 524–532.

Green, M. F., Horan, W. P., & Lee, J. (2015). Social cognition in schizophrenia. *Nature Reviews Neuroscience*, *16*(10), 620–631.

Green, M. F., Penn, D. L., Bentall, R., Carpenter, W. T., Gaebel, W., Gur, R. C., … Heinssen, R. (2008). Social cognition in schizophrenia: an NIMH workshop on definitions, assessment, and research opportunities. *Schizophrenia Bulletin*, *34*(6), 1211–1220.

Groppe, S. E., Gossen, A., Rademacher, L., Hahn, A., Westphal, L., Gründer, G., & Spreckelmeyer, K. N. (2013). Oxytocin influences processing of socially relevant cues in the ventral tegmental area of the human brain. *Biological Psychiatry*, *74*(3), 172–179.

Haber, S. N., & Knutson, B. (2010). The reward circuit: linking primate anatomy and human imaging. *Neuropsychopharmacology*, *35*(1), 4–26.

Howes, O. D., Kambeitz, J., Kim, E., Stahl, D., Slifstein, M., Abi-Dargham, A., & Kapur, S. (2012). The nature of dopamine dysfunction in schizophrenia and what this means for treatment: meta-analysis of imaging studies. *Archives of General Psychiatry*, *69*(8), 776–786.

Howes, O. D., Montgomery, A. J., Asselin, M. -C., Murray, R. M., Valli, I., Tabraham, P., … Broome, M. (2009). Elevated striatal dopamine function linked to prodromal signs of schizophrenia. *Archives of General Psychiatry*, *66*(1), 13–20.

Hu, J., Qi, S., Becker, B., Luo, L., Gao, S., Gong, Q., … Kendrick, K. M. (2015). Oxytocin selectively facilitates learning with social feedback and increases activity and functional connectivity in emotional memory and reward processing regions. *Human Brain Mapping*, *36*(6), 2132–2146.

Hung, L. W., Neuner, S., Polepalli, J. S., Beier, K. T., Wright, M., Walsh, J. J., … Dölen, G. (2017). Gating of social reward by oxytocin in the ventral tegmental area. *Science*, *357*(6358), 1406–1411.

Jones, R. M., Somerville, L. H., Li, J., Ruberry, E. J., Libby, V., Glover, G., … Casey, B. (2011). Behavioral and neural properties of social reinforcement learning. *Journal of Neuroscience*, *31*(37), 13039–13045.

Kennerley, S. W., Walton, M. E., Behrens, T. E., Buckley, M. J., & Rushworth, M. F. (2006). Optimal decision making and the anterior cingulate cortex. *Nature Neuroscience*, *9*(7), 940–947.

Klucharev, V., Hytönen, K., Rijpkema, M., Smidts, A., & Fernández, G. (2009). Reinforcement learning signal predicts social conformity. *Neuron, 61*(1), 140–151.
Knutson, B., Westdorp, A., Kaiser, E., & Hommer, D. (2000). FMRI visualization of brain activity during a monetary incentive delay task. *NeuroImage, 12*(1), 20–27.
Kraepelin, E. (1919). *Dementia praecox and paraphenia*. (M. Barclay, Trans.) (Vol. Original work published 1913). Chicago, IL: Chicago Medical Book Company.
Kring, A. M., & Moran, E. K. (2008). Emotional response deficits in schizophrenia: insights from affective science. *Schizophrenia Bulletin, 34*(5), 819–834.
Kring, A. M., & Neale, J. M. (1996). Do schizophrenic patients show a disjunctive relationship among expressive, experiential, and psychophysiological components of emotion? *Journal of Abnormal Psychology, 105*(2), 249–257.
Kring, A. M., Siegel, E. H., & Barrett, L. F. (2014). Unseen affective faces influence person perception judgments in schizophrenia. *Clinical Psychological Science, 2*(4), 443–454.
Kwapil, T. R. (1998). Social anhedonia as a predictor of the development of schizophrenia-spectrum disorders. *Journal of Abnormal Psychology, 107*(4), 558–565.
Lee, J., Altshuler, L., Glahn, D. C., Miklowitz, D. J., Ochsner, K., & Green, M. F. (2013). Social and nonsocial cognition in bipolar disorder and schizophrenia: relative levels of impairment. *American Journal of Psychiatry, 170*(3), 334–341.
Lewandowski, K. E., Whitton, A. E., Pizzagalli, D. A., Norris, L. A., Ongur, D., & Hall, M. H. (2016). Reward learning, neurocognition, social cognition, and symptomatology in psychosis. *Frontiers in Psychiatry, 7*, 100.
Maia, T. V., & Frank, M. J. (2017). An integrative perspective on the role of dopamine in schizophrenia. *Biological Psychiatry, 81*(1), 52–66.
Manduca, A., Servadio, M., Damsteegt, R., Campolongo, P., Vanderschuren, L. J., & Trezza, V. (2016). Dopaminergic neurotransmission in the nucleus accumbens modulates social play behavior in rats. *Neuropsychopharmacology, 41*(9), 2215–2223.
McCarthy, J. M., Bradshaw, K. R., Catalano, L. T., Garcia, C. P., Malik, A., Bennett, M. E., & Blanchard, J. J. (2018). Negative symptoms and the formation of social affiliative bonds in schizophrenia. *Schizophrenia Research, 193*, 225–231.
Moran, E. K., & Kring, A. M. (2018). Anticipatory emotion in schizophrenia. *Clinical Psychological Science, 6*(1), 63–75.
Morris, S. E., Heerey, E. A., Gold, J. M., & Holroyd, C. B. (2008). Learning-related changes in brain activity following errors and performance feedback in schizophrenia. *Schizophrenia Research, 99*(1–3), 274–285.
Mote, J., Minzenberg, M. J., Carter, C. S., & Kring, A. M. (2014). Deficits in anticipatory but not consummatory pleasure in people with recent-onset schizophrenia spectrum disorders. *Schizophrenia Research, 159*(1), 76–79.
Mote, J., Stuart, B. K., & Kring, A. M. (2014). Diminished emotion expressivity but not experience in men and women with schizophrenia. *Journal of Abnormal Psychology, 123*(4), 796–801.
Myin-Germeys, I., Delespaul, P. A., & DeVries, M. W. (2000). Schizophrenia patients are more emotionally active than is assumed based on their behavior. *Schizophrenia Bulletin, 26*(4), 847–854.
Nakagami, E., Xie, B., Hoe, M., & Brekke, J. S. (2008). Intrinsic motivation, neurocognition and psychosocial functioning in schizophrenia: testing mediator and moderator effects. *Schizophrenia Research, 105*(1–3), 95–104.
Norman, R. M., Manchanda, R., Harricharan, R., & Northcott, S. (2015). The course of negative symptoms over the first five years of treatment: data from an early intervention program for psychosis. *Schizophrenia Research, 169*(1–3), 412–417.
O'Doherty, J. P., Deichmann, R., Critchley, H. D., & Dolan, R. J. (2002). Neural responses during anticipation of a primary taste reward. *Neuron, 33*(5), 815–826.

Oldham, S., Murawski, C., Fornito, A., Youssef, G., Yücel, M., & Lorenzetti, V. (2018). The anticipation and outcome phases of reward and loss processing: a neuroimaging meta-analysis of the monetary incentive delay task. *Human Brain Mapping*, *39*, 3398–3418.

Oliveira, S. E., Esteves, F., & Carvalho, H. (2015). Clinical profiles of stigma experiences, self-esteem and social relationships among people with schizophrenia, depressive, and bipolar disorders. *Psychiatry Research*, *229*(1–2), 167–173.

Pecina, S., & Berridge, K. C. (2005). Hedonic hot spot in nucleus accumbens shell: where do μ-opioids cause increased hedonic impact of sweetness? *Journal of Neuroscience*, *25*(50), 11777–11786.

Pizzagalli, D. A., Goetz, E., Ostacher, M., Iosifescu, D. V., & Perlis, R. H. (2008). Euthymic patients with bipolar disorder show decreased reward learning in a probabilistic reward task. *Biological Psychiatry*, *64*(2), 162–168.

Rademacher, L., Krach, S., Kohls, G., Irmak, A., Gründer, G., & Spreckelmeyer, K. N. (2010). Dissociation of neural networks for anticipation and consumption of monetary and social rewards. *NeuroImage*, *49*(4), 3276–3285.

Radua, J., Schmidt, A., Borgwardt, S., Heinz, A., Schlagenhauf, F., McGuire, P., & Fusar-Poli, P. (2015). Ventral striatal activation during reward processing in psychosis: a neurofunctional meta-analysis. *JAMA Psychiatry*, *72*(12), 1243–1251.

Reddy, L. F., Horan, W. P., Barch, D. M., Buchanan, R. W., Dunayevich, E., Gold, J. M., … Wynn, J. K. (2015). Effort-based decision-making paradigms for clinical trials in schizophrenia: Part 1—Psychometric characteristics of 5 paradigms. *Schizophrenia Bulletin*, *41*(5), 1045–1054.

Rescorla, R. A., & Solomon, R. L. (1967). Two-process learning theory: relationships between Pavlovian conditioning and instrumental learning. *Psychological Review*, *74*(3), 151.

Rosenfeld, A. J., Lieberman, J. A., & Jarskog, L. F. (2010). Oxytocin, dopamine, and the amygdala: a neurofunctional model of social cognitive deficits in schizophrenia. *Schizophrenia Bulletin*, *37*(5), 1077–1087.

Samamé, C., Martino, D., & Strejilevich, S. (2012). Social cognition in euthymic bipolar disorder: systematic review and meta-analytic approach. *Acta Psychiatrica Scandinavica*, *125*(4), 266–280.

Schlagenhauf, F., Huys, Q. J., Deserno, L., Rapp, M. A., Beck, A., Heinze, H. -J., … Heinz, A. (2014). Striatal dysfunction during reversal learning in unmedicated schizophrenia patients. *NeuroImage*, *89*, 171–180.

Schlosser, D. A., Fisher, M., Gard, D., Fulford, D., Loewy, R. L., & Vinogradov, S. (2014). Motivational deficits in individuals at-risk for psychosis and across the course of schizophrenia. *Schizophrenia Research*, *158*(1–3), 52–57.

Scholten, M. R., van Honk, J., Aleman, A., & Kahn, R. S. (2006). Behavioral inhibition system (BIS), behavioral activation system (BAS) and schizophrenia: relationship with psychopathology and physiology. *Journal of Psychiatric Research*, *40*(7), 638–645.

Schultz, W. (2016). Dopamine reward prediction-error signalling: a two-component response. *Nature Reviews Neuroscience*, *17*(3), 183.

Shamay-Tsoory, S. G., & Abu-Akel, A. (2016). The social salience hypothesis of oxytocin. *Biological Psychiatry*, *79*(3), 194–202.

Shankar, A., Rafnsson, S. B., & Steptoe, A. (2015). Longitudinal associations between social connections and subjective wellbeing in the English Longitudinal Study of Ageing. *Psychology & Health*, *30*(6), 686–698.

Smith, K. S., Tindell, A. J., Aldridge, J. W., & Berridge, K. C. (2009). Ventral pallidum roles in reward and motivation. *Behavioural Brain Research*, *196*(2), 155–167.

Strauss, G. P., Waltz, J. A., & Gold, J. M. (2013). A review of reward processing and motivational impairment in schizophrenia. *Schizophrenia Bulletin*, *40*(Suppl. 2), S107–S116.

Treadway, M. T., Buckholtz, J. W., Schwartzman, A. N., Lambert, W. E., & Zald, D. H. (2009). Worth the 'EEfRT'? The effort expenditure for rewards task as an objective measure of motivation and anhedonia. *PLoS One*, *4*(8), e6598.

Treadway, M. T., Peterman, J. S., Zald, D. H., & Park, S. (2015). Impaired effort allocation in patients with schizophrenia. *Schizophrenia Research*, *161*(2–3), 382–385.

Treadway, M. T., & Zald, D. H. (2013). Parsing anhedonia: translational models of reward-processing deficits in psychopathology. *Current Directions in Psychological Science*, *22*(3), 244–249.

Tso, I. F., Grove, T. B., & Taylor, S. F. (2014). Differential hedonic experience and behavioral activation in schizophrenia and bipolar disorder. *Psychiatry Research*, *219*(3), 470–476.

Waltz, J. A., Frank, M. J., Robinson, B. M., & Gold, J. M. (2007). Selective reinforcement learning deficits in schizophrenia support predictions from computational models of striatal-cortical dysfunction. *Biological Psychiatry*, *62*(7), 756–764.

Waltz, J. A., & Gold, J. M. (2007). Probabilistic reversal learning impairments in schizophrenia: further evidence of orbitofrontal dysfunction. *Schizophrenia Research*, *93*(1–3), 296–303.

Weinstein, J. J., Chohan, M. O., Slifstein, M., Kegeles, L. S., Moore, H., & Abi-Dargham, A. (2017). Pathway-specific dopamine abnormalities in schizophrenia. *Biological Psychiatry*, *81*(1), 31–42.

Whitton, A. E., Treadway, M. T., & Pizzagalli, D. A. (2015). Reward processing dysfunction in major depression, bipolar disorder and schizophrenia. *Current Opinion in Psychiatry*, *28*(1), 7.

Willuhn, I., Tose, A., Wanat, M. J., Hart, A. S., Hollon, N. G., Phillips, P. E., ... Wöhr, M. (2014). Phasic dopamine release in the nucleus accumbens in response to pro-social 50 kHz ultrasonic vocalizations in rats. *Journal of Neuroscience*, *34*(32), 10616–10623.

CHAPTER EIGHT

# Substance misuse and social cognition on the psychosis-spectrum: A bottom-up framework

**Jacob J. Crouse*, Daniel S. Quintana[†], Daniel F. Hermens*[,‡]**

*Youth Mental Health Team, Brain & Mind Centre, The University of Sydney, Camperdown, NSW, Australia
[†]NORMENT, KG Jebsen Centre for Psychosis Research, Division of Mental Health and Addiction, University of Oslo and Oslo University Hospital, Oslo, Norway
[‡]Sunshine Coast Mind and Neuroscience—Thompson Institute, University of the Sunshine Coast, Birtinya, QLD, Australia

## Introduction

The relationship between substance use and psychosis has been long-documented and is relevant to the etiologies, expression, course, and outcomes of patients with psychosis-spectrum illnesses (Barnett et al., 2007; Hambrecht & Hafner, 1996, 2000; Lambert et al., 2005; Warnock, 1903). Indeed, the high degree of comorbidity between substance misuse and psychosis is suggestive of a shared liability or related neurobiological diathesis (Chambers, Krystal, & Self, 2001). While other psychiatric illnesses are associated with high rates of co-occurring substance use, psychotic illnesses are particularly overrepresented at this juncture (Addington & Addington, 2007; Volkow, 2009). The markedly high rates of substance use observed among patients with psychosis is concerning, not only with respect to the cardiovascular and associated physical health problems associated with substance misuse (Brown, Inskip, & Barraclough, 2000; Laursen, Munk-Olsen, & Vestergaard, 2012), but also in that comorbid use may lead to poorer mental health and functional outcomes (van Os et al., 2002; Wade, Harrigan, McGorry, Burgess, & Whelan, 2007).

Social cognition is a common illness feature in psychosis, which has been linked in some studies to substance use. While definitions differ, social cognition will be operationalized here as "*the mental operations underlying social interactions, which include the ability to perceive the intentions and dispositions of others*" (Brothers, 1990). Deficits in social cognitive processes are well-

*Social Cognition in Psychosis*
https://doi.org/10.1016/B978-0-12-815315-4.00008-2

established in individuals with psychotic disorders (e.g., schizophrenia), including problems with cue detection and interpretation (e.g., tone of voice and facial expression), inference of others' mental states (i.e., theory of mind), and hostile attributional style, among others (Javitt, 2010; Javitt, Shelley, & Ritter, 2000; Leitman et al., 2005; Mancuso, Horan, Kern, & Green, 2011; Penn, Sanna, & Roberts, 2008). In the following sections, we will summarize the literature showing direct links between substance use and social cognitive deficits, and then argue that substance use may disrupt low-level processing underpinning higher-order social cognition. Our efforts will dovetail with bottom-up models of disturbed cognition in schizophrenia (Javitt, 2009).

## Direct links between substance use and social cognition in psychotic disorders

### Tobacco and social cognition

Tobacco smoking is highly prevalent among individuals with psychotic disorders such as schizophrenia and bipolar disorder (Brown et al., 2000). The high degree of comorbidity has been suggested to arise from a shared neurobiological diathesis between psychosis and reward sensitivity, or instead as a means of self-medicating cognitive and other illness features sensitive to nicotinic receptor stimulation. The literature has been mixed on whether nicotinic administration improves social cognitive functions such as facial affect recognition. For example, Drusch et al. (2013) investigated the effects of nicotine on social cognition and social competence in patients with schizophrenia and healthy participants. There were no significant effects of nicotine use on a task of facial affect recognition or greater expressive behavior in patients. However, there were also no differences in group performance on the facial recognition task between patients and controls, possibly indicating a more socially-intact patient group for whom a ceiling effect of additional nicotinic stimulation may hide an intervention effect (Drusch et al., 2013). A double-blind placebo-controlled randomized controlled trial (RCT)tested whether oromucosal spray (i.e., directed toward mucous surfaces of the mouth) would improve general and social cognitive tasks in patients with schizophrenia. While a modest effect was observed for improving attention in the smoking group ($n = 16$), no enhancing effects were observed for social cognitive functioning (Quisenaerts et al., 2014). A separate analysis of the same RCT, however, observed an enhancing effect of acute nicotinic administration on a social decision-making ultimatum game in patients with schizophrenia (Quisenaerts et al., 2013). A recent study assessed both social and nonsocial cognitive performance in 335 patients with first-episode psychosis, and 253 healthy controls. The authors compared nontobacco-using patients and tobacco-using patients, and observed no significant differences in social or nonsocial cognition between the groups (Sánchez-Gutiérrez et al., 2018).

The importance of considering history of psychosis with respect to social cognition in bipolar disorder is becoming increasingly recognized (Thaler, Allen, Sutton, Vertinski, & Ringdahl, 2013). Ospina et al. (2016) examined social and general cognition in a group of euthymic (i.e., stable mood) patients with bipolar disorder ($n = 105$), some of whom had a history of psychosis. The authors observed a main effect of cigarette smoking group (i.e., smoker versus nonsmoker) on social processing, particularly for the Emotion Recognition Task (ERT), whereby nonsmokers outperformed smokers on the percentage of correctly identified fearful and surprised faces. Moreover, there were several significant smoking status × psychosis history interactions on measures of the ERT and Mayer-Salovey-Caruso Emotional Intelligence Test (MSCEIT). Specifically, patients with a psychosis history who were nonsmokers outperformed smokers on the total MSCEIT and the ERT when asked to identify sad, fearful, and surprised faces (Ospina et al., 2016).

Further research is needed to discern whether neurobiological or other factors explain variance in response to nicotine administration. Moreover, understanding why individuals with psychotic disorders are prone to tobacco smoking is critical for improving cessation programs targeting this group.

## Ketamine and social cognition

As glutamatergic neurotransmission is robustly implicated in schizophrenia-related cognitive impairments, the *N*-methyl-D-aspartate antagonist ketamine has been widely used as a pharmacological model of psychosis. Ebert, Haussleiter, Juckel, Brune, and Roser (2012) tested whether ketamine administration would cause social cognitive dysfunction indexed by the Ekman 60 Faces test (EK-60F). Using 18 healthy male participants tested on one occasion with ketamine and one without, the authors observed a nonsignificant deterioration of global recognition abilities, and a significant reduction in the ability to correctly identify sad facial expressions, independent of attention and psychotic symptoms (Ebert et al., 2012).

## Alcohol and social cognition

Facial emotion recognition is impaired both in patients with schizophrenia and those with alcohol and substance use disorders (AUD/SUD). Carrà and colleagues (2017) tested whether the comorbidity between these disorders would lead to even poorer facial emotion recognition. They additionally examined the role of the catechol-O-methyltransferase (COMT) Val158Met polymorphism in this effect. Sixty-seven patients with schizophrenia were recruited and administered the Ekman 60 Faces Test (EK-60F). Contrary to expectations, patients with a comorbid AUD/SUD did not perform significantly worse on the EK-60F than those with schizophrenia alone. Moreover, the beneficial impact of the COMT Val158Met polymorphism was observed only in those without a comorbid AUD/SUD. A meta-analysis synthesizing data from eight studies also suggests that theory of mind is impaired in individuals with AUDs (Onuoha, Quintana, Lyvers, & Guastella, 2016). Whether theory of mind deficits in psychotic disorders are further impaired in those with an AUD remains to be established.

## Cannabis and social cognition

Associations between cannabis use and cognition with respect to psychotic disorders are complex are highly contentious. The vast majority of this research has focused on general neurocognitive function, with little attention paid to social cognition. Sanchez-Torres et al. (2013) recruited 42 patients with a schizophrenia-spectrum disorder and 35 unaffected twins, following them for 10 years. The authors observed a negative effect of continued cannabis use on performance on the Managing Emotions section of the MSCEIT in patients, with evidence of possible confounding when lifetime tobacco use was added to the model (Sanchez-Torres et al., 2013). Some research has observed that better social cognition is predictive of recency of cannabis use in individuals with first-episode psychosis, which may be explained by socially savvy patients being better able to acquire illicit substances than their peers with greater social and general neurocognition (Arnold, Allott, Farhall, Killackey, & Cotton, 2015). However, while this may be true in countries where cannabis is illegal, it would not explain this paradoxical association in countries where the drug is legal and easily accessible. Accordingly, Helle, Loberg, Gjestad, Schnakenberg Martin, and Lysaker (2017) modeled whether the positive link between executive function and lifetime cannabis use in patients with schizophrenia is explained by superior levels of social cognition. Eighty-seven patients with schizophrenia were administered measures of executive function and social cognition. Using structural equation modeling, the authors found that cannabis use was positively associated with executive function, negatively related to affect recognition, and had no relationship with theory of mind (Helle et al., 2017). Meijer et al. (2012) observed in a cross-sectional study of 956 patients with nonaffective psychosis, 953 unaffected siblings, and 554 healthy control participants that lifetime cannabis use was associated with better performance on facial affect recognition and face identity recognition relative to never-using patients, albeit with small effect sizes. These results have been taken to be evidence for cannabis-using patients representing a subgroup that is less vulnerable to developing schizophrenia than individuals who have no lifetime exposure to cannabis (Mueser, Drake, & Wallach, 1998; Zubin & Spring, 1977). In this way, cannabis may act to precipitate a psychotic illness in individuals with sufficient biological vulnerability. However, this group may have better premorbid functioning and cognitive reserve than the nonusing group, who develop psychosis without cannabis use as a function of greater vulnerability (e.g., aberrant brain development or genetic factors) (Meijer et al., 2012).

## Amphetamines and social cognition

Studies of the link between social cognition and amphetamine-type stimulant drugs (e.g., methamphetamine, MDMA) in the context of psychotic disorders are sparse. One published study by Uhlmann, Ipser, Wilson, and Stein (2018) assessed social cognitive impairment and its association with aggression in three groups: individuals with methamphetamine dependence (MD; $n=21$), methamphetamine-associated psychosis (MAP, $n=20$), and healthy nonusing controls ($n=21$). The groups completed a facial morphing ERT for anger, fear, happiness, and sadness, the reading the mind in the eyes task (RMET), and an aggression questionnaire. Both methamphetamine groups showed poorer performance on the RMET relative to controls, with the MAP group performing worse than the MD group. The MD group also had difficulty recognizing emotional expressions of anger, with a generalized impairment across all four emotions evident in the MAP group, and no associations between social cognition and aggressive behavior (Uhlmann et al., 2018). This is an important area for future research to expand on, as aggressive and violent acts are common among heavy methamphetamine users (Stretesky, 2008), and substance use is a predictor of violence in individuals with psychotic disorders (Large & Nielssen, 2011). The role of social cognitive processes (e.g., inferring emotional state of others) may be important with respect to substance-related violence during psychotic episodes.

As can be seen, research directly testing associations between substance misuse and social cognition in psychotic disorders is limited, and many questions remain unaddressed. Accordingly, we will present several possible indirect paths from substance use to disturbed social cognition that may be useful in generating hypotheses for future studies.

# Indirect links between substance misuse and social cognition

## The impact of substance use on preattentional functioning, and links with social cognition

An increasingly common method of studying the pathophysiology of psychotic disorders is to gather electrophysiological recordings from the brains of patients to noninvasively monitor brain activity. Such measurements have proven useful in uncovering disruptions to early sensory processing as an important disturbance in psychotic illnesses with possible links to social cognitive function. Similar efforts have been applied to understanding the effects of substance use on the brain, including in substance-using patients with psychosis. In the following, we will summarize the literature relevant at this intersection, and highlight ways in which neurophysiological disturbances attributable to substance use may disturb perception and social cognition.

## Mismatch negativity and related components

The mismatch negativity (MMN) is a key neurophysiological index that has been the subject of particular attention in psychosis research. The MMN is a negative component of the event-related potential (ERP), which marks a change-discrimination process underlying the useful ability to detect a discriminable deviance in one's environment (e.g., the breaking of a stick beneath a predator's foot) (Näätänen, 1995). Importantly, the MMN is an unconscious, preattentional index of processing, meaning that it can be utilized to probe sensory processing in disorders accompanied by attentional dysfunction, as is common in psychotic disorders. Evidence for a compromised MMN signal in schizophrenia has been long documented (Shelley et al., 1991), with more recent work extending samples to early illness stages across a broader psychosis-spectrum, observing comparable deficits (Kaur et al., 2011, 2012).

The MMN is thought to mark an important temporal window representing a transitional "gateway" from early sensory-based processing to engagement of higher-order networks required for cognitive and psychosocial functioning. Some have suggested that disruptions to early sensorial processing may lead to flow-on effects to higher-order processing (e.g., tone-of-voice interpretation) (Rissling et al., 2013). Indeed, individuals with schizophrenia are observed to have deficits in emotional interpretation based on prosody (Leitman et al., 2005, 2007), with such deficits thought to map onto functional difficulties (Brekke, Kay, Lee, & Green, 2005). It is thought that these impairments may arise due to an underlying failure to utilize sensory cues associated with preattentional dysfunction in auditory cortices (Javitt, 2010).

Light and Braff (2005) were the first to demonstrate an association of MMN deficits and poorer global functioning in patients with schizophrenia. This association with psychosocial functioning was later extended to healthy participants (Light, Swerdlow, & Braff, 2007). Wynn and colleagues (2010) observed in patients with schizophrenia that greater MMN amplitudes recorded at fronto-central sites of the scalp were associated with better performance on a task indexing response to perceptual cues associated with social interaction (Profile of Nonverbal Sensitivity task). Hermens and colleagues (2010) observed significant associations between the MMN and a positive component of the ERP (P3a) and psychosocial functioning, with specific associations between fronto-central MMN amplitudes and the WHO Quality of Life Environment subscale (which includes social factors such as social care and participation). Further, preliminary research suggests that MMN is predictive of social skills acquisition in people with schizophrenia (Kawakubo et al., 2007), which is an important line for future research. As can be seen, there appears to be an important relationship between preattentional processing and higher-order social cognition and psychosocial functioning. The following sections will focus on studies demonstrating an impact of substance use on preattentional processing, which may be relevant to substance-using patients with psychosis.

### *Cannabis and preattentional functioning*

Cannabis has been shown to be associated with a compromised MMN in both clinical and nonclinical samples. To begin with, Greenwood and colleagues (2014) compared a group of chronic cannabis users and nonusing healthy controls in an MMN paradigm, observing a reduction in MMN amplitude among chronic cannabis users relative to controls. Subgroup analyses revealed that relative to both controls and short-term users, long-term users had reduced MMN durations that were associated with increased duration of cannabis exposure and increased psychosis-like experiences while intoxicated. Impey and colleagues (2015) compared MMN of a group of tobacco-naive cannabis users and individuals without a history of tobacco or cannabis use. The authors observed that cannabis users had impaired MMN relative to nonusers, with subgroup analyses showing MMN impairments in both chronic/heavy users as well as nonchronic/light users, however, with a dose-dependent effect of greater cannabis use. Pesa and colleagues (2012) recruited 21 healthy participants and 44 patients with early psychosis, 22 of whom were cannabis users and 22 who were nonusers. Using a two-tone, deviance-detection paradigm, the authors observed that both patient groups showed marked reductions in MMN/P3a amplitudes compared with controls. However, the cannabis-using patient group displayed further impairments in frontal MMN/P3a latencies. It has been suggested that patients and controls may exhibit differential alterations as a consequence of preexisting pathophysiological differences, such as in the endocannabinoid system (Rentzsch et al., 2011).

### *Ketamine and preattentional functioning*

As the MMN is generated by glutamatergic neurotransmission, NMDAR antagonists such as ketamine robustly perturb its generation (Heekeren et al., 2008; Rosburg & Kreitschmann-Andermahr, 2016). Indeed, a number of studies have demonstrated that the MMN is dose-dependently reduced by ketamine (Ehrlichman, Maxwell, Majumdar, & Siegel, 2008; Umbricht et al., 2000; Umbricht, Koller, Vollenweider, & Schmid, 2002), with phenomena resembling negative symptoms of schizophrenia apparent following ketamine administration (Heekeren et al., 2008). A recent preliminary study observed smaller MMN amplitudes associated with more negative symptoms, suggestive of an interrelationship between MMN, negative symptomology, and glutamatergic neurotransmission (Thiebes et al., 2017). It has been suggested that dysfunction of the NMDAR in schizophrenia-related psychoses may result in difficulty forming and utilizing transient memory traces at multiple levels of information processing (Umbricht et al., 2000), which may be important with respect to bottom-up disruption of higher-order cognition involved in social interaction (e.g., utilizing information about facial expression to guide social behavior). Unfortunately, no available studies have examined social cognitive effects following ketamine administration related to neurophysiological measures.

### *Alcohol and preattentional functioning*

Alcohol is another important substance with effects at the NMDAR associated with MMN alterations. Acute ingestion of alcohol has been shown to suppress the MMN (He et al., 2013; Hirvonen, Jaaskelainen, Näätänen, & Sillanaukee, 2000; Jaaskelainen et al., 1995), whereby alcohol increases the effects of gamma-aminobutyric acid (GABA) on $GABA_A$ receptors, blocking the NMDAR (Boyce-Rustay & Holmes, 2005; Deitrich, Dunwiddie, Harris, & Erwin, 1989).

Research in patients with preexisting disruptions to the NMDA system, such as those with psychotic and bipolar disorders, has demonstrated that misuse of alcohol may have an additive effect on NMDA dysregulation and MMN generation. For example, Chitty and colleagues (2011) observed that young psychotic patients with higher alcohol use demonstrated reduced temporal MMN amplitudes relative to both patients with low use and controls. The same group observed similar associations between risky alcohol use and temporal MMN impairments in patients with bipolar disorder (Chitty, Kaur, Lagopoulos, Hickie, & Hermens, 2014). More work is required to test whether these additive effects of alcohol map onto social and functional deficits.

## The impact of substance use on heart rate variability and links to social cognition

Heart rate variability (HRV) refers to the modification of the heart rate over time via the interplay of autonomic, circulatory, respiratory, endocrine, and mechanical influences. Reductions in HRV, which are indicative of reduced parasympathetic activity (Akselrod et al., 1981), have been associated with several cardiovascular illnesses (Kemp & Quintana, 2013) and immune system dysfunction (Tracey, 2002). HRV is reduced in patients with psychosis-spectrum illnesses (Bar et al., 2008), and these reductions in HRV have also been shown to be independent of medication or body mass index effects (Quintana et al., 2016). Although reductions in HRV have also been observed in mood disorders (Kemp et al., 2010) and anxiety disorders (Chalmers, Quintana, Abbott, & Kemp, 2014), a recent meta-analysis of HRV across psychiatric disorders revealed that HRV reductions are largest in psychosis-spectrum illnesses (Alvares, Quintana, Hickie, & Guastella, 2016).

The Polyvagal theory proposes that the autonomic nervous system evolved to support social engagement and disengagement, among its other functions (Porges, 2007); however, see Grossman & Taylor (2007) for caveats regarding vagal tone interpretation. In line with the Polyvagal theory, healthy individuals with higher resting-state HRV have been shown to perform better on a theory of mind task (Lischke, Lemke, Neubert, Hamm, & Lotze, 2017; Quintana, Guastella, Outhred, Hickie, & Kemp, 2012). A positive relationship between HRV and theory of mind performance has also been shown in children with autism spectrum disorders, a diagnosis that is characterized by social dysfunction (Bal et al., 2010). Not only does HRV appear to be related to the processing of emotions, but also the projection of emotion as it has been reported to positively predict skill in expressing some facial emotions (Tuck, Grant, Sollers, Booth, & Consedine, 2016). While these studies point to the role of autonomic system regulation in social cognition, their correlational design limits any conclusions regarding causality. Vagus nerve stimulation (VNS) has been shown to increase HRV (Zhang et al., 2009); however, this requires invasive surgery, rendering this approach impractical for research. The relatively recent development of transcutaneous VNS offers a noninvasive means of experimentally increasing HRV by stimulating the nerve fibers of the auricular branch of the vagus nerve (ABVN) in the left ear (Shiozawa et al., 2014). As it is relatively easy to provide a sham stimulation condition by placing the stimulating electrodes on parts of the ear that are not innervated by the AVBN, this approach is well suited for experimental research. Research suggests that experimentally increasing HRV via transcutaneous VNS improves theory of mind performance compared with sham stimulation (Colzato, Sellaro, & Beste, 2017; Sellaro, de Gelder, Finisguerra, & Colzato, 2018).

Individuals with alcohol dependence also demonstrate reduced HRV compared with controls, with a meta-analysis of published studies yielding a summary hedges' *g* effect size of 0.61 (Quintana, McGregor, Guastella, Malhi, & Kemp, 2013). Acute alcohol consumption is reported to reduce HRV via the blocking of vagal inhibitory processes (Reed, Porges, & Newlin, 1999). As direct administration of alcohol to the sinoatrial node of dogs has little effect on the heart rate (James & Bear, 1967), these changes are likely to be centrally mediated. Sustained attenuation of HRV via the blocking of vagal inhibitory processes from chronic alcohol consumption may then contribute to the reductions in resting-state HRV observed in individuals with alcohol dependence. However, the relationship between alcohol consumption and HRV might be J-shaped, as moderate alcohol consumption has been associated with higher HRV compared with abstainers and heavy drinkers (Karpyak, Romanowicz, Schmidt, Lewis, & Bostwick, 2014; Quintana, Guastella, McGregor, Hickie, & Kemp, 2013). The effects of other substances of abuse on HRV have been less studied than alcohol. Research suggests that cannabis use is associated with increased HRV in males (Schmid, Schonlebe, Drexler, & Mueck-Weymann, 2010), and that methamphetamine use is associated with reduced HRV (Henry, Minassian, & Perry, 2012). Tobacco use is also associated with reduced HRV, with increases in HRV observed after smoking cessation (Harte & Meston, 2014; Hayano et al., 1990). While it is likely that reductions in HRV caused by chronic alcohol intake contribute to the deficits in social cognition observed in alcohol misuse, it is also conceivable that reductions in HRV from other causes (e.g., co-morbid psychosis-spectrum disorders) can increase the likelihood of alcohol misuse due to impairments in behavioral inhibition and self-regulation (Thayer, Hansen, Saus-Rose, & Johnsen, 2009).

## Concluding remarks and future directions

Substance misuse is an important and worthwhile factor to consider in patients with psychotic disorders, with important relevance to social cognition both directly and indirectly, operating through pathways involving disrupted early sensory processing and autonomic nervous system activity (see Fig. 1). Unfortunately, research examining direct associations between substance use and social cognition in people with psychosis is limited, and may represent an important target for intervention, through which social cognitive impairments may be improved by reducing substance use. We illustrate in the Fig. 1 a framework in which use of various substances are directly associated with deficits in social cognitive processes. We additionally show two "indirect" paths through which substance use may act on "lower-level" processes, which may in turn affect "higher-order" cognitive processes in a bottom-up fashion. These paths are currently speculative, but may indeed partially represent the biological mechanisms of reported associations between substance use and social cognition. Research is needed to tease apart the precise associations between specific substances and specific social cognitive tasks, determine mediating and moderating factors of these effects, and establish mechanisms governing them.

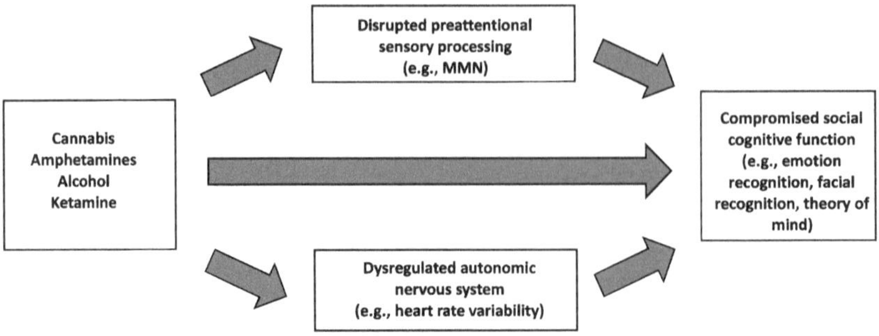


**Fig. 1** Direct and indirect impacts of substance use on social cognition.

Future research should directly test relationships between social cognition and substance use in healthy individuals, as well as those with psychotic disorders and associated social cognitive impairments. For example, researchers should test relationships between MMN integrity and specific social cognitive tasks, examining the impact of NMDAR modulation by way of experimental administration of $\Delta^9$-THC, alcohol, and ketamine. Further, it may be useful to manipulate ratios of $\Delta^9$-THC-to-CBD in such experiments, as concentrations in favor of higher CBD-to-$\Delta^9$-THC may better preserve cognitive skills required for social interaction. This may be a particularly important question in light of the growing prevalence of commercially-available cannabinoid products around the world, as harm minimization approaches should seek to moderate the risk associated with propsychotic compounds that may arise as a function of cultivation processes associated with wide-scale production (Potter, 2014). As social disturbances are common to many individuals with psychotic illnesses, research determining the potential impact of substance use on social cognition directly is warranted. Further, as comorbid substance use is a modifiable factor amenable to treatment, it is a worthwhile endeavor to understand the totality of its effects on psychotic disorders, so that treatment may be optimized and individually tailored.

## References


Addington, J., & Addington, D. (2007). Patterns, predictors and impact ofsubstance use in early psychosis: a longitudinal study. *Acta Psychiatrica Scandinavica*, *115*, 304–309.

Akselrod, S., Gordon, D., Ubel, F. A., Shannon, D. C., Berger, A. C., & Cohen, R. J. (1981). Power spectrum analysis of heart rate fluctuation: a quantitative probe of beat-to-beat cardiovascular control. *Science*, *213*, 220–222.

Alvares, G. A., Quintana, D. S., Hickie, I. B., & Guastella, A. J. (2016). Autonomic nervous system dysfunction in psychiatric disorders and the impact of psychotropic medications: a systematic review and meta-analysis. *Journal of Psychiatry & Neuroscience*, *41*, 89–104.

Arnold, C., Allott, K., Farhall, J., Killackey, E., & Cotton, S. (2015). Neurocognitive and social cognitive predictors of cannabis use in first-episode psychosis. *Schizophrenia Research, 168*, 231–237.

Bal, E., Harden, E., Lamb, D., Van Hecke, A. V., Denver, J. W., & Porges, S. W. (2010). Emotion recognition in children with autism spectrum disorders: relations to eye gaze and autonomic state. *Journal of Autism and Developmental Disorders, 40*, 358–370.

Bar, K. J., Wernich, K., Boettger, S., Cordes, J., Boettger, M. K., Loffler, S., ... Agelink, M. W. (2008). Relationship between cardiovagal modulation and psychotic state in patients with paranoid schizophrenia. *Psychiatry Research, 157*, 255–257.

Barnett, J. H., Werners, U., Secher, S. M., Hill, K. E., Brazil, R., Masson, K., ... Jones, P. B. (2007). Substance use in a population-based clinic sample of people with first-episode psychosis. *The British Journal of Psychiatry, 190*, 515–520.

Boyce-Rustay, J. M., & Holmes, A. (2005). Functional roles of NMDA receptor NR2A and NR2B subunits in the acute intoxicating effects of ethanol in mice. *Synapse, 56*, 222–225.

Brekke, J., Kay, D. D., Lee, K. S., & Green, M. F. (2005). Biosocial pathways to functional outcome in schizophrenia. *Schizophrenia Research, 80*, 213–225.

Brothers, L. (1990). The social brain: a project for integrating primate behavior and neurophysiology in a new domain. *Concepts in Neuroscience, 1*, 27–51.

Brown, S., Inskip, H., & Barraclough, B. (2000). Causes of the excess mortality of schizophrenia. *The British Journal of Psychiatry, 177*, 212–217.

Carrà, G., Nicolini, G., Lax, A., Bartoli, F., Castellano, F., Chiorazzi, A., & Papagno, C. (2017). Facial emotion recognition in schizophrenia: an exploratory study on the role of comorbid alcohol and substance use disorders and COMT Val158Met. *Human Psychopharmacology, 32*(6), 1–7.

Chalmers, J. A., Quintana, D. S., Abbott, M. J., & Kemp, A. H. (2014). Anxiety disorders are associated with reduced heart rate variability: a meta-analysis. *Frontiers in Psychiatry, 5*, 80.

Chambers, R. A., Krystal, J. H., & Self, D. W. (2001). A neurobiological basis for substance abuse comorbidity in schizophrenia. *Biological Psychiatry, 50*, 71–83.

Chitty, K. M., Kaur, M., Lagopoulos, J., Hickie, I. B., & Hermens, D. F. (2011). *Alcohol use and mismatch negativity in young patients with psychotic disorder Neuroreport, 22*(17), 918–922.

Chitty, K. M., Kaur, M., Lagopoulos, J., Hickie, I. B., & Hermens, D. F. (2014). Risky alcohol use predicts temporal mismatch negativity impairments in young people with bipolar disorder. *Biological Psychology, 99*, 60–68.

Colzato, L. S., Sellaro, R., & Beste, C. (2017). Darwin revisited: the vagus nerve is a causal element in controlling recognition of other's emotions. *Cortex, 92*, 95–102.

Deitrich, R. A., Dunwiddie, T. V., Harris, R. A., & Erwin, V. G. (1989). Mechanism of action of ethanol: initial central nervous system actions. *Pharmacological Reviews, 41*, 489–537.

Drusch, K., Lowe, A., Fisahn, K., Brinkmeyer, J., Musso, F., Mobascher, A., ... Wolwer, W. (2013). Effects of nicotine on social cognition, social competence and self-reported stress in schizophrenia patients and healthy controls. *European Archives of Psychiatry and Clinical Neuroscience, 263*, 519–527.

Ebert, A., Haussleiter, I. S., Juckel, G., Brune, M., & Roser, P. (2012). Impaired facial emotion recognition in a ketamine model of psychosis. *Psychiatry Research, 200*, 724–727.

Ehrlichman, R. S., Maxwell, C. R., Majumdar, S., & Siegel, S. J. (2008). Deviance-elicited changes in event-related potentials are attenuated by ketamine in mice. *Journal of Cognitive Neuroscience, 20*, 1403–1414.

Greenwood, L. M., Broyd, S. J., Croft, R., Todd, J., Michie, P. T., Johnstone, S., & Solowij, N. (2014). Chronic effects of cannabis use on the auditory mismatch negativity. *Biological Psychiatry, 75*(6), 449–458.

Grossman, P., & Taylor, E. W. (2007). Toward understanding respiratory sinus arrhythmia: relations to cardiac vagal tone, evolution and biobehavioral functions. *Biological Psychology, 74*, 263–285.

Hambrecht, M., & Hafner, H. (1996). Substance abuse and the onset of schizophrenia. *Biological Psychiatry, 40,* 1155–1163.
Hambrecht, M., & Hafner, H. (2000). Cannabis, vulnerability, and the onset of schizophrenia: an epidemiological perspective. *The Australian and New Zealand Journal of Psychiatry, 34,* 468–475.
Harte, C. B., & Meston, C. M. (2014). Effects of smoking cessation on heart rate variability among long-term male smokers. *International Journal of Behavioral Medicine, 21,* 302–309.
Hayano, J., Yamada, M., Sakakibara, Y., Fujinami, T., Yokoyama, K., Watanabe, Y., & Takata, K. (1990). Short- and long-term effects of cigarette smoking on heart rate variability. *The American Journal of Cardiology, 65,* 84–88.
He, J., Li, B., Guo, Y., Näätänen, R., Pakarinen, S., & Luo, Y. J. (2013). Effects of alcohol on auditory pre-attentive processing of four sound features: evidence from mismatch negativity. *Psychopharmacology, 225,* 353–360.
Heekeren, K., Daumann, J., Neukirch, A., Stock, C., Kawohl, W., Norra, C., … Gouzoulis-Mayfrank, E. (2008). Mismatch negativity generation in the human 5HT2A agonist and NMDA antagonist model of psychosis. *Psychopharmacology, 199,* 77–88.
Helle, S., Loberg, E. M., Gjestad, R., Schnakenberg Martin, A. M., & Lysaker, P. H. (2017). The positive link between executive function and lifetime cannabis use in schizophrenia is not explained by current levels of superior social cognition. *Psychiatry Research, 250,* 92–98.
Henry, B. L., Minassian, A., & Perry, W. (2012). Effect of methamphetamine dependence on heart rate variability. *Addiction Biology, 17,* 648–658.
Hermens, D. F., Ward, P. B., Hodge, M. A., Kaur, M., Naismith, S. L., & Hickie, I. B. (2010). Impaired MMN/P3a complex in first-episode psychosis: cognitive and psychosocial associations. Progress in Neuro-Psychopharmacology &amp. *Biological Psychiatry, 34*(6), 822–829.
Hirvonen, J., Jaaskelainen, I. P., Näätänen, R., & Sillanaukee, P. (2000). Adenosine A1/A2a receptors mediate suppression of mismatch negativity by ethanol in humans. *Neuroscience Letters, 278,* 57–60.
Impey, D., El-Marj, N., Parks, A., Choueiry, J., Fisher, D., & Knott, V. J. (2015). Mismatch negativity in tobacco-naïve cannabis users and its alteration with acute nicotine administration. *Pharmacology, Biochemistry, and Behavior, 136,* 73–81.
Jaaskelainen, I. P., Lehtokoski, A., Alho, K., Kujala, T., Pekkonen, E., Sinclair, J. D., … Sillanaukee, P. (1995). Low dose of ethanol suppresses mismatch negativity of auditory event-related potentials. *Alcoholism, Clinical and Experimental Research, 19,* 607–610.
James, T. N., & Bear, E. S. (1967). Effects of ethanol and acetaldehyde on the heart. *American Heart Journal, 74,* 243–255.
Javitt, D. C. (2009). When doors of perception close: bottom-up models of disrupted cognition in schizophrenia. *Annual Review of Clinical Psychology, 5,* 249–275.
Javitt, D. C. (2010). Glutamatergic theories of schizophrenia. *The Israel Journal of Psychiatry and Related Sciences, 47,* 4–16.
Javitt, D. C., Shelley, A., & Ritter, W. (2000). Associated deficits in mismatch negativity generation and tone matching in schizophrenia. *Clinical Neurophysiology, 111,* 1733–1737.
Karpyak, V. M., Romanowicz, M., Schmidt, J. E., Lewis, K. A., & Bostwick, J. M. (2014). Characteristics of heart rate variability in alcohol-dependent subjects and nondependent chronic alcohol users. *Alcoholism, Clinical and Experimental Research, 38,* 9–26.
Kaur, M., Battisti, R. A., Ward, P. B., Ahmed, A., Hickie, I. B., & Hermens, D. F. (2011). MMN/P3a deficits in first episode psychosis: comparing schizophrenia-spectrum and affective-spectrum subgroups. *Schizophrenia Research, 130,* 203–209.
Kaur, M., Lagopoulos, J., Ward, P. B., Watson, T. L., Naismith, S. L., Hickie, I. B., & Hermens, D. F. (2012). Mismatch negativity/P3a complex in young people with psychiatric disorders: a cluster analysis. *PLoS One,* 7, e51871.

Kawakubo, Y., Kamio, S., Nose, T., Iwanami, A., Nakagome, K., Fukuda, M., … Kasai, K. (2007). Phonetic mismatch negativity predicts social skills acquisition in schizophrenia. *Psychiatry Research, 152*, 261–265.

Kemp, A. H., & Quintana, D. S. (2013). The relationship between mental and physical health: insights from the study of heart rate variability. *International Journal of Psychophysiology, 89*, 288–296.

Kemp, A. H., Quintana, D. S., Gray, M. A., Felmingham, K. L., Brown, K., & Gatt, J. M. (2010). Impact of depression and antidepressant treatment on heart rate variability: a review and meta-analysis. *Biological Psychiatry, 67*, 1067–1074.

Lambert, M., Conus, P., Lubman, D. I., Wade, D., Yuen, H., Moritz, S., … Schimmelmann, B. G. (2005). The impact of substance use disorders on clinical outcome in 643 patients with first-episode psychosis. *Acta Psychiatrica Scandinavica, 112*, 141–148.

Large, M. M., & Nielssen, O. (2011). Violence in first-episode psychosis: a systematic review and meta-analysis. *Schizophrenia Research, 125(2–3)*, 209–220.

Laursen, T. M., Munk-Olsen, T., & Vestergaard, M. (2012). Life expectancy and cardiovascular mortality in persons with schizophrenia. *Current Opinion in Psychiatry, 25*, 83–88.

Leitman, D. I., Foxe, J. J., Butler, P. D., Saperstein, A., Revheim, N., & Javitt, D. C. (2005). Sensory contributions to impaired prosodic processing in schizophrenia. *Biological Psychiatry, 58*, 56–61.

Leitman, D. I., Hoptman, M. J., Foxe, J. J., Saccente, E., Wylie, G. R., Nierenberg, J., … Javitt, D. C. (2007). The neural substrates of impaired prosodic detection in schizophrenia and its sensorial antecedents. *The American Journal of Psychiatry, 164*, 474–482.

Light, G. A., & Braff, D. L. (2005). Mismatch negativity deficits are associated with poor functioning in schizophrenia patients. *Archives of General Psychiatry, 62*(2), 127–136.

Light, G. A., Swerdlow, N. R., & Braff, D. L. (2007). Preattentive sensory processing as indexed by the MMN and P3a brain responses is associated with cognitive and psychosocial functioning in healthy adults. *Journal of Cognitive Neuroscience, 19*, 1624–1632.

Lischke, A., Lemke, D., Neubert, J., Hamm, A. O., & Lotze, M. (2017). Inter-individual differences in heart rate variability are associated with inter-individual differences in mind-reading. *Scientific Reports*, 7, 11557.

Mancuso, F., Horan, W. P., Kern, R. S., & Green, M. F. (2011). Social cognition in psychosis: multidimensional structure, clinical correlates, and relationship with functional outcome. *Schizophrenia Research, 125*, 143–151.

Meijer, J. H., Dekker, N., Koeter, M. W., Quee, P. J., van Beveren, N. J., & Meijer, C. J. (2012). Cannabis and cognitive performance in psychosis: a cross-sectional study in patients with non-affective psychotic illness and their unaffected siblings. *Psychological Medicine, 42*, 705–716.

Mueser, K. T., Drake, R. E., & Wallach, M. A. (1998). Dual diagnosis: a review of etiological theories. *Addictive Behaviors, 23*, 717–734.

Näätänen, R. (1995). The mismatch negativity: a powerful tool for cognitive neuroscience. *Ear and Hearing, 16*(1), 6–18.

Onuoha, R. C., Quintana, D. S., Lyvers, M., & Guastella, A. J. (2016). A meta-analysis of theory of mind in alcohol use disorders. *Alcohol and Alcoholism, 51*, 410–415.

Ospina, L. H., Russo, M., Nitzburg, G. M., Cuesta-Diaz, A., Shanahan, M., Perez-Rodriguez, M. M., … Burdick, K. E. (2016). The effects of cigarette smoking behavior and psychosis history on general and social cognition in bipolar disorder. *Bipolar Disorders, 18*, 528–538.

Penn, D. L., Sanna, L. J., & Roberts, D. L. (2008). Social cognition in schizophrenia: an overview. *Schizophrenia Bulletin, 34*, 408–411.

Pesa, N., Hermens, D. F., Battisti, R. A., Kaur, M., Hickie, I. B., & Solowij, N. (2012). Delayed preattentional functioning in early psychosis patients with cannabis use. *Psychopharmacology, 222*(3), 507–518.

Porges, S. W. (2007). The polyvagal perspective. *Biological Psychology, 74*, 116–143.
Potter, D. J. (2014). A review of the cultivation and processing of cannabis (*Cannabis sativa* L.) for production of prescription medicines in the UK. *Drug Testing and Analysis, 6*, 31–38.
Quintana, D. S., Guastella, A. J., McGregor, I. S., Hickie, I. B., & Kemp, A. H. (2013). Moderate alcohol intake is related to increased heart rate variability in young adults: implications for health and well-being. *Psychophysiology, 50*, 1202–1208.
Quintana, D. S., Guastella, A. J., Outhred, T., Hickie, I. B., & Kemp, A. H. (2012). Heart rate variability is associated with emotion recognition: direct evidence for a relationship between the autonomic nervous system and social cognition. *International Journal of Psychophysiology, 86*, 168–172.
Quintana, D. S., McGregor, I. S., Guastella, A. J., Malhi, G. S., & Kemp, A. H. (2013). A meta-analysis on the impact of alcohol dependence on short-term resting-state heart rate variability: implications for cardiovascular risk. *Alcoholism, Clinical and Experimental Research, 37*(Suppl. 1), E23–E29.
Quintana, D. S., Westlye, L. T., Kaufmann, T., Rustan, O. G., Brandt, C. L., Haatveit, B., … Andreassen, O. A. (2016). Reduced heart rate variability in schizophrenia and bipolar disorder compared to healthy controls. *Acta Psychiatrica Scandinavica, 133*, 44–52.
Quisenaerts, C., Morrens, M., Hulstijn, W., de Boer, P., Timmers, M., Sabbe, B., & de Bruijn, E. R. (2013). Acute nicotine improves social decision-making in non-smoking but not in smoking schizophrenia patients. *Frontiers in Neuroscience*, 7, 197.
Quisenaerts, C., Morrens, M., Hulstijn, W., de Bruijn, E., Timmers, M., Streffer, J., … Sabbe, B. (2014). The nicotinergic receptor as a target for cognitive enhancement in schizophrenia: barking up the wrong tree? *Psychopharmacology, 231*, 543–550.
Reed, S. F., Porges, S. W., & Newlin, D. B. (1999). Effect of alcohol on vagal regulation of cardiovascular function: contributions of the polyvagal theory to the psychophysiology of alcohol. *Experimental and Clinical Psychopharmacology*, 7, 484–492.
Rentzsch, J., Buntebart, E., Stadelmeier, A., Gallinat, J., & Jockers-Scherübl, M. C. (2011). *Differential effects of chronic cannabis use on preattentional cognitive functioning in abstinent schizophrenic patients and healthy subjects Schizophrenia Research, 130*(1–3), 222–227.
Rissling, A. J., Park, S. H., Young, J. W., Rissling, M. B., Sugar, C. A., Sprock, J., & Light, G. A. (2013). Demand and modality of directed attention modulate "pre-attentive" sensory processes in schizophrenia patients and nonpsychiatric controls. *Schizophrenia Research, 146*(1–3), 326–335.
Rosburg, T., & Kreitschmann-Andermahr, I. (2016). The effects of ketamine on the mismatch negativity (MMN) in humans—a meta-analysis. *Clinical Neurophysiology, 127*, 1387–1394.
Sánchez-Gutiérrez, T., García-Portilla, M. P., Parellada, M., Bobes, J., Calvo, A., Moreno-Izco, L., … Balanzá, V. (2018). Smoking does not impact social and non-social cognition in patients with first episode psychosis. *Schizophrenia Research, 199*, 64–74.
Sanchez-Torres, A. M., Basterra, V., Rosa, A., Fananas, L., Zarzuela, A., Ibanez, B., … Cuesta, M. J. (2013). Lifetime cannabis use and cognition in patients with schizophrenia spectrum disorders and their unaffected siblings. *European Archives of Psychiatry and Clinical Neuroscience, 263*, 643–653.
Schmid, K., Schonlebe, J., Drexler, H., & Mueck-Weymann, M. (2010). The effects of cannabis on heart rate variability and well-being in young men. *Pharmacopsychiatry, 43*, 147–150.
Sellaro, R., de Gelder, B., Finisguerra, A., & Colzato, L. S. (2018). Transcutaneous vagus nerve stimulation (tVNS) enhances recognition of emotions in faces but not bodies. *Cortex, 99*, 213–223.

Shelley, A. M., Ward, P. B., Catts, S. V., Michie, P. T., Andrews, S., & McConaghy, N. (1991). Mismatch negativity: an index of a preattentive processing deficit in schizophrenia. *Biological Psychiatry*, *30*, 1059–1062.

Shiozawa, P., Silva, M. E., Carvalho, T. C., Cordeiro, Q., Brunoni, A. R., & Fregni, F. (2014). Transcutaneous vagus and trigeminal nerve stimulation for neuropsychiatric disorders: a systematic review. *Arquivos de Neuro-Psiquiatria*, *72*, 542–547.

Stretesky, P. B. (2008). National case-control study of homicide offending and methamphetamine use. *Journal of Interpersonal Violence*, *24*(6), 911–924.

Thaler, N. S., Allen, D. N., Sutton, G. P., Vertinski, M., & Ringdahl, E. N. (2013). Differential impairment of social cognition factors in bipolar disorder with and without psychotic features and schizophrenia. *Journal of Psychiatric Research*, *47*, 2004–2010.

Thayer, J. F., Hansen, A. L., Saus-Rose, E., & Johnsen, B. H. (2009). Heart rate variability, prefrontal neural function, and cognitive performance: the neurovisceral integration perspective on self-regulation, adaptation, and health. *Annals of Behavioral Medicine*, *37*, 141–153.

Thiebes, S., Leicht, G., Curic, S., Steinmann, S., Polomac, N., Andreou, C., … Mulert, C. (2017). Glutamatergic deficit and schizophrenia-like negative symptoms: new evidence from ketamine-induced mismatch negativity alterations in healthy male humans. *Journal of Psychiatry & Neuroscience*, *42*, 273–283.

Tracey, K. J. (2002). The inflammatory reflex. *Nature*, *420*, 853–859.

Tuck, N. L., Grant, R. C. I., Sollers, J. J., Booth, R. J., & Consedine, N. S. (2016). Higher resting heart rate variability predicts skill in expressing some emotions. *Psychophysiology*, *53*, 1852–1857.

Uhlmann, A., Ipser, J. C., Wilson, D., & Stein, D. J. (2018). Social cognition and aggression in methamphetamine dependence with and without a history of psychosis. *Metabolic Brain Disease*, *33*, 559–568.

Umbricht, D., Koller, R., Vollenweider, F. X., & Schmid, L. (2002). Mismatch negativity predicts psychotic experiences induced by NMDA receptor antagonist in healthy volunteers. *Biological Psychiatry*, *51*, 400–406.

Umbricht, D., Schmid, L., Koller, R., Vollenweider, F. X., Hell, D., & Javitt, D. C. (2000). Ketamine-induced deficits in auditory and visual context-dependent processing in healthy volunteers: implications for models of cognitive deficits in schizophrenia. *Archives of General Psychiatry*, *57*, 1139–1147.

van Os, J., Bak, M., Hanssen, M., Bijl, R. V., de Graaf, R., & Verdoux, H. (2002). Cannabis use and psychosis: a longitudinal population-based study. *American Journal of Epidemiology*, *156*, 319–327.

Volkow, N. D. (2009). Substance use disorders in schizophrenia—clinical implications of comorbidity. *Schizophrenia Bulletin*, *35*, 469–472.

Wade, D., Harrigan, S., McGorry, P. D., Burgess, P. M., & Whelan, G. (2007). Impact of severity of substance use disorder on symptomatic and functional outcome in young individuals with first-episode psychosis. *The Journal of Clinical Psychiatry*, *68*, 767–774.

Warnock, J. (1903). Insanity from hasheesh. *Journal of Mental Science*, *49*, 96–110.

Zhang, Y., Popovic, Z. B., Bibevski, S., Fakhry, I., Sica, D. A., Van Wagoner, D. R., & Mazgalev, T. N. (2009). Chronic vagus nerve stimulation improves autonomic control and attenuates systemic inflammation and heart failure progression in a canine high-rate pacing model. *Circulation. Heart Failure*, *2*, 692–699.

Wynn, J. K., Sugar, C., Horan, W. P., Kern, R., & Green, M. F. (2010). Mismatch negativity, social cognition, and functioning in schizophrenia patients. *Biological Psychiatry*, *67* (10), 940–947.

Zubin, J., & Spring, B. (1977). Vulnerability: a new view of schizophrenia. *Journal of Abnormal Psychology*, *86*, 103–126.

CHAPTER NINE

# Assessment of social cognition

**Juliete M. Silberstein*, Philip D. Harvey*,†**
*University of Miami Miller School of Medicine, Miami, FL, United States
†Bruce W. Carter VA Medical Center, Miami, FL, United States

Cognition has emerged as a consistent, and perhaps the strongest, predictor of the everyday deficits in individuals living with schizophrenia. Efforts to understand the relationship between cognition and functional outcomes in schizophrenia have primarily focused on neurocognition. Nearly three-quarters of individuals with schizophrenia perform below the general population on important neurocognitive domains such as processing speed, attention, learning, problem-solving, and working memory. Recently, social cognition has emerged as a critical research area. Evidence suggests that social cognition explains unique variances in real-world outcomes, and appears to be a stronger predictor of social outcomes (i.e., interpersonal relationships and other socially focused behaviors) than neurocognitive ability. Importantly, treating social cognitive deficits leads to improvements in real-world social outcomes (Fiszdon, 2013; Kurtz & Richardson, 2012), including social adjustment (Eack, Greenwald, Hogarty, & Keshavan, 2010), social functioning (Lindenmayer et al., 2013; Mazza et al., 2010; Tas, Danaci, Cubukcuoglu, & Brune, 2012), social relationships (Combs et al., 2007; Sachs et al., 2012), aggressive incidents (Combs et al., 2007), and social skills (Roberts & Penn, 2009).

These factors have led to social cognition being recognized as an essential treatment target; yet, problems in consensus regarding the best measures to assess this critical domain, and concerns about quality of the measures available, have restricted the utility of social cognition in clinical care and outcome in treatment trials (Green et al., 2008; Green, Olivier, Crawley, Penn, & Silverstein, 2005). Several consensus initiatives have attempted to address the importance of social cognition, as well as these measurement limitations, and to develop well-validated measures suitable for both clinical trials and treatment. This chapter aims to provide an overview of the current evidence surrounding assessment of social cognition in schizophrenia. First, we will discuss the challenges that have arisen in using social cognition as an endpoint. Second, we will provide an overview of the psychometric strengths and

*Social Cognition in Psychosis*
https://doi.org/10.1016/B978-0-12-815315-4.00009-4

limitations of measures that have shown to be the most well-validated. We will next examine alternative strategies for performance-based measures of social cognition, such as observational and self-assessment strategies. Finally, we will discuss future areas of work necessary to guarantee that clinical trials can rigorously assess social cognition in schizophrenia and related disorders with an eventual goal of developing replicated and targeted treatments.

## Shared challenges of assessment: Neurocognition and social cognition

While measures of social cognition in schizophrenia pose challenges distinct from challenges in assessing neurocognitive ability, such as defining the domains of importance and choosing live, video, or static assessments, the shared characteristics of performance-based measures of neurocognition and social cognition underscore the importance of developing high-quality and clinically appropriate measures. In contrast to assessment strategies that could be employed in less-impaired populations, assessment of neurocognition, social cognition, functional capacity, and everyday functioning pose particular challenges in people with schizophrenia. Self-report and, more surprisingly, family or friend-reported assessments of an individual with schizophrenia's cognitive ability and everyday functioning had yielded limited association to objective outcomes as measured on performance-based evaluations or achievement of functional milestones (Bowie et al., 2007; Durand et al., 2015; Harvey et al., 2012; Keefe, Poe, Walker, Kang, & Harvey, 2006; McKibbin, Patterson, & Jeste, 2004; Patterson et al., 1997; Sabbag et al., 2011). Significantly, high-contact clinician's or caregiver's evaluations and ratings have been shown to be highly aligned with performance measures, and have shown strong correlations across domains of neurocognition, social cognition, and everyday functioning (Durand et al., 2015; Sabbag et al., 2011). Collectively, these findings accentuate the need for reliable and clinically-feasible performance-based or high-contact informant measures.

## Challenges in social cognitive measurement

The putative multifactorial nature of social cognition has historically resulted in numerous construct definitions, and has complicated the determination of the key domains and their most appropriate assessments. In 2006, the National Institute of Mental Health (NIMH) convened a consensus panel aimed at streamlining inconsistent terminology and means of measuring social cognition in schizophrenia. The NIMH Workshop of Social Cognition in Schizophrenia defined social cognition as "the mental operations that underlie social interactions, including perceiving, interpreting, and generating responses to the intentions, dispositions, and behaviors of others" (Green et al., 2008). However, the NIMH Workshop cited conceptual and measurement-related overlap within subdomains, and as such, reported a lack of expert consensus on which abilities defined social cognition in schizophrenia (Green et al., 2008). The Cognitive Neuroscience for Treatment Research to Improve Cognition in Schizophrenia (CNTRICS), another NIMH initiative, subsequently attempted to subdivide social cognition based on neuronal processes (Carter et al., 2009). CNTRICS ultimately recommended a measurement development focus on emotion identification and responding, which they defined as the "ability to detect, recognize, and judge the affective value of both linguistic (e.g., seen or spoken words and their prosodic contour) and nonlinguistic (e.g., images of people, facial expressions, eye gaze, scenes) stimuli" (Carter et al., 2009). Last, the few factor analyses that were completed suggest social cognition may be best parsed according to the level of information processing (i.e., perception versus inferential and regulatory processing) rather than the domain of social information (i.e., emotion versus mental state) (Lin, Wynn, Hellemann, & Green, 2012; Mancuso, Horan, Kern, & Green, 2011). However, some other factor analyses suggest that with current measures, social cognition may actually be largely unifactorial (Browne et al., 2016). Ultimately, the ambiguity of how to partition social cognition into defined end-points produced a barrier in the evaluation or development of valid measurements.

A further challenge in the assessment of social cognition has been its relationship with and independence from neurocognition. Previous studies and reviews have generally concluded that they reflect related, but separable domains (Mehta et al., 2013; Sergi et al., 2007). However, it has also been argued that social cognition and neurocognition have independent effects on social outcomes (Fett et al., 2011). An additional important consideration is whether global deficits on cognitive tests are also associated with challenges in the performance of social cognitive tests, leading to exaggerated social cognitive deficits.

Only recently experts have systematically attempted to reach an agreement on the core domains of social cognition. The Social Cognition Psychometric Evaluation (SCOPE) study was a five-phase project designed to improve measurement of social cognition in schizophrenia by systematically surveying the field on the current state of the art of social cognition assessment, developing systematic domain definitions, and evaluating the psychometric properties of the most widely used measures (Pinkham et al., 2014; Pinkham, Harvey, & Penn, 2018; Pinkham, Penn, Green, & Harvey, 2016). Phase 1 was an extensive survey to identify the core domains of social cognition in schizophrenia. Roughly 60 experts in the field were engaged in two rounds of survey and discussion. The initial phase garnered more than 150 nominations for different social cognitive domains (100 of which were unique terms). After combining terms referring to the same general process (e.g., affect recognition and emotion identification), a compilation of nominations revealed that 80% of the submissions of varied domains converged on six domains: emotion processing, social perception, theory of mind/mental state attribution, attributional style/bias, social metacognition, and social reciprocity (Pinkham et al., 2014). The second round of surveys supported the value and validity in the existing research literature of four of the nominated domains: emotion processing, social perception, theory of mind/mental state attribution, and attributional style/bias. The domains of social metacognition and social reciprocity were dropped during the second round (Pinkham et al., 2014), largely because of a lack of systematic research in the field. While they were viewed as valid core domains of social cognition, continued development in these areas was encouraged because of the lack of clear measures. Later stages of the SCOPE project systematically collected preliminary data on social metacognition using novel applications of existing methods. Definitions of the four core social cognitive domains ([1] emotion processing; [2] social perception; [3] theory of mind/mental state attribution; [4] attributional style/bias) are summarized as follows.

*Emotion processing*: This domain is broadly defined as perceiving and using emotions (Green et al., 2008). It subsumes three subdomains that represent both lower-level and higher-level processes. At a lower perceptual level is the first subdomain—emotion perception/recognition (identifying and recognizing emotional displays from facial expressions and/or nonface cues such as voice)—and at a higher level are the two subdomains of understanding emotions (i.e., comprehending others' emotional displays) and managing emotions (i.e., correctly reacting to emotional displays).

*Social perception*: Social perception refers to decoding and interpreting social cues in others (Penn, Ritchie, Francis, Combs, & Martin, 2002; Sergi & Green, 2003; Toomey, Schuldberg, Corrigan, & Green, 2002).

It includes social context processing and social knowledge, which can be defined as knowing social rules, roles, and goals (RRGs), utilizing those RRGs, and understanding how such RRGs may influence others' behaviors (Addington, Saeedi, & Addington, 2006; Corrigan & Green, 1993).

*Theory of mind/mental state attribution*: This domain is defined as the ability to comprehend and represent the mental states of others, including the inference of intentions, dispositions, and/or beliefs (Frith, 1992; Penn, Addington, & Pinkham, 2006). Theory of mind is also referred to as mentalizing, mental state attribution, or cognitive empathy (Shamay-Tsoory, 2011).

*Attributional style/bias*: Attributional style describes the way in which individuals explain or make sense of the cause of social events or interactions (Green et al., 2008; Penn et al., 2006).

## Social Cognition Psychometric Evaluation: An overview

Multiple measures have often been used to assess the same domain of social cognition. For example, a prior metaanalysis of theory of mind concluded that the previous inconsistency of reported theory of mind deficits among populations of individuals with schizophrenia stemmed from the "heterogeneity of the methods" used to assess theory of mind (Bora, Yucel, & Pantelis, 2009). Further, the majority of social cognitive measures have little available psychometric data in terms of understanding of normative performance, test-retest reliability, and construct validity. The previous lack of psychometric data related to tasks of social cognitive ability has made evaluating the adequacy of tests difficult, if not impossible (Bora et al., 2009; Yager & Ehmann, 2006). The relative scarcity of studies implementing multiple measures of social cognition with sample sizes large enough to allow for either exploratory or confirmatory factor analyses also contributes to the limitations in the construct structure present in research of social cognition in schizophrenia. The variability in quality among assessments may impede the utilization of social cognition as a treatment target, and muddy the relationships between social cognition, neurocognition, and functional outcome. These quality concerns included psychometric problems such as floor or ceiling effects, poor test-retest reliability, and poorly related alternative forms. Further, some of the stimulus materials appear too complicated for use with cognitively impaired patients. As such, the SCOPE trial also aimed to address the lack of consensus about which measures provided the best index for a given social cognitive domain, and to resolve previous methodological problems with an empirical psychometric study, followed by modifications of some tasks, and a subsequent final analysis of psychometrics.

The ultimate aim of the SCOPE study was to identify the best current tasks of social cognition with a systematic evaluation and consensus development process (Pinkham et al., 2014, 2016, 2018). Experts nominated the most suitable tests in each core domain during the initial phase of the study (Pinkham et al., 2014) according to the following criteria. Nominations were then reduced according to criteria based on the availability and quality of data in the literature, and all available published and otherwise available psychometric information about these measures was summarized and provided to panelists for evaluation. Phase 2 used methods similar to other NIMH measurement initiatives, in which a carefully selected group of expert panelists appraised the identified tasks utilizing a consensus procedure, the RAND Appropriateness Method (Pinkham et al., 2014).

Panelists rated the quality of each measure on (1) reliability—test-retest and interrater reliability as applicable, as well as internal consistency, (2) distributions—floor and/or ceiling effects and normality of distributions, (3) utility as a repeated measure—stability over time in the absence of intervention or sensitivity to intervention associated change, (4) convergent and discriminant validity—relationship to social cognitive measures relative to other abilities and constructs, (5) criterion validity—correlations with real-world social outcomes, (6) practicality for administration, and (7) tolerability for patients. Ultimately, diverging ratings were discussed at the in-person meeting to obtain consensus (Pinkham et al., 2014).

The most promising tasks, defined by the highest ratings, underwent further evaluation and development. In Phases 3–5, candidate measures were administered to large samples of healthy controls and patients with schizophrenia to assess the reliability and validity characteristics of each task (Pinkham et al., 2016, 2018). Phase 3 evaluated the first eight assessments selected with the RAND appropriateness procedures. These measures of social cognition were administered to 179 individuals with schizophrenia and 104 healthy controls in Miami, Florida and Dallas, Texas. All participants completed each task twice, approximately 2 weeks apart. The results were then reported to a subset of the previous panel of experts to obtain consensus on which tasks could be recommended for use in clinical trials (Pinkham et al., 2016). Measures showing adequate test-test reliability, small practice effects, and significant associations with outcomes were re-examined in Phase 5 of SCOPE. Two hundred and eighteen individuals with schizophrenia and 154 healthy controls completed all acceptable Phase 3 tasks twice, replicating the procedure of task administration at two time-points (Pinkham et al., 2018). We will discuss psychometric properties of tasks deemed acceptable at the culmination of the SCOPE studies later in the chapter. Fig. 1 depicts the five phases of the SCOPE study, along with their goals.

An additional feature of Phase 4 was the evaluation of promising, but apparently limited, tasks that were modified and evaluated for feasibility. Expansions of the measures focused on increasing predictive utility, and included instructions for speeded performance and the collection of response times for each item, and ratings of confidence in the correctness of each response (Cornacchio, Pinkham, Penn, & Harvey, 2017). Our previous work has indicated that response time for social cognitive decisions can be a strong individual predictor of functional outcomes (Pinkham & Penn, 2006), and that introspective accuracy, or the awareness of one's abilities, might be a more potent predictor of outcomes than task performance (Gould et al., 2015; Harvey & Pinkham, 2015). As previously discussed, the concept of social metacognition, which overlaps with introspective accuracy, was identified as a potential core domain of social cognition in the earliest phase of the SCOPE study, but was ultimately eliminated because of lack of suitable instrumentation (Pinkham et al., 2014). Social metacognition is broadly defined as evaluating thinking, including both discrete acts, such as assessing the correctness of a response, and synthetic acts, such as integrating thoughts and feelings into complex representations (Lysaker & Dimaggio, 2014). Thus, confidence ratings and response times were integrated into

| Phase | Title | Task |
|---|---|---|
| 1 | World wide expert survey | Solicit nominations on social cognitive domains and tasks to measure those domains.<br>Explicitly avoid any quality evaluation for the measures |
| 2 | RAND appropriateness panel | Assemble the worldwide database on each nominated measure<br>Nominate a wide-ranging expert panel<br>Provide data to panelists and obtain preliminary ratings<br>Conduct in person meeting to obtain ratings and agree on tasks |
| 3 | Initial psychometric study | Examine healthy controls and schizophrenia patients on selected tasks<br>Analyze data and re-present to a subgroup of the RAND panel<br>Identify tasks as adequate, requiring modification, or unsuitable |
| 4 | Modification of selected tasks | Modify tasks to improve their performance<br>Preliminarily examine new tests developed since the RAND panel |
| 5 | Final psychometric study | Examine healthy controls and schizophrenia patients on selected tasks<br>Analyze data and hold investigator consensus meeting<br>Identify tasks as adequate, requiring modification, or unsuitable<br>Make general recommendations to the field |

**Fig. 1** Phases and goals of the SCOPE project.

modified tasks because these endpoints could provide a potentially useful method of evaluating social metacognition within the context of the other core domains.

## Social Cognition Psychometric Evaluation: Results

At the conclusion of all five phases in the SCOPE study, three tasks from the domains of emotion processing and mental state attribution were deemed suitable for immediate use in clinical trials: The Bell Lysaker Emotion Recognition Task (BLERT), the Hinting task, and the Penn Emotion Recognition Task (ER-40) (Bryson, Bell, & Lysaker, 1997; Corcoran, Mercer, & Frith, 1995; Kohler et al., 2003). Three other measures showed adequate psychometric properties, but with limitations: Reading the Mind in the Eyes (Eyes), The Awareness of Social Inferences Task (TASIT), and Intentionally Bias Task (IBT) (Baron-Cohen, Wheelwright, Hill, Raste, & Plumb, 2001; McDonald, Flanagan, Rollins, & Kinch, 2003; Rosset, 2008). All of the acceptable social cognitive indices showed significant correlations with at least one outcome measure, which included indices of real-world functioning, cognitive performance, and functional capacity. The magnitude of these relations ranged from small to medium (all $r = 0.18$–$0.44$). Test-retest reliability was adequate for all tests, except IBT (Pinkham et al., 2016, 2018). For patients, these measures approached an acceptable level of internal consistency ($\alpha = 0.80$) for accuracy of performance (Nunnally, 1967). The psychometric properties of these tasks allowed our team to recommend them for use in clinical trials, particularly studies seeking to improve those aspects of social cognition that have strong links to functioning. The subsequent discussion will provide an overview of the strengths and limitations of the measures deemed appropriate for use as end-points in clinical trials of social cognition in schizophrenia.

## Measures suitable for immediate use in clinical trials

### The Bell Lysaker Emotion Recognition Task (Bryson et al., 1997)

An assessment of emotional processing, BLERT, asks participants to identify emotion from short video clips (Bryson et al., 1997). In both trials, BLERT showed good test-retest reliability, limited potential for floor/ceiling effects, sensitivity to group differences, and high practicality and tolerability. The strength of psychometric properties differed between Phase 3 and Phase

5. BLERT showed greater practice effects in Phase 5 (Cohen's $d = 0.40$) than Phase 3 (Cohen's $d = 0.22$) (Pinkham et al., 2016, 2018). Notably, the BLERT was the only uniquely significant predictor of real-world outcomes in Phase 3, as rated by high-quality informants on the Specific Level of Functioning Scale (SLOF). BLERT also showed a small-to-moderate relation to functional outcomes ($r = 0.26$, $P < .001$) (Pinkham et al., 2016). However, the Phase 5 study failed to replicate the measure's significant correlation to functional outcomes (Pinkham et al., 2018). In light of the strong psychometric properties shown in Phase 3, we continued to recommend this task for use in clinical trials, despite relatively weaker findings in Phase 5, which may have been affected by the changes that were made to tasks in Phase 4 in an attempt to add social metacognitive demands and to increase the difficulty.

For the latter stages of the SCOPE study, the BLERT task was expanded to accommodate evaluation of social metacognition within the core cognitive domain of emotional processing. Participants were asked to perform the task as rapidly as possible, and to rate their confidence in the accuracy of their responses after each item, and the time to complete each item was recorded (Pinkham et al., 2018). Significantly, BLERT confidence ratings were correlated with functional outcomes as rated by high-contact clinicians. In fact, higher confidence ratings on BLERT on the part of patients were the strongest significant correlate of more impaired social functioning in the Phase 5 study ($r = -0.41$, $P < .001$). This correlation was higher than the association between social outcomes and performance on any neurocognitive or social cognitive test (all $r < 0.35$, $P < .05$) (Pinkham et al., 2018). A subsequent evaluation of the sample showed that both patients and healthy controls were less confident in, and spent more time on, incorrect items. However, individuals with schizophrenia did not adjust their effort as a function of personal difficulty, as indexed by scores on correct and incorrect items to the same degree as healthy individuals ($P < .001$) (Cornacchio et al., 2017). While both patients and healthy controls accurately (albeit to a variable extent) recognized when a question was more challenging to them, patients did not adjust their effort appropriately to task demands (Cornacchio et al., 2017). These results suggest that even when individuals with schizophrenia show generally intact social metacognition, patients failed to utilize their performance to change their strategy. An interesting finding from the modification of the BLERT was that confidence ratings collected on a trial-by-trial basis were not correlated with performance on measures of social cognition. This result is noteworthy regarding the independence of the social metacognitive construct (Pinkham et al., 2018). Confidence ratings and response time were easily implemented modifications of the BLERT, and our findings support that their inclusion could have a unique relation to real-world deficits.

## The hinting task (hinting) (Corcoran et al., 1995)

The Hinting Task displayed strong psychometric properties in all SCOPE trials (Pinkham et al., 2016, 2018). The measure assesses mental state attribution/theory of mind by examining the ability of individuals to infer the true intent of indirect speech in short passages that end in a character dropping a hint (Corcoran et al., 1995). Hinting showed adequate test-retest reliability, small practice effects, and strong relations to functional outcomes, including uniquely accounting for variance in outcomes while controlling for other social cognitive tasks. Hinting was the only measure that showed significant incremental validity in the prediction of functional capacity and social competence in both Phase 3 and Phase 4. The task could be administered quickly, was liked by patients, and distinguished patient and control performance (Pinkham et al., 2016, 2018). Ceiling effects have previously been reported (Marjoram et al., 2006; Roberts & Penn, 2009; Versmissen et al., 2008); however, they were not evident in Phase 3 (7%) or in Phase 5 (8%) (Pinkham et al., 2016, 2018). The implications of this limitation in future studies is small, but potentially a proportion of patients may reach the highest possible score, which could restrict evaluation of whether the person's true level of functioning had been accurately measured. Overall, the psychometric data were uniformly strong in both SCOPE clinical trials, and Hinting's association to real-world deficits advocates for the usage of Hinting in future research.

## The Penn Emotion Recognition Task (Kohler et al., 2003)

The ER-40 has participants identify the emotion presented on 40 color photographs of static faces (Kohler et al., 2003). Similar to the BLERT, ER-40 also measures emotional processing, and performed differently in Phase 3 than Phase 5 (Pinkham et al., 2016, 2018). Unlike the BLERT, which had weaker psychometric results in Phase 5, the psychometric properties of ER-40 strengthened the latter study. The Phase 3 expert panel reported uncertainty about whether the ER-40 offered a unique contribution beyond the BLERT because the two assessments showed strong overlap ($r = 0.59$, $P < .001$). Ultimately, the Phase 3 panel decided that the BLERT was more suitable for use because it predicted real-world functional outcome, whereas the ER-40 did not (Pinkham et al., 2016). In Phase 5, the psychometric properties of ER-40 were reexamined with a new sample of 218 individuals with schizophrenia. We also modified the task to include response time and confidence ratings of each answer (Pinkham et al., 2018). Both phases found that ER-40 correlated with some functional outcomes, but did not emerge as a uniquely significant predictor when other measures were considered (Pinkham et al., 2016, 2018). Notably, in Phase 5, the total ER-40 score and the time spent on each question emerged as uniquely significant predictors of social competence (e.g., social skills), even when controlling for all other cognitive and social cognitive variables (Pinkham et al., 2018). As the ER-40 showed only limited relations to functional outcomes in the Phase 3 study, the Phase 5 findings suggest our modifications have the potential to increase the functional utility of this measure. Further, the ER-40 is very easy to administer, and has multiple formats for delivery of the stimuli, increasing its feasibility.

## Measures showing adequate psychometric properties, but with limitations

### Reading the Mind in the Eyes test (Baron-Cohen et al., 2001)

Eyes evaluates mental state attribution/theory of mind by asking participants to discriminate the mental state (thoughts/feelings) of others from photos of the eye region (Baron-Cohen et al., 2001). Eyes was rated as acceptable, but concerns included the potential dependence of performance on vocabulary, and the somewhat limited relation with outcomes in both SCOPE trial phases (Pinkham et al., 2016, 2018). Phase 5 modifications to the Eyes focused on reducing the dependence of performance on vocabulary, and encouraging the use of a glossary of terms that is provided with the task. Therefore, a version of the task was created that includes embedded definitions, which can be viewed on the same screen as the stimuli (Pinkham et al., 2018). These modifications did not successfully reduce this confounding relationship. Eyes failed to offer any unique contribution to the prediction of outcomes, and task performance was strongly correlated with Wechsler abbreviated scale of intelligence Vocabulary (WASI; Wechsler, 1999) scores in patients ($r = 0.63$, $P < .001$) and controls ($r = 0.47$, $P < .001$) (Pinkham et al., 2018). Previous studies report a correlation of 0.49 between WASI Vocabulary and Eyes performance in healthy individuals (Peterson &

Miller, 2012), which suggests the dependence on vocabulary remained a confounder, despite the modifications. The large amount of shared variance raises questions as to whether this is an intelligence test or a social cognition measure.

## The Awareness of Social Inferences Test, part III (McDonald et al., 2003)

Similar to Hinting and Eyes, the TASIT evaluated the theory of mind domain by asking participants to detect lies and sarcasm in short clips of everyday interactions (McDonald et al., 2003). TASIT showed acceptable psychometric properties, but also had limitations that should be considered carefully. TASIT showed small relations to functional outcomes, and had the longest administration time across both phases, which may be impractical for use in some clinical trials where a detailed assessment might be desired (Pinkham et al., 2016, 2018). Specifically, the Phase 3 expert panel had concerns about alternative forms of the TASIT, and whether the differences between performance at visits 1 and 2 were due to interference from previous administration, or nonequivalence between the two task forms (Pinkham et al., 2016). Counterbalancing form administration in Phase 5 did appear to reduce the discrepancy, and thus, we recommend counterbalancing be implemented when using both forms (Pinkham et al., 2018). Overall, additional research should be conducted before the acceptability of the TASIT can be reported.

## The Intentional Bias Task (Rosset, 2008)

The IBT assesses the Attributional Style/Bias domain, specifically the tendency to attribute intentionality to the actions of others. Participants indicated whether 24 brief descriptions of actions (e.g., "He broke the window") occurred "on purpose" or "by accident" (Rosset, 2008). We did not include the IBT in the early stages of the SCOPE study based on the RAND panel suggestions, and the measure was added to Phase 5 because no task evaluating attributional style/bias had previously shown psychometric promise. The major limitations of IBT included the lowest test-retest reliability ($r = 0.59$, $P < .001$) of all tasks, and increased missing data due to the IBT procedure of limiting response times. Importantly, IBT uniquely accounted for variance in both functional capacity and real-world functional outcome. The relationship IBT showed with everyday deficits supported IBT as a potentially useful measure of attributional style/bias, and more

**Table 1** Characteristics of the final tasks in the SCOPE study.

| Task | Psychometrics | Correlation with outcomes | Practicality/ tolerability |
|---|---|---|---|
| ***Currently acceptable*** | | | |
| BLERT | XXX | XX | XXX |
| ER-40 | XXX | XX | XXX |
| Hinting | XXX | XXX | XXX |
| ***Acceptable with concerns*** | | | |
| Eyes | XX | X | XXX |
| IBT | XX | X | XXX |
| TASIT | X | X | – |
| ***Not acceptable*** | | | |
| Mini-PONS | – | X | – |
| SAT-MC | – | X | XXX |

Excellent: XXX.
Good: XX.
Adequate: X.
Not currently adequate: –.

detailed analyses (e.g., examination of relations to symptoms, consideration of condition effects, etc.) appear warranted (Pinkham et al., 2018).

The overall summary of the core findings of the of the SCOPE project is presented in Table 1. In this table, we evaluate each of the final tasks for their usefulness in clinical trials, and highlight any areas where the task's performance was limited in some way.

## Extension of Social Cognition Psychometric Evaluation: Informant rating measures of social cognition

Multiple reliable and valid interview-based measures of neurocognition that incorporate informant reports have strong correlations with functional outcomes, quite comparable to objective cognitive performance (Keefe et al., 2006; Ventura et al., 2010; Ventura, Cienfuegos, Boxer, & Bilder, 2008). Currently, only one informant-based rating scale exists for social cognition. The Observable Social Cognition Rating Scale (OSCARS), the first informant-based interview scale for social cognition, has shown internal consistency and test-retest reliability (Healey et al., 2015). The OSCARS is comprised of eight questions probing different social cognitive domains (theory of mind, emotional perception, cognitive rigidity, jumping to conclusions, and attributional style) followed by general example behaviors that reflect impairment in that domain. Participants rank each item on a seven-point scale, with higher ratings indicating greater impairment. An additional ninth question can be utilized to assess global social cognitive impairment. The global rating uses a 10-point scale; again higher ratings designate more severe impairment (range 1–10). The global score has shown significant correlations to all eight items ($P < .05$), and a strong relationship to the mean total rating ($r = 0.67$, $P < .05$) (Silberstein, Pinkham, Penn, & Harvey, 2018).

Evidence suggests the scores on the OSCARS might be a stronger predictor of social outcomes than information collected from performance-based social cognitive tasks. Healey et al. (2015) found that the OSCARS predicted real-world outcomes beyond objective social cognitive measures with a magnitude of nearly double the variance contributed by objective-based measures of social cognition, as long as ratings were generated by high contact informants and not the patients themselves. Similarly, analyses based on the fifth phase of the SCOPE trial has suggested that informant-based ratings on the OSCARS might be a stronger predictor of functional outcomes than performance on social cognitive performance-based assessments (Silberstein et al., 2018). Regression analysis was used to evaluate the relative importance of social cognitive performance-based tasks (BLERT, ER-40, Eyes, Hinting, TASIT, IBT) and high-contact informant OSCARS ratings in predicting three domains of social outcomes (interpersonal relationships, social acceptability, employment). OSCARS ratings accounted for 25%–36% of the variance in social outcomes, but as a group, performance-based social cognitive measures failed to predict any of the three functional outcomes significantly (all $P > .08$). However, as previously discussed, several of the individual tasks significantly correlated with functional outcomes in the initial SCOPE analysis (Pinkham et al., 2016, 2018). Additionally, social outcomes were defined by informant reports, which might conflate the correlation results between informant assessments of patients' social cognitive ability and everyday disability. Despite limitations, these findings support that the short, eight-question OSCARS might be an effective proxy for extensive performance-based testing, and may aid in determining factors producing a patients' social impairments.

A further finding of the Silberstein et al. study was that discrepancies between self-reports and informant reports of social cognitive competence on the OSCARS were consistently correlated with functional outcomes.

Patients whose self-reports of their social cognitive competence overestimated their ability compared with informant reports consistently manifested greater impairment in everyday functioning. Much like the confidence ratings from SCOPE 5 and the results of the Gould et al. (2015) study of neurocognition and functional outcomes, reporting greater confidence in your ability, particularly in reference to informant observations, is associated with impairment in everyday functioning. Thus, the direct comparison of self-assessment in social and social cognitive domains is an assessment of social metacognition, and impairments in these domains can be seen to reflect reduced awareness of illness and its associated features.

## Implication of Social Cognition Psychometric Evaluation results

Overall, six social cognitive measures (BLERT, ER-40, Eyes, Hinting, TASIT, and IBT) displayed acceptable or acceptable with limitations psychometric characteristics. The BLERT, Hinting, and ER-40 tasks were the strongest of the six acceptable tasks (Pinkham et al., 2014, 2016, 2018). Hinting provided a suitable measure addressing the domain of mental state attribution/theory of mind, and the modified BLERT and ER-40 offered good representation for emotion processing. Phase 5 of SCOPE identified the IBT as a potentially promising measure of attributional bias/style, and the OSCARS emerged as a potentially robust measure of global social cognition (Healey et al., 2015). The OSCARS has been shown to be correlated to the performance-based measures of social cognition. Importantly, our most recent findings suggest that the OSCARS might be a stronger predictor of social outcome than performance-based social cognitive tests (Silberstein et al., 2018). The SCOPE results lay the groundwork for developing an array of valid performance- and rating-based measures that will increase the rigor and reliability of future research in social cognition in schizophrenia.

The SCOPE results also underscore the challenges that continue to face the emerging field of social cognitive research in schizophrenia. More than 100 domains and more than 100 measures were initially nominated, and 21 were forwarded to the RAND panel for consideration. Less than a quarter of the measures received average ratings from the RAND panel in the good range, with the remainder rated as only fair (Pinkham et al., 2014). On the positive side, the large number of nominated tasks provide strong evidence for the health, breadth, and diversity of social cognitive research.

From a more negative viewpoint, however, the use of so many different measures can result in the problems noted previously; namely, the lack of standardized measures and potential for discrepant findings between studies that may be due merely to measurement variation rather than true differences. Across all five phases, SCOPE identified more measures that should not be used (or used cautiously), rather than those that should be used. This was likely a reflection of the relative youth of social cognition as a research topic (e.g., as compared to neurocognition), but made clear that developing new measures will be required to generate a reliable, valid battery for social cognition (Pinkham et al., 2014, 2016, 2018).

Unfortunately, no measures of social perception were recommended. Given that the initial expert survey in Phase 1 identified social perception as a core domain, this shortcoming underscores the need to either improve existing measures, or develop new ones (Pinkham et al., 2014). The exclusion of the social perception domain additionally raises questions about how best to assess the full range of social cognitive processes in a clinical trial. The failure to identify a well-validated task corresponding to each of the four-core social cognitive domains in schizophrenia made the creation of an assessment battery infeasible. In comparison, a similar consensus effort for neurocognition in schizophrenia successfully developed a well-validated assessment battery that was the conclusion of the expert panels and clinical trials (Nuechterlein et al., 2008). This limitation emphasizes the importance of assessment creation, particularly in the domain of social perception, to bring the research standards of social cognition in schizophrenia on par with those of neurocognition.

## Social cognitive assessments and functional outcomes

The results from the SCOPE trials support the association between social cognitive measures and functional outcomes, but found the strength of the relationship to be small to medium. Half of the social cognitive tasks assessed showed significant, yet limited, correlations. Social cognitive performance accounted for less than 5% of the variance in social and nonsocial outcomes (Pinkham et al., 2016). A metaanalysis of 52 studies supported SCOPE's findings (Fett et al., 2011). The metaanalysis found that social cognitive performance-based scores had a medium-strength correlation with functional outcomes (all $r = 0.31–0.48$, all $P < .004$). A subsequent analysis of the data from the studies showed that social cognitive performance had a stronger relationship to social everyday outcomes and nonsocial outcomes. Social cognition remains an important small- to moderate-strength predictor of functional outcomes, and future use of social cognitive tasks appears to be particularly indicated in studies evaluating real-world social deficits.

Social cognition and neurocognition have emerged as two of the most reliable predictors of functional outcomes in schizophrenia, with evidence of some specificity of prediction across social and nonsocial functional domains. Yet studies of the determinants of everyday functional deficits in schizophrenia have stalled at accounting for 50% or less of the variance in real-world functioning (Bowie et al., 2008, 2010; Bowie, Reichenberg, Patterson, Heaton, & Harvey, 2006; Harvey et al., 2011). The attempts to identify factors to account for additional association neurocognition and outcome have been limited in their success. The CNTRACS Consortium identified a number of cognitive abilities that were reliably measured with performance-based assessments and linked to specific elements of brain functioning. However, the consortium was not as successful at identifying predictive measures that added to the understanding of the causes of everyday disability, despite using highly selective neuroscience-oriented tests with clear neurobiological correlates (Gold et al., 2012). On the surface, these findings seem disheartening, but there are at least two reasons for this outcome. First, the more cognitive domains that a specific neuropsychological test predicts, the more globally sensitive it is to outcomes. The classic example is that of various coding tests, which require intact functioning in several domains to be performed efficiently, including visual perception, problem-solving, working memory, and speed (Knowles et al., 2015). Thus, these tasks routinely account for more than 60% of the variance in composite measures of neurocognition (Keefe et al., 2006), and are routinely found to be the strongest predictors of everyday disability (McClure et al., 2007). Therefore, tasks that measure a single highly selective aspect of cognitive performance (either social cognitive or neurocognitive) seem unlikely to add to the prediction of the everyday disability by broad and nonspecific tests already capturing multiple cognitive domains.

Second, a growing body of research has shown a divergence of neurocognition and social cognition for the prediction of different domains of real-world outcomes, such that social cognition predicts social outcomes, but not everyday activities (e.g., living independently), and neurocognition provides a minimal prediction of social outcomes (e.g., interpersonal relationships) (Depp et al., 2012; Fett et al., 2011). As part of the SCOPE trials, neurocognition was included to evaluate the comparative strength of prediction between social cognitive ability and neurocognitive ability. The combination of neurocognitive and social cognitive tasks accounted for 13% of the variance in functional outcomes, as rated by high-quality informants. When the outcomes domains were differentiated, social cognitive tasks accounted for 10%–16% of the variance in social outcomes (e.g., interpersonal relationships and employment), compared with 9% of the variance in nonsocial outcomes (e.g., everyday activities) (Pinkham et al., 2016). These findings suggest that future usage of social cognitive tasks in clinical research is most indicated in studies attempting to understand social outcomes.

## Social competence assessments

In going beyond social cognitive performance, it is also important to consider social interaction ability. The ability to perform the social skills required for activities of daily living or social communications in performance-based assessments is referred to as social competence (social skills) (Green et al., 2011; McKibbin et al., 2004). Social competence requires a combination of cognition, skill, and motivation. As such, social competence has emerged as a mediator between social cognition and real-world outcomes (Couture, Granholm, & Fish, 2011), with social competence related to social cognition and possibly neurocognition.

## Social skills performance assessment

One easy method for assessing social competence is through the Social Skills Performance Assessment (SSPA) (Patterson, Moscona, McKibbin, Davidson, & Jeste, 2001). The SSPA is an abbreviation and adaptation of the role-play components of the Maryland Assessment of Social Competence (MASC: Sayers, Bellack, Wade, Bennett, & Fong, 1995). Two role-play scenarios are presented after a brief role practice (i.e., greeting a new neighbor and calling a landlord to request the repair of a leak that remains unrepaired despite a previous request); the participants are then expected to start and maintain a conversation for 3 minutes per role-play. Assessments are commonly administered and audiotaped by raters trained in test administration, but untrained in scoring, thus reducing potential biases in the administration of the test. Studies have demonstrated that the SSPA has high test-test reliability ($r=0.91$) over 1 week (Patterson et al., 2001), but has displayed decreased test-test reliability over a more extended retest period ($r = 0.49$–$0.79$; Leifker, Patterson, Bowie, Mausbach, & Harvey, 2010). The SSPA has previously been shown to be sensitive to the effects of antipsychotic treatment (Harvey, Patterson, Potter, Zhong, & Brecher, 2006), and importantly, the SSPA has shown correlations to functional outcomes and quality of life (Patterson et al., 2001; Pinkham et al., 2016, 2018).

While social cognition and social competence are interrelated, evidence suggests that they are separable constructs, and that they have independent effects on everyday outcomes. A previous study that included no social cognition tests (Robertson et al., 2014) found that performance-based measures of social competence accounted for 2% of the variance in everyday social outcomes, as rated by high-contact clinicians. In contrast, negative symptoms accounted for 27% of the variance in those same outcomes. Analyses we conducted utilizing the SCOPE Phase 3 sample ($n = 179$) found that social cognition tasks accounted for 19% of the variance in performance-based measures of social competence (Kalin et al., 2015). The addition of negative symptoms (e.g., social and emotional withdrawal and avoidance) increased the percent of the variance in social competence, which accounted for to 32%, but did not reduce the impact of social cognition on social competence. Additionally, social competence and social cognition both uniquely contributed 5%–9% of the variance in functional outcomes (Kalin et al., 2015). These findings suggest separable influences of social competence and social cognition, and that efforts to improve social outcomes should focus on both social cognition and social competence.

## The importance of reduced social motivation on social outcomes

The elephant in the room in studies of social cognition, social competence, and social outcomes is the commonly seen reduced social motivation in schizophrenia. In the studies noted herein, the predictive strength of social cognition and social competence for the prediction of social outcomes was overshadowed by the correlations between reduced social motivation and impairments in functional outcomes. In the Robertson et al. study, social outcomes did not differ between patients with perfect scores on the SSPA and patients with reduced performance. In both the Robertson et al. and Kalin et al. studies, conducted on separate samples, the same two negative symptoms, active social avoidance and passive-apathetic social withdrawal, combined to account for 27% and 17% of the variance. Research to date has not examined the impact of social cognition on social outcomes in patients with low levels of social motivation, but this would clearly be an interesting target for later research.

## Other considerations: Social cognition in other psychiatric conditions

While the majority of mental health research on social cognition has been done in people with schizophrenia, studies have been conducted in people with bipolar disorder and major depression as well. In a recent study, Ospina et al. (2018) reported that patients with bipolar disorder manifested deficits in both global social cognition and theory of mind compared with healthy controls. In this sample, cognitive impairments did not predict impairments in everyday outcomes, but when social cognition was examined as a mediating factor, the correlations became significant. In a previous study with a smaller sample (Lee et al., 2013, but with a schizophrenia control group, people with bipolar disorder did not differ from healthy controls on any test in a detailed social cognition battery. The authors note that the profile of social cognition in bipolar disorder was also quite similar to that seen in the healthy controls, and notably different from schizophrenia as well. In a metaanalysis of social cognition and schizophrenia and bipolar disorder, Bora and Pantelis (2016) found that schizophrenia patients had modestly greater deficits than bipolar patients. They commented that the level of difference was not that dissimilar from differences in levels of neurocognitive impairments across the samples. Adding interest to this area, Bora and Ozerdem (2017) published a metaanalysis of social cognitive deficits in first-degree relatives of people with bipolar disorder. They reported modest deficits on the part of these relatives compared with healthy people, suggesting that social cognitive deficits may be a meaningful endophenotype for bipolar illness.

The databases for social cognition in bipolar disorder are not as large or detailed compared with schizophrenia. Further, bipolar disorder has a more variable clinical course and truly meaningful subtype differences, compared with schizophrenia. Clearly this is an area where much more research will be needed, hopefully, starting from the outset, with high methodological standards.

## Future research areas

We recommend future research into social metacognition. Social metacognition was selected as a key social cognitive domain in the initial-round of SCOPE's consensus procedure, but was ultimately dropped as a core social cognitive domain due to lack of available data. A newly defined component of the larger metacognition construct addresses just how well individuals evaluate their abilities and performance (Koren, Seidman, Goldsmith, & Harvey, 2006), and we refer to this type of self-awareness as "introspective accuracy." Results from the modified BLERT task in Phases 4 and 5 of SCOPE support social metacognition as an important future research area. Confidence ratings of the BLERT emerged as the largest unique predictor of social functioning in Phase 5 ($r=0.41$, $P<.001$; Pinkham et al., 2018). The preliminary data from a subsequent analysis of the Phase 5 sample (Silberstein et al., 2018) showed that introspective accuracy accounted for a unique variance of 6%–8% in the prediction of social outcomes beyond social cognitive performance, as measured by acceptable SCOPE tasks (BLERT, ER-40, Eyes, TASIT, IBT, Hinting) or clinician ratings (OSCARS). Introspective accuracy also displayed a significant association with social outcomes (all $r=0.32$–$0.45$) that was comparable or stronger than any performance-based measure (all $r=0.20$–$0.35$). Importantly, deficits in introspective accuracy that present across cognition, capacity, and outcome domains have shown relate to medication adherence, suicidality, everyday activities, vocational functioning, and social outcomes (Green et al., 2011; Holshausen, Bowie, Mausbach, Patterson, & Harvey, 2014; McKibbin et al., 2004; Patterson et al., 1997). The current body of research suggests that impaired introspective accuracy of ability might be an independent predictor of disability for individuals with schizophrenia, and an important future research area.

Assessment of social metacognition in schizophrenia implies that self-ratings of social cognitive ability add value, despite the lack of correlation with real-world outcomes. Self-reports lie at the crux of clinically assessing introspective accuracy. Impairments in introspective accuracy can be defined by discrepancies between how one rates one's abilities and achievements, and his/her actual performance or a high-contact informant's rating of his/her ability. For example, introspective accuracy can easily be assessed by subtracting a patient's self-reported OSCARS rating from a high-contact informant's rating. On performance-based measures, such as the BLERT, a person's confidence rating of their accuracy, overall, on each question, or in terms of confidence when correct versus incorrect could be utilized. As such, patients may overestimate or underestimate their ability, both of which can challenge functional outcomes. For example, a person who overestimates their ability may incorrectly believe they can utilize maps and public transportation schedules to locate an appointment, but get lost. A person that underestimates their ability may incorrectly believe they lack social skills to engage in a conversation, and may choose to withdrawal socially, even if their true limitation is anxiety, and not competence. Overall, more work is needed to define and understand the significance of introspective accuracy and the potential subdomains of the construct, but recent findings suggest that impaired social metacognition, and in particular, introspective accuracy, may factor into our understanding of the determinants of functional outcomes among individuals with schizophrenia. Treatment of social metacognitive deficits may hold promise in improving functional outcomes in social and other domains.

A second major recommendation is the development of more adequate measures of social perception. This is the main consensus-validated domain in SCOPE where there was no success in identifying a useful measure. Many of the measures that were tested failed because of their complexity. It would seem like a reasonable strategy, if possible, to develop measures of social perception that share some of the characteristics of other successful tests of emotion recognition and theory of mind. The tests that worked had stimuli presented for abbreviated periods (20 seconds or less) and had instructions that were easy to understand. Also, those tests that were successful did not have high-level intellectual demands, requiring neither complex instructions in an attempt to establish a testing context, or requiring participants to process and recall extensive information in order to perform the tasks. This seems to be a major area of need, and the field would be well served by better tests. This is particularly true because social perception seems, at face value, to be a critical factor in interactive social skills.

## Summary

Social cognition in schizophrenia continues to be a rapidly emerging area of study. The youth of the field, combined with the relative paucity of well-validated measures and confounding of subdomains, has limited the evaluation of social cognition in schizophrenia. A number of consensus measure (NIHM Workshop, CNTRICS/CNTRACS, SCOPE) have made progress in these areas, particularly in determining unified definitions and psychometrically robust tasks. Social cognition has now been defined as the "mental operations that underlie social interactions, including perceiving, interpreting, and generating responses to the intentions, dispositions, and behaviors of others" (Green et al., 2008). Further, consensus agreement has broken social cognition in schizophrenia down into four domains: emotion processing, social perception, attributional bias, and theory of mind (Pinkham et al., 2014). This organizing framework has provided a needed foundation to develop and critique assessments.

Our work in the SCOPE study underscores that continued effort is necessary to build a well-validated assessment battery. Because the majority of tasks had weak psychometric properties, the SCOPE results highlight that clinical researchers must use caution in task selection. The bulk of the roughly 100 measures nominated had poor test-test reliability, showed large practice effects, and poor internal consistency (Pinkham et al., 2014, 2016, 2018). Six measures ultimately displayed strong psychometric properties, and were selected as acceptable (or acceptable with caution) for clinical trials (Pinkham et al., 2018). These tasks assess three out of four core domains of social cognition in schizophrenia: emotional processing (BLERT, ER-40); mental state-attribution/theory of mind (Hinting, TASIT, Eyes); attributional Bias/Style (IBT). Despite a large number of nominated tasks, no task assessing the social perception domain was found to have psychometrically sound properties. The dearth of well-validated social perception measures stresses the substantial difficulties associated with social cognitive research in schizophrenia, and the need for continued measurement development.

Assessment strategies in clinical trials and clinical practice fall in several different domains. The BLERT, ER-40, Eyes, Hinting, TASIT, and IBT represent performance-based assessments. Administration time was less than 10 minutes for the all of the tasks. TASIT was a notable exception, taking 18 minutes, on average, for both patients and controls. As such, the selected measures were well tolerated, and participants rated all tasks as pleasant (Pinkham et al., 2018). The ability of task administration might be somewhat more limited in the clinical context because of technology, training, and time constraints. Unfortunately, the lack of validity of self-ratings of cognitive and functional ability has enhanced the need for reliable and clinically-feasible measures to examine cognitive abilities (Bowie et al., 2007; Durand et al., 2015; Keefe et al., 2006; McKibbin et al., 2004; Patterson et al., 1997; Sabbag et al., 2011). Fortunately, high-contact informant ratings have been shown to be highly accurate in terms of predicting objective indices of functional ability (Sabbag et al., 2011). The OSCARS rating scale has emerged as a potential proxy for extensive performance-based social cognitive testing. Recent evidence suggests that social cognition OSCARS ratings completed by informants who indicated knowing the patient "very well" might be a stronger predictor of outcome than social cognitive ability based on performance-based tasks (Healey et al., 2015; Silberstein et al., 2018). Collectively, the implementation of performance-based and high-contact informant ratings can help clinicians create more targeted treatment goals.

## References

Addington, J., Saeedi, H., & Addington, D. (2006). Influence of social perception and social knowledge on cognitive and social functioning in early psychosis. *The British Journal of Psychiatry*, *189*, 373–378. https://doi.org/10.1192/bjp.bp.105.021022.

Baron-Cohen, S., Wheelwright, S., Hill, J., Raste, Y., & Plumb, I. (2001). The 'Reading the mind in the eyes' Test revised version: a study with normal adults, and adults with Asperger syndrome or high-functioning autism. *Journal of Child Psychology and Psychiatry*, *42*(2), 241–251. https://doi.org/10.1111/1469-7610.00715.

Bora, E., & Pantelis, C. (2016). Social cognition in schizophrenia in comparison to bipolar disorder: a meta-analysis. *Schizophrenia Research*, *175*, 72–78.

Bora, E., & Ozerdem, A. (2017). Social cognition in first-degree relatives of patients with bipolar disorder: a meta-analysis. *European Neuropsychopharmacology*, *27*, 293–300.

Bora, E., Yucel, M., & Pantelis, C. (2009). Theory of mind impairment in schizophrenia: meta-analysis. *Schizophrenia Research*, *109*(1–3), 1–9. https://doi.org/10.1016/j.schres.2008.12.020.

Bowie, C. R., Depp, C., McGrath, J. A., Wolyniec, P., Mausbach, B. T., Thornquist, M. H., … Pulver, A. E. (2010). Prediction of real-world functional disability in chronic mental disorders: a comparison of schizophrenia and bipolar disorder. *The American Journal of Psychiatry*, *167*(9), 1116–1124. https://doi.org/10.1176/appi.ajp.2010.09101406.

Bowie, C. R., Leung, W. W., Reichenberg, A., McClure, M. M., Patterson, T. L., Heaton, R. K., & Harvey, P. D. (2008). Predicting schizophrenia patients' real-world behavior with specific neuropsychological and functional capacity measures. *Biological Psychiatry*, *63*(5), 505–511. https://doi.org/10.1016/j.biopsych.2007.05.022.

Bowie, C. R., Reichenberg, A., Patterson, T. L., Heaton, R. K., & Harvey, P. D. (2006). Determinants of real-world functional performance in schizophrenia subjects: correlations with cognition, functional capacity, and symptoms. *The American Journal of Psychiatry*, *163*(3), 418–425. https://doi.org/10.1176/appi.ajp.163.3.418.

Bowie, C. R., Twamley, E. W., Anderson, H., Halpern, B., Patterson, T. L., & Harvey, P. D. (2007). Self-assessment of functional status in schizophrenia. *Journal of Psychiatric Research*, *41*(12), 1012–1018. https://doi.org/10.1016/j.jpsychires.2006.08.003.

Browne, J., Penn, D. L., Raykov, T., Pinkham, A. E., Kelsven, S., Buck, B. A., & Harvey, P. D. (2016). Social cognition in schizophrenia: factor structure of emotion processing and theory of mind. *Psychiatry Research*, *242*, 150–156.

Bryson, G., Bell, M., & Lysaker, P. (1997). Affect recognition in schizophrenia: a function of global impairment or a specific cognitive deficit. *Psychiatry Research*, *71*(2), 105–113.

Carter, C. S., Barch, D. M., Gur, R., Gur, R., Pinkham, A., & Ochsner, K. (2009). CNTRICS final task selection: social cognitive and affective neuroscience-based measures. *Schizophrenia Bulletin*, *35*(1), 153–162. https://doi.org/10.1093/schbul/sbn157.

Combs, D. R., Adams, S. D., Penn, D. L., Roberts, D., Tiegreen, J., & Stem, P. (2007). Social Cognition and Interaction Training (SCIT) for inpatients with schizophrenia spectrum disorders: preliminary findings. *Schizophrenia Research, 91*(1–3), 112–116. https://doi.org/10.1016/j.schres.2006.12.010.

Corcoran, R., Mercer, G., & Frith, C. D. (1995). Schizophrenia, symptomatology and social inference: investigating "theory of mind" in people with schizophrenia. *Schizophrenia Research, 17*(1), 5–13.

Cornacchio, D., Pinkham, A. E., Penn, D. L., & Harvey, P. D. (2017). Self-assessment of social cognitive ability in individuals with schizophrenia: appraising task difficulty and allocation of effort. *Schizophrenia Research, 179*, 85–90.

Corrigan, P. W., & Green, M. F. (1993). Schizophrenic patients' sensitivity to social cues: the role of abstraction. *The American Journal of Psychiatry, 150*(4), 589–594. https://doi.org/10.1176/ajp.150.4.589.

Couture, S. M., Granholm, E. L., & Fish, S. C. (2011). A path model investigation of neurocognition, theory of mind, social competence, negative symptoms and real-world functioning in schizophrenia. *Schizophrenia Research, 125*(2–3), 152–160. https://doi.org/10.1016/j.schres.2010.09.020.

Depp, C. A., Mausbach, B. T., Harmell, A. L., Savla, G. N., Bowie, C. R., Harvey, P. D., & Patterson, T. L. (2012). Meta-analysis of the association between cognitive abilities and everyday functioning in bipolar disorder. *Bipolar Disorders, 14*(3), 217–226. https://doi.org/10.1111/j.1399-5618.2012.01011.x.

Durand, D., Strassnig, M., Sabbag, S., Gould, F., Twamley, E. W., Patterson, T. L., & Harvey, P. D. (2015). Factors influencing self-assessment of cognition and functioning in schizophrenia: implications for treatment studies. *European Neuropsychopharmacology, 25*(2), 185–191. https://doi.org/10.1016/j.euroneuro.2014.07.008.

Eack, S. M., Greenwald, D. P., Hogarty, S. S., & Keshavan, M. S. (2010). One-year durability of the effects of cognitive enhancement therapy on functional outcome in early schizophrenia. *Schizophrenia Research, 120*(1–3), 210–216. https://doi.org/10.1016/j.schres.2010.03.042.

Fett, A. K., Viechtbauer, W., Dominguez, M. D., Penn, D. L., van Os, J., & Krabbendam, L. (2011). The relationship between neurocognition and social cognition with functional outcomes in schizophrenia: a meta-analysis. *Neuroscience and Biobehavioral Reviews, 35*(3), 573–588. https://doi.org/10.1016/j.neubiorev.2010.07.001.

Fiszdon, J. M. (2013). Introduction to social cognitive treatment approaches for schizophrenia. In *Social cognition in schizophrenia: From evidence to treatment* (pp. 285–310). New York, NY: Oxford University Press.

Frith, C. (1992). *The cognitive neuropsychology of schizophrenia.* East Sussex: Psychology Press.

Gold, J. M., Barch, D. M., Carter, C. S., Dakin, S., Luck, S. J., MacDonald, A. W., 3rd, … Strauss, M. (2012). Clinical, functional, and intertask correlations of measures developed by the Cognitive Neuroscience Test Reliability and Clinical Applications for Schizophrenia Consortium. *Schizophrenia Bulletin, 38*(1), 144–152. https://doi.org/10.1093/schbul/sbr142.

Gould, F., McGuire, L. S., Durand, D., Sabbag, S., Larrauri, C., Patterson, T. L., … Harvey, P. D. (2015). Self-assessment in schizophrenia: accuracy of evaluation of cognition and everyday functioning. *Neuropsychology, 29*(5), 675–682. https://doi.org/10.1037/neu0000175.

Green, M. F., Olivier, B., Crawley, J. N., Penn, D. L., & Silverstein, S. (2005). Social cognition in schizophrenia: recommendations from the measurement and treatment research to improve cognition in schizophrenia new approaches conference. *Schizophrenia Bulletin, 31*(4), 882–887. https://doi.org/10.1093/schbul/sbi049.

Green, M. F., Penn, D. L., Bentall, R., Carpenter, W. T., Gaebel, W., Gur, R. C., … Heinssen, S. (2008). Social cognition in schizophrenia: an NIMH workshop on definitions, assessment, and research opportunities. *Schizophrenia Bulletin*, *34*(6), 1211–1220. https://doi.org/10.1093/schbul/sbm145.

Green, M. F., Schooler, N. R., Kern, R. S., Frese, F. J., Granberry, W., Harvey, P. D., … Marder, S. R. (2011). Evaluation of functionally meaningful measures for clinical trials of cognition enhancement in schizophrenia. *The American Journal of Psychiatry*, *168*(4), 400–407. https://doi.org/10.1176/appi.ajp.2010.10030414.

Harvey, P. D., & Pinkham, A. (2015). Impaired self-assessment in schizophrenia: why patients misjudge their cognition and functioning. *Current Psychiatry*, *14*, 6.

Harvey, P. D., Patterson, T. L., Potter, L. S., Zhong, K., & Brecher, M. (2006). Improvement in social competence with short-term atypical antipsychotic treatment: a randomized, double-blind comparison of quetiapine versus risperidone for social competence, social cognition, and neuropsychological functioning. *The American Journal of Psychiatry*, *163*(11), 1918–1925. https://doi.org/10.1176/ajp.2006.163.11.1918.

Harvey, P. D., Raykov, T., Twamley, E. W., Vella, L., Heaton, R. K., & Patterson, T. L. (2011). Validating the measurement of real-world functional outcomes: phase I results of the VALERO study. *The American Journal of Psychiatry*, *168*(11), 1195–1201. https://doi.org/10.1176/appi.ajp.2011.10121723.

Harvey, P. D., Sabbag, S., Prestia, D., Durand, D., Twamley, E. W., & Patterson, T. L. (2012). Functional milestones and clinician ratings of everyday functioning in people with schizophrenia: overlap between milestones and specificity of ratings. *Journal of Psychiatric Research*, *46*(12), 1546–1552. https://doi.org/10.1016/j.jpsychires.2012.08.018.

Healey, K. M., Combs, D. R., Gibson, C. M., Keefe, R. S., Roberts, D. L., & Penn, D. L. (2015). Observable Social Cognition—A Rating Scale: an interview-based assessment for schizophrenia. *Cognitive Neuropsychiatry*, *20*(3), 198–221. https://doi.org/10.1080/13546805.2014.999915.

Holshausen, K., Bowie, C. R., Mausbach, B. T., Patterson, T. L., & Harvey, P. D. (2014). Neurocognition, functional capacity, and functional outcomes: the cost of inexperience. *Schizophrenia Research*, *152*(2–3), 430–434. https://doi.org/10.1016/j.schres.2013.08.004.

Kalin, M., Kaplan, S., Gould, F., Pinkham, A., Penn, D., & Harvey, P. D. (2015). Social cognition, social competence, negative symptoms and social outcomes: interrelationships in people with schizophrenia. *Journal of Psychiatric Research*, *68*, 254–260. https://doi.org/10.1016/j.jpsychires.2015.07.008.

Keefe, R. S., Poe, M., Walker, T. M., Kang, J. W., & Harvey, P. D. (2006). The Schizophrenia Cognition Rating Scale: an interview-based assessment and its relationship to cognition, real-world functioning, and functional capacity. *The American Journal of Psychiatry*, *163*(3), 426–432. https://doi.org/10.1176/appi.ajp.163.3.426.

Knowles, E. E., Weiser, M., David, A. S., Glahn, D. C., Davidson, M., & Reichenberg, A. (2015). The puzzle of processing speed, memory, and executive function impairments in schizophrenia: fitting the pieces together. *Biological Psychiatry*, *78*(11), 786–793. https://doi.org/10.1016/j.biopsych.2015.01.018.

Kohler, C. G., Turner, T. H., Bilker, W. B., Brensinger, C. M., Siegel, S. J., Kanes, S. J., … Gur, R. C. (2003). Facial emotion recognition in schizophrenia: intensity effects and error pattern. *The American Journal of Psychiatry*, *160*(10), 1768–1774. https://doi.org/10.1176/appi.ajp.160.10.1768.

Koren, D., Seidman, L. J., Goldsmith, M., & Harvey, P. D. (2006). Real-world cognitive—and metacognitive—dysfunction in schizophrenia: a new approach for measuring (and remediating) more "right stuff" *Schizophrenia Bulletin*, *32*(2), 310–326. https://doi.org/10.1093/schbul/sbj035.

Kurtz, M. M., & Richardson, C. L. (2012). Social cognitive training for schizophrenia: a meta-analytic investigation of controlled research. *Schizophrenia Bulletin*, *38*(5), 1092–1104. https://doi.org/10.1093/schbul/sbr036.

Lee, J., Altshuler, L., Glahn, D. C., Miklowitz, D. J., Ochsner, K., & Green, M. F. (2013). Social and nonsocial cognition in bipolar disorder and schizophrenia: relative levels of impairment. *The American Journal of Psychiatry, 170*(3), 334–341. https://doi.org/10.1176/appi.ajp.2012.12040490.

Leifker, F. R., Patterson, T. L., Bowie, C. R., Mausbach, B. T., & Harvey, P. D. (2010). Psychometric properties of performance-based measurements of functional capacity: test-retest reliability, practice effects, and potential sensitivity to change. *Schizophrenia Research, 119*(1–3), 246–252. https://doi.org/10.1016/j.schres.2010.03.021.

Lin, Y. C., Wynn, J. K., Hellemann, G., & Green, M. F. (2012). Factor structure of emotional intelligence in schizophrenia. *Schizophrenia Research, 139*(1–3), 78–81. https://doi.org/10.1016/j.schres.2012.04.013.

Lindenmayer, J. P., McGurk, S. R., Khan, A., Kaushik, S., Thanju, A., Hoffman, L., ... Herrmann, E. (2013). Improving social cognition in schizophrenia: a pilot intervention combining computerized social cognition training with cognitive remediation. *Schizophrenia Bulletin, 39*(3), 507–517. https://doi.org/10.1093/schbul/sbs120.

Lysaker, P. H., & Dimaggio, G. (2014). Metacognitive capacities for reflection in schizophrenia: implications for developing treatments. *Schizophrenia Bulletin, 40*(3), 487–491. https://doi.org/10.1093/schbul/sbu038.

Mancuso, F., Horan, W. P., Kern, R. S., & Green, M. F. (2011). Social cognition in psychosis: multidimensional structure, clinical correlates, and relationship with functional outcome. *Schizophrenia Research, 125*(2–3), 143–151. https://doi.org/10.1016/j.schres.2010.11.007.

Marjoram, D., Miller, P., McIntosh, A. M., Cunningham Owens, D. G., Johnstone, E. C., & Lawrie, S. (2006). A neuropsychological investigation into 'Theory of Mind' and enhanced risk of schizophrenia. *Psychiatry Research, 144*(1), 29–37. https://doi.org/10.1016/j.psychres.2006.01.008.

Mazza, M., Lucci, G., Pacitti, F., Pino, M. C., Mariano, M., Casacchia, M., & Roncone, R. (2010). Could schizophrenic subjects improve their social cognition abilities only with observation and imitation of social situations? *Neuropsychological Rehabilitation, 20*(5), 675–703. https://doi.org/10.1080/09602011.2010.486284.

McClure, M. M., Bowie, C. R., Patterson, T. L., Heaton, R. K., Weaver, C., Anderson, H., & Harvey, P. D. (2007). Correlations of functional capacity and neuropsychological performance in older patients with schizophrenia: evidence for specificity of relationships? *Schizophrenia Research, 89*(1–3), 330–338. https://doi.org/10.1016/j.schres.2006.07.024.

McDonald, S., Flanagan, S., Rollins, J., & Kinch, J. (2003). TASIT: a new clinical tool for assessing social perception after traumatic brain injury. *The Journal of Head Trauma Rehabilitation, 18*(3), 219–238.

McKibbin, C., Patterson, T. L., & Jeste, D. V. (2004). Assessing disability in older patients with schizophrenia: results from the WHODAS-II. *The Journal of Nervous and Mental Disease, 192*(6), 405–413.

Mehta, U. M., Thirthalli, J., Subbakrishna, D. K., Gangadhar, B. N., Eack, S. M., & Keshavan, M. S. (2013). Social and neuro-cognition as distinct cognitive factors in schizophrenia: a systematic review. *Schizophrenia Research, 148*(1–3), 3–11.

Nuechterlein, K. H., Green, M. F., Kern, R. S., Baade, L. E., Barch, D. M., Cohen, J. D., ... Marder, S. R. (2008). The MATRICS Consensus Cognitive Battery, part 1: test selection, reliability, and validity. *The American Journal of Psychiatry, 165*(2), 203–213. https://doi.org/10.1176/appi.ajp.2007.07010042.

Nunnally, J. (1967). *Psychometric theory*. New York: McGraw-Hill.

Ospina, L. H., Nitzburg, G. C., Shanahan, M., Perez-Rodriguez, M. M., Larsen, E., Latifoglu, A., & Burdick, K. E. (2018). Social cognition moderates the relationship between neurocognition and community functioning in bipolar disorder. *Journal of Affective Disorders, 235*, 7–14.

Patterson, T. L., Moscona, S., McKibbin, C. L., Davidson, K., & Jeste, D. V. (2001). Social skills performance assessment among older patients with schizophrenia. *Schizophrenia Research, 48*(2–3), 351–360.

Patterson, T. L., Semple, S. J., Shaw, W. S., Halpain, M., Moscona, S., Grant, I., & Jeste, D. V. (1997). Self-reported social functioning among older patients with schizophrenia. *Schizophrenia Research, 27*(2–3), 199–210. https://doi.org/10.1016/S0920-9964(97)00078-9.

Penn, D., Addington, J., & Pinkham, A. (2006). Social cognitive impairments. In T. S. Stroup, J. A. Lieberman, & D. O. Perkins (Eds.), *The American Psychiatric Publishing textbook of schizophrenia* (pp. 261–274). Arlington, VA: American Psychiatric Publishing, Inc.

Penn, D. L., Ritchie, M., Francis, J., Combs, D., & Martin, J. (2002). Social perception in schizophrenia: the role of context. *Psychiatry Research, 109*(2), 149–159.

Peterson, E., & Miller, S. F. (2012). The eyes test as a measure of individual differences: how much of the variance reflects verbal IQ? *Frontiers in Psychology, 3*, 220. https://doi.org/10.3389/fpsyg.2012.00220.

Pinkham, A. E., Harvey, P. D., & Penn, D. L. (2018). Social cognition psychometric evaluation: results of the final validation study. *Schizophrenia Bulletin, 44*(4), 737–748. https://doi.org/10.1093/schbul/sbx117.

Pinkham, A. E., & Penn, D. L. (2006). Neurocognitive and social cognitive predictors of interpersonal skill in schizophrenia. *Psychiatry Research, 143*(2–3), 167–178. https://doi.org/10.1016/j.psychres.2005.09.005.

Pinkham, A. E., Penn, D. L., Green, M. F., Buck, B., Healey, K., & Harvey, P. D. (2014). The social cognition psychometric evaluation study: results of the expert survey and RAND panel. *Schizophrenia Bulletin, 40*(4), 813–823. https://doi.org/10.1093/schbul/sbt081.

Pinkham, A. E., Penn, D. L., Green, M. F., & Harvey, P. D. (2016). Social cognition psychometric evaluation: results of the initial psychometric study. *Schizophrenia Bulletin, 42*(2), 494–504. https://doi.org/10.1093/schbul/sbv056.

Roberts, D. L., & Penn, D. L. (2009). Social cognition and interaction training (SCIT) for outpatients with schizophrenia: a preliminary study. *Psychiatry Research, 166*(2–3), 141–147. https://doi.org/10.1016/j.psychres.2008.02.007.

Robertson, B. R., Prestia, D., Twamley, E. W., Patterson, T. L., Bowie, C. R., & Harvey, P. D. (2014). Social competence versus negative symptoms as predictors of real world social functioning in schizophrenia. *Schizophrenia Research, 160*, 136–141. https://doi.org/10.1016/j.schres.2014.10.037.

Rosset, E. (2008). It's no accident: our bias for intentional explanations. *Cognition, 108*(3), 771–780.

Sabbag, S., Twamley, E. M., Vella, L., Heaton, R. K., Patterson, T. L., & Harvey, P. D. (2011). Assessing everyday functioning in schizophrenia: not all informants seem equally informative. *Schizophrenia Research, 131*(1–3), 250–255. https://doi.org/10.1016/j.schres.2011.05.003.

Sachs, G., Winklbaur, B., Jagsch, R., Lasser, I., Kryspin-Exner, I., Frommann, N., & Wolwer, W. (2012). Training of affect recognition (TAR) in schizophrenia–impact on functional outcome. *Schizophrenia Research, 138*(2–3), 262–267. https://doi.org/10.1016/j.schres.2012.03.005.

Sayers, M. D., Bellack, A. S., Wade, J. H., Bennett, M. E., & Fong, P. (1995). An empirical method for assessing social problem solving in schizophrenia. *Behavior Modification, 19*(3), 267–289. https://doi.org/10.1177/01454455950193001.

Sergi, M. J., & Green, M. F. (2003). Social perception and early visual processing in schizophrenia. *Schizophrenia Research, 59*(2–3), 233–241.

Sergi, M. J., Rassovsky, Y., Widmark, C., Reist, C., Erhart, S., Braff, D. L., ... Green, M. F. (2007). Social cognition in schizophrenia: relationships with neurocognition and negative symptoms. *Schizophrenia Research, 90*(1–3), 316–324.

Shamay-Tsoory, S. G. (2011). The neural bases for empathy. *The Neuroscientist, 17*(1), 18–24. https://doi.org/10.1177/1073858410379268.

Silberstein, J., Pinkham, A., Penn, D. L., & Harvey, P. D. (2018). Self assessment of social cognitive ability in schizophrenia: association with social cognitive test performance, informant assessments of social cognitive ability, and everyday outcomes. *Schizophrenia Research, 199*, 75–82.

Tas, C., Danaci, A. E., Cubukcuoglu, Z., & Brune, M. (2012). Impact of family involvement on social cognition training in clinically stable outpatients with schizophrenia—a randomized pilot study. *Psychiatry Research, 195*(1–2), 32–38. https://doi.org/10.1016/j.psychres.2011.07.031.

Toomey, R., Schuldberg, D., Corrigan, P., & Green, M. F. (2002). Nonverbal social perception and symptomatology in schizophrenia. *Schizophrenia Research, 53*(1–2), 83–91.

Ventura, J., Cienfuegos, A., Boxer, O., & Bilder, R. (2008). Clinical global impression of cognition in schizophrenia (CGI-CogS): reliability and validity of a co-primary measure of cognition. *Schizophrenia Research, 106*(1), 59–69. https://doi.org/10.1016/j.schres.2007.07.025.

Ventura, J., Reise, S. P., Keefe, R. S., Baade, L. E., Gold, J. M., Green, M. F., ... Bilder, R. M. (2010). The Cognitive Assessment Interview (CAI): development and validation of an empirically derived, brief interview-based measure of cognition. *Schizophrenia Research, 121*(1–3), 24–31. https://doi.org/10.1016/j.schres.2010.04.016.

Versmissen, D., Janssen, I., Myin-Germeys, I., Mengelers, R., Campo, J. A., van Os, J., & Krabbendam, L. (2008). Evidence for a relationship between mentalising deficits and paranoia over the psychosis continuum. *Schizophrenia Research, 99*(1–3), 103–110. https://doi.org/10.1016/j.schres.2007.09.024.

Wechsler, D. (1999). *Wechsler abbreviated scale of intelligence*. Psychological Corporation.

Yager, J. A., & Ehmann, T. S. (2006). Untangling social function and social cognition: a review of concepts and measurement. *Psychiatry, 69*(1), 47–68. https://doi.org/10.1521/psyc.2006.69.1.47.

CHAPTER TEN

# Improving ecological validity in research on social cognition

**Karlijn Hermans*[,a], Robin Achterhof*[,a], Inez Myin-Germeys*, Zuzana Kasanova*, Olivia Kirtley*, Maude Schneider*[,†]**

*KU Leuven, Department of Neurosciences, Research Group Psychiatry, Center for Contextual Psychiatry, Leuven, Belgium

[†]University of Geneva, Developmental Imaging and Psychopathology Lab, Department of Psychiatry, Faculty of Medicine, Geneva, Switzerland

Social cognition has been conceptualized as a set of psychological processes involved in social behavior and social interaction. It allows an individual to process social information within an interaction by identifying, interpreting, and reacting to others' thoughts and emotions. Traditional research on social cognition has primarily used cognitive tasks in laboratory settings to investigate distinct aspects of social cognition, mainly emotion perception, Theory of Mind (ToM), and attributional style (Penn, Sanna, & Roberts, 2008). Research in patients with psychosis has shown a clear deficit in social cognition across its multiple domains (Savla, Vella, Armstrong, Penn, & Twamley, 2013). It has also been hypothesized that the traditional concept of social cognition plays a role in the etiology of both positive and negative psychotic symptoms, due to the assumed impact social cognition has on how people interpret and make sense of their social world. For example, experiences of paranoia have been related to deficits in emotion recognition and attribution biases, that is, misattributing neutral stimuli to be negative (Lahera et al., 2015; Premkumar et al., 2008). Studies on negative symptoms in psychosis have mainly found links to emotion perception and processing, which is linked to social behavior. For example, one study found that 39% of variance in asocial behavior (such as social isolation and social anhedonia) can be explained by social cognition variables (Lincoln, Mehl, Kesting, & Rief, 2011).

The relevance of the social cognition concept lies in the tacit assumption that adequate social cognition is required for successful social functioning. For patients with psychosis, the concept is meaningful insofar as it explains

[a] These authors contributed equally to this work

*Social Cognition in Psychosis*
https://doi.org/10.1016/B978-0-12-815315-4.00010-0

social impairments in their day-to-day lives. In other words, social cognition necessitates a degree of ecological validity. This can be investigated by assessing its link with social functioning. The capacity to adapt one's behavior to changing social contingencies and cues, that is, social cognition, is thought to be essential for optimal social functioning in general, and by extension, an important indicator of mental health (Green et al., 2008; Green, Horan, & Lee, 2015). Unequivocally, psychosis is associated with social impairments. When assessed in terms of family situation and employment, prevalent deficits are observed in participants with psychotic disorders compared with controls (Bebbington et al., 2005). Furthermore, higher rates of functional disability were reported in patients with psychosis than in controls (Green et al., 2008, 2015), such as diminished social connections, lower employment rates, and fewer independent living situations (Green et al., 2018). These studies indicate that social impairments are crucial targets for prevention and intervention. However, it is not evident that social cognition deficits actually drive these impairments. A complicating factor in the assessment of the ecological validity of social cognition lies in the ambiguity of the definition of social functioning. Despite being a much-researched construct, there is a lack of consensus on the definition of social functioning. Definitions are typically used interchangeably across research domains, employing different perspectives to describe social functioning (Brissos, Molodynski, Videira Dias, & Luisa Figueira, 2011; Burns & Patrick, 2007; Kakela et al., 2014; Priebe, 2007). This ranges from objective assessment of the frequency of engagement in social interaction and fulfillment of social roles (Social Functioning Scale [SFS], Birchwood, Smith, Cochrane, Wetton, & Copestake, 1990; Global Assessment Scale [GAS], Endicott, Spitzer, Fleiss, & Cohen, 1976), to measuring subjective appraisals (feelings and thoughts), or an evaluation of social situations, such as social stress. Many authors argue that the concept of social functioning should be approached from both an objective and a subjective perspective (Brissos et al., 2011; Burns & Patrick, 2007; Schneider et al., 2017). In practice, however, social functioning has predominantly been described with objective, or primarily observer-based indices, instead of self-report. In this chapter, we will offer evidence that a subjective, individual perspective on social functioning is ideally suited to advance our understanding of social impairments and the role of social cognition.

For optimal assessment of social impairments and understanding of the factors underlying social impairments, the link between social cognition and real-life social functioning should be firmly established. Currently, however, the evidence is equivocal and, despite thorough general psychometric validations of social cognition tasks (Pinkham, Harvey, & Penn, 2018; Pinkham, Penn, Green, & Harvey, 2016), there is a dearth of evidence specifically investigating this link. This, therefore, raises the fundamental question of whether social cognition is a prerequisite for social functioning. In light of this question, this chapter discusses the link between social cognition and traditional social functioning, before discussing a broader framework of ecological validity of social cognition. The chapter then focuses on two approaches with high ecological validity to investigate social cognition and social functioning in a wider sense—virtual reality (VR) and the experience sampling methododology (ESM)—and finally, discusses considerations and implications for future research and clinical practice.

## Social cognition: A prerequisite for social functioning?

Individual studies have reported on correlations between social functioning and specific domains of social cognition, such as ToM (Sprong, Schothorst, Vos, Hox, & Van Engeland, 2007), emotion recognition (Henry, Cowan, Lee, & Sachdev, 2015; Hooker & Park, 2002), emotion perception (Irani, Seligman, Kamath, Kohler, & Gur, 2012), and attribution bias (Lahera et al., 2015). Overall, the link between social cognition and social functioning has mainly been based on correlational studies, which precludes conclusions about directionality (Couture, Penn, & Roberts, 2006; Fett et al., 2011). Moreover, a meta-analysis by Fett et al. (2011) shows that only 16% of variance in social functioning can be explained by social cognition. This percentage is surprisingly small, given the widely held assumption that social cognition is a strong determinant of real-life social functioning (Green et al., 2015). A recent study also did not observe any significant association between social cognition and self-reported social functioning (Simons, Bartels-Velthuis, & Pijnenborg, 2016). One possible explanation for these inconsistencies is that the relationship between social cognition and social functioning is mainly accounted for by other factors. For example, one study showed that the association between measures of social cognition and social functioning disappeared when neurocognition was controlled for (Woolverton, Bell, Moe, Harrison-Monroe, & Breitborde, 2017). Based on these findings, it is argued that, although a link between traditionally measured social functioning and social cognition was established, the lack of a subjective, ecologically valid perspective prevents it from being clinically and mechanistically relevant.

Based on the assumption that deficits in social cognition strongly contribute to social functioning impairments, and play a role in the emergence of positive and negative symptoms, several interventions targeting social cognition have been developed. These aim to improve social functioning and reduce symptoms in patients with psychosis. In particular, several programs of cognitive remediation therapy focus on social cognition (e.g., CIRCuiTS; Reeder et al., 2016; Wykes et al., 2018), and include daily functioning as one of the main outcome variables (Wykes, Huddy, Cellard, McGurk, & Czobor, 2011). Although an initial meta-analysis showed a medium effect size in different cognitive domains (McGurk, Twamley, Sitzer, McHugo, & Mueser, 2007), a more methodologically rigorous meta-analysis showed a small-to-medium effect size of cognitive remediation therapy on functioning outcomes (Wykes et al., 2011). The authors concluded that the improved or newly acquired cognitive skills should be complemented with strategies for their implementation in daily life. That is, the authors pointed out that personalization for specific participants and their environments was critical for the effectiveness of the treatment (Fett et al., 2011). These and other social cognitive treatments have, thus far, provided limited evidence for a direct relationship between social cognition deficits and social functioning in daily life (Fiszdon & Reddy, 2012).

In sum, the basic assumption of theories of social cognition is that these deficits result in alterations in social behavior, thereby having a negative impact on social functioning. The main empirical support for this association, however, comes from laboratory studies that show weak correlations between social cognition and social functioning. Additionally, treatments specifically aimed at improving social cognition do not show large improvements in daily social functioning. Therefore, it is uncertain whether social cognition is, in fact, a prerequisite for social functioning. Other authors have gone beyond the traditional idea that social cognition drives impairments in social functioning. They focused on alternative factors driving social functioning that should be investigated within this link as well, such as effort allocation, prediction error processing, expected value-driven learning, and action selection (Barch, Pagliaccio, & Luking, 2016). An alternative way of conceptualizing the link between social cognition and social functioning, however, is the Social Interaction First (SIF) hypothesis, in which social cognition has been posited as the result of more basic social interactive abilities (Schneider, Myin, & Myin-Germeys, n.d.). This is in contrast to the Social Cognition First (SCF) hypothesis, which assumes that social cognition is a prerequisite for social functioning (Carruthers, 2009; Schonherr, 2017), and therefore questions whether social cognition explains and predicts social difficulties in patients with psychosis. Particularly in clinical research, an alternative line of study would be to focus on mechanisms underlying social interactive abilities in daily life. The next section will further discuss the concept of ecological validity with regard to social cognition.

## Ecological validity

Cognitive psychology has seen a surge of attention devoted to increasing ecological validity of cognitive concepts in general (Burgess et al., 2006), and social cognition in particular (Zaki & Ochsner, 2009). Conceptually, Burgess et al. (2006) advocated that the focus of research should shift from a "construct-driven" to a "function-led" approach in order to improve ecological validity. From this perspective, the cognitive construct (e.g., emotion perception) is not the central focus of attention, but rather its function in observable behavior (e.g., being empathic to someone). Related to this view are the concepts of *representativeness* and *generalizability*, which have been proposed as the requirements for experimental tasks to be considered ecologically valid (analogous to the terms *verisimilitude* and *veridicality*; Franzen & Wilhelm, 1996). Representativeness refers to the similarity in content and experience of an experimental task with the real world. For example, a paradigm in which participants have to perform a complex shopping task is more representative of potential limitations in the real world than, for example, a Stroop task (MacLeod, 1991; Stroop, 1935). Generalizability, on the other hand, refers to how well an experimental task is actually predictive of its associated behavior or functioning in real life. A generalizable task is not necessarily "real-world"-like. For example, the result of a pen and paper driving test could be more predictive of actual driving capacity than performance on a highly realistic racing video game. In this example, the game might be more representative, but the test would be more generalizable.

As highlighted in the previous section of the chapter, several studies observed no significant association between tasks of social cognition and actual daily-life social functioning (Simons et al., 2016; Woolverton et al., 2017), which provides arguments against the generalizability of such tasks. In addition, social cognition and its related concepts have mainly been investigated and developed in controlled laboratory settings. Because traditional social cognition tasks (involving for instance emotion recognition of photographed faces, or identifying beliefs of characters in written scenarios) are highly experimental in nature, they also do not meet the representativeness requirement. The link between social cognition concepts and real-world social behavior is therefore often indirect and artificial. Compared with real-life social situations, experimental tasks designed to assess aspects of social cognition tend to lack at least three real-world attributes of interacting partners: (1) multimodality (visual, semantic, and prosodic elements); (2) dynamic change; and (3) a context of previous experiences with that partner (Zaki & Ochsner, 2009). In experimental tasks of social cognition, the social situation is thus usually discrete, static, and isolated. Additionally, the interactive nature of real-life social processes may involve qualities that extend beyond the sum of individual social capacities. Within the field of social cognition, two main approaches can be identified, a "third-person" and a "first-person" approach (Keysers & Gazzola, 2007; Schilbach et al., 2013; Zaki & Ochsner, 2009). In the third-person approach, one is a somewhat detached observer of the other, inferring at a cognitive level about another person's mind (as in ToM). In the first-person approach, however, emotions or mental states of the other are more directly experienced, as if they were one's own (as in emotion recognition). Neuroimaging studies based on either the "third-person" or the "first-person" approach have found mostly nonoverlapping neural networks, even though they both aim to target and explain the same concepts (i.e., social cognition and social behavior). This implies that these traditional views of social cognition have insufficient explanatory value, and that an alternative paradigm is necessary (Keysers & Gazzola, 2007; Pfeiffer, Timmermans, Vogeley, Frith, & Schilbach, 2013; Schilbach et al., 2013; Zaki & Ochsner, 2009). As argued by Schilbach et al. (2013), social cognition research would benefit from a "second-person" approach, where the interactive element of social behavior is taken into account. Their perspective builds upon the idea of "embodied cognition," in which an individual is not just a spectator of their environment and the people in it, but is in a state of constant interaction. Social cognition then arises when someone is both emotionally engaged and interacting with another person, as opposed to being detached and merely observant.

In sum, by not meeting the requirements of representativeness and generalizability, traditional social cognition research has limited ecological validity. In an attempt to address this issue, some investigations into social cognition have employed more naturalistic assessment methods. For example, participants were asked to infer the thoughts and feelings of an interactive partner during a secretly videotaped interaction from moments before (Ickes, 1997; Ickes, Stinson, Bissonnette, & Garcia, 1990). The accuracy of these assessments was obtained by independently asking the interactive partner to provide an account of their thoughts and feelings. This method is, however, still a somewhat indirect way of tapping into everyday social interactions. For one, participants have to recall their (meta-)thoughts, feelings, and inferences at a later point, which makes them prone to recall bias. Additionally, the social situations that can be assessed using these sorts of paradigms still take place in laboratory settings, leaving the question of generalizability to real-world situations unanswered.

Alternative methods for assessing social cognition and social functioning that meet the conditions of representativeness and generalizability are available. In the next sections, we will review existing research using VR and the ESM in patients with psychosis. In VR, virtual environments are becoming increasingly similar to real life, invoking experiences that are more representative of real-life situations. In ESM, on the other hand, measurements are drawn directly from daily life, thereby allowing for the assessment of the generalizability of findings obtained in experimental settings. Accordingly, we will discuss the contribution of these methods to a more naturalistic understanding of social cognition.

## Virtual reality

In the advent of the technological revolution of the past two decades, VR techniques have become increasingly widespread, available, and sophisticated (Valmaggia, 2017). VR is typically defined as the presentation of a computer-generated, three-dimensional environment, in which the user receives visual (and sometimes auditory or haptic) input that is directly influenced by that user's movements[1] (Eichenberg & Wolters, 2012). Different types of VR technology are currently available, with the most well-known application being a head-mounted display that links head movements in real time to the virtual environment that it presents. The aim of VR is to present a virtual world that is immersive and is experienced as being authentic.

As such, VR has also been adapted for use in the field of mental health, with a specific focus on the assessment of social processes (Brunet-Gouet, Oker, Martin, Grynszpan, & Jackson, 2016; Freeman et al., 2017; Parsons, 2015). As discussed, social cognition and related social processes

[1]Other definitions of VR include simpler virtual environments (as in e.g., Froese, Iizuka, & Ikegami, 2015), but in this chapter only immersive VR applications are discussed.

are difficult to assess with traditional experimental tasks, because such tasks usually do not take into account the multimodal, contextual, and dynamic aspects of social interaction. Furthermore, through the standardized presentation of a virtual environment, and the possibility of controlling and manipulating every aspect of that environment, it becomes possible to differentiate between objective context characteristics and personal interpretations of that context. Following the "function-led" approach to ecologically validity proposed by Burgess et al. (2006), VR most directly meets the criterion of representativeness, defined as the extent to which a task relates to a situation outside of the lab in form and environment.

VR techniques seem to be acceptable and safe to use for people with psychotic symptoms (for further discussion of this, see Rus-Calafell, Garety, Sason, Craig, & Valmaggia, 2017). Furthermore, the feasibility of VR technology to assess aspects of social cognition in individuals with psychotic symptoms has also been tested in a number of studies. In an initial feasibility study by Ku et al. (2003), eleven patients with schizophrenia engaged in a VR task by controlling a joystick, which enabled them to move toward or away from another person embodied as an avatar. The VR was not immersive, as the virtual environment was projected on a screen. The avatar had breathing motions and blinked, but showed no facial emotions, did not make any gesture, and had a very basic appearance. Outcome variables assessed were the virtual interpersonal distance, and verbal response time in a simple scripted conversation. The authors found that interpersonal distance was negatively related to the severity of negative symptoms. Head-mounted displays were used to present the same virtual environment in a follow-up study (Park et al., 2009). Interpersonal distance to the avatar was found to be significantly larger for 30 patients with schizophrenia than for 30 controls. In another study from the same research group (Kim et al., 2007), several aspects of social perception were investigated, although a clear description of the VR procedure is lacking. Overall, significant differences between schizophrenia patients and controls were observed on a number of social perception variables, involving emotion recognition and interpretation of environmental cues. Two other studies investigated the feasibility of VR to assess emotion recognition through a task where participants with schizophrenia or controls had to indicate the basic emotions of virtual faces (Dyck, Winbeck, Leiberg, Chen, & Mathiak, 2010; Gutiérrez-Maldonado, Rus-Calafell, Márquezrejón, & Ribas-Sabaté, 2012). These authors found similar emotion recognition impairments as those reported in studies using photographs of real faces.

Although the preceding studies employed rather unsophisticated VR techniques, virtual environments have quickly become more immersive and convincing. However, realistic and immersive VR does not necessarily imply high ecological validity. As noted by Parsons (2015), many early VR studies incorporated traditional neurocognitive tests (e.g., a Stroop task) into a virtual environment (e.g., Henry, Joyal, & Nolin, 2012). Similarly, the emotion recognition and social perception VR studies focused on specific components of social cognition (Dyck et al., 2010; Gutiérrez-Maldonado et al., 2012; Kim et al., 2007; Ku et al., 2003; Park et al., 2009). Because such studies target specific cognitive modalities, they do not enrich our understanding of how such modalities might be involved in the complexities and dynamics of actual behavior in daily life.

A function-led approach would be better suited to gain more insight into the underpinnings of social processes that have actual relevance for day-to-day interactions. Through focusing on behaviors and the natural context in which they occur, more information can be gathered about the entire spectrum of factors involved in function versus dysfunction. An example is a study by Greenwood et al. (2016), in which 43 participants with schizophrenia engaged in both a VR and real-life supermarket shopping task. The participants also completed a comprehensive test battery of traditional neurocognitive tests aimed at capturing the wide range of cognitive performance involved in daily functioning. The authors found that after taking into account these cognitive measures, the virtual shopping task still significantly predicted real-life performance. This implies that the abilities needed to perform functions in everyday life exceed the joint sum of traditionally proposed cognitive modalities, and that VR techniques are able to capture these additional capacities.

Within a function-led approach, it is necessary to select the functions, or behaviors, to target. With the aim of finding mechanisms underlying the social aspects of psychosis, it might be worthwhile to assess relevant symptoms, as expressed in a naturalistic environment. As opposed to, for example, assessing a context-free impairment in emotion recognition, it may be more informative to track (mal)adjustment in a realistic situation in which a subject may be required to understand another person's feelings or intentions. With respect to studies on social dysfunction in psychosis, this would imply a shift in focus toward assessing psychotic symptoms in a naturalistic social space. Most VR studies targeting mechanisms of psychosis do so by assessing paranoid experiences in representative and appropriate virtual environments (Valmaggia, Day, & Rus-Calafell, 2016). For example, some studies investigated paranoia in naturalistic virtual environments (such as a train ride, or a bar) with ambiguous social cues (e.g., Freeman et al., 2008; Veling, Pot-Kolder, Counotte, Van Os, & Van Der Gaag, 2016). Such studies have the benefit of both full control over the experimental set-up and high real-life representation. In this way, they are able to assess the relation between known contextual and social factors and paranoid experiences with high ecological validity.

One limitation of contemporary VR is that it presents virtual avatars with which the user has had no previous contact. As such, the embeddedness of a social interaction in previous experiences (a key feature of real-life social situations) is not considered in VR. An additional important point is that realistic social interactions are not yet possible to render in VR (Pan & Hamilton, 2018). In the first place, this is because avatars that have been developed so far are not truly life-like, neither in appearance nor in behavior. Second, although participants in VR studies interpret potential emotions and thoughts of virtual avatars, they never actually interact with them. This is an especially glaring limitation from a "second-person," or embodied, perspective of social cognition, which states that actual interaction is key to understanding social cognition, over and above any individual capacities.

Technological advances in artificial intelligence might provide a potential solution to these issues, by developing truly life-like virtual avatars. Still, it would be necessary to meet the generalizability criterion of ecological validity, and to assess whether results from VR studies do actually translate to real life. In order to learn more about social cognition or social functioning in the context of daily life, a method is needed that is able to most directly assess daily-life processes. A particularly well-suited candidate for achieving this is ESM.

## Experience sampling methodology

ESM, or ecological momentary assessment (EMA) (Shiffman, Stone, & Hufford, 2008), is another methodology that has been increasingly used because of its high ecological validity (Hektner, Schmidt, & Csikszentmihalyi, 2007; Myin-Germeys et al., 2018). Via booklets, dedicated devices, or smartphone applications, this structured diary technique prompts participants to complete brief questionnaires during the day for several consecutive days in order to map experience and behavior in real-world and individual contexts. These brief questionnaires include questions about individuals' current social context (i.e., "Right now, who is with you?") and their appraisals of this context in the moment (i.e., "Right now, I would rather be alone"). This method avoids recall bias, as participants are requested to refer to their thoughts and feelings as they occur in the moment. While laboratory settings typically use actors or standardized expressive faces to add a social component (Green et al., 2015), ESM inquires about the actual social contexts of participants, which can differ throughout the day and between individuals. Additionally, the naturalistic setting of ESM allows the assessment of multimodal information about emotions across *different* physical contexts (Myin-Germeys et al., 2009), instead of an isolated assessment of either facial expression or auditory cues in the *same* physical context (Henry et al., 2015).

Although several links between laboratory measures of social cognition and daily-life measures of social functioning have been implied, few studies have directly compared them. Janssens et al. (2012) compared a traditional computer task assessing emotion recognition with daily-life reports on social context and participants' appraisal of this situation as an indicator of social functioning. No association was found between performance on the emotion recognition task and different aspects of social functioning in daily life (amount of time spent in these social contexts and the appraisals of these contexts). The authors argued that emotion recognition tasks in laboratory settings do not capture the complexity of daily life, which is in line with aforementioned critiques on these types of laboratory tasks (Zaki & Ochsner, 2009). Furthermore, Schneider et al. (n.d.) have not found any significant association between traditional ToM tasks and social functioning as measured in daily life. They argue that more basic interactive abilities than social cognition are at the core of social interaction, providing evidence for the SIF hypothesis, as opposed to the SCF hypothesis, as referred to earlier. Although comparative research is scarce, these few studies that have directly compared social cognition tasks in the lab with social functioning in daily life show that they are poorly related. An alternative view would therefore be to focus on mechanisms underlying social interactive abilities and measure these directly in daily life.

Although objective measurements with the ESM (e.g., amount of time spent alone) provide finer-grained information about social situations throughout the day than questionnaire measures, they do not provide insight regarding potential mechanisms of social impairment. In order to study how social interactions are altered in psychosis, it is crucial to explore what drives social behavior, and which thoughts and emotions are associated with this behavior. For example, current explanatory models of negative symptoms highlight the importance of impaired effort-based decision making, resulting in individuals with psychosis to engage less in rewarding situations (Barch et al., 2016). This has primarily been investigated with traditional tasks in laboratory settings, such as the spatial orienting task (i.e., motivated game measuring response time to rewarding or punishing simple circular targets; Derryberry & Reed, 1997) used in a study by Vrijen, Hartman, and Oldehinkel (2018). In order to map subtle states and changes in behavior though, more insight could be gained from studies exploring subjective indicators in daily life.

In mapping transient affective states (Myin-Germeys et al., 2009, 2018; Palmier-Claus et al., 2011), ESM has shed new light on the conceptualization of negative symptoms in psychosis, especially asociality and social anhedonia (i.e., diminished pleasure experienced in company). For example, momentary experiences of positive affect in company seemed intact in patients compared with controls (Oorschot et al., 2013). Further, while some studies have found a reduced interest for social contact in individuals scoring high on a social anhedonia scale (Brown, Silvia, Myin-Germeys, & Kwapil, 2007; Kwapil, 1998), recent work has challenged and amended the concept of asociality (Kasanova, Oorschot, & Myin-Germeys, 2018). The latter showed that patients who, as a group tended to be unemployed and single, were less often engaged in social contexts corresponding to work and family obligations (i.e., structured goal-directed contexts), but equally often in unstructured social situations such as conversations and visits. These studies therefore challenge the idea that patients have limited interest in social interactions in general (true asociality), but rather argue to look into specific social contexts and appraisals thereof in daily life.

ESM studies have also explored several mechanisms related to the development of positive psychotic symptoms in daily life. One of them is enhanced threat anticipation, a cognitive-affective bias toward threat, linked to attribution bias and information processing (Freeman et al., 2013). Reininghaus, Kempton, et al. (2016) found support for this mechanism in daily life, showing an association between threat anticipation (likelihood of personal negative events in the future) and intensity of psychotic experiences. This study did not investigate whether threat anticipation was mostly related to other individuals, so direct inferences on the social component could not be made. However, enhanced threat anticipation could inform the way individuals appraise their (social) environment. This would therefore be a fruitful avenue for future ESM studies in elucidating how positive symptoms, such as paranoia, can develop from threatening social situations.

ESM studies focusing on the mechanisms implicated in the development of psychotic symptoms have shown that a fine-grained mapping of everyday social life can provide detailed, individual information on social functioning, which is highly relevant for clinical research.

In sum, we have shown that the link between social cognition and real-life social functioning in psychosis is not as strong as would be expected, given traditional views of social cognition. We highlighted several potential explanations for the difference between findings from traditional social cognitive tasks in a laboratory environment and findings from assessments of social functioning in daily life. From a methodological point of view, we argue that laboratory tasks lack ecological validity, whereas ESM uses individuals' actual real-world reports with personal connections and backgrounds in different physical environments. However, ESM also has some methodological limitations that should be considered when using this method for assessment of social functioning. First, there is currently no consistency in the ESM items that are used across research groups. As a result, a different operationalization has been used across ESM studies to measure the same concepts (e.g., social functioning). Second, appraisals are often addressed with a single item (i.e., "I like this company"), whereas questionnaires usually group together several items assessing the same construct to improve validity of the measure. Both aspects could be improved by conducting methodological studies with the aim to establish a validated composition of items that correlates with self-report questionnaires. Nevertheless, through individual moment-to-moment measurements, ESM still has many added benefits. For example, it allows for more reliable individual scores on actual social interactive capacities, and for variability measures throughout days or weeks. Last, we have shown that ESM can provide new insights into how we can conceptualize symptoms in psychosis, and elucidate their occurrence in daily life.

## Conclusion and implications

In this chapter, we discussed the usefulness of the concept of social cognition within the context of daily life, with a specific focus on psychotic disorder. In order to do this, the ecological validity of traditional social cognitive tasks has been compared with two methods having high ecological validity. We argued that clinical research would benefit most from focusing on social functioning, which has shown to be weakly correlated with social cognition. Hence, VR and ESM have been proposed as ecologically valid alternatives to study social impairments in psychosis. We have also raised the question whether social cognition is a prerequisite for social functioning. The few studies comparing social cognition with social functioning in daily life showed weak correlations between the two ways of measurement, yet only a few studies are available. If further investigations fail to support the assumed link between social cognition and social functioning, a paradigm shift would be needed. In this regard, an interesting alternative would be to explore the SIF approach.

Moreover, VR and ESM can be used to target identified mechanisms. Research has started to use ecological momentary interventions (EMI) (Myin-Germeys, Klippel, Steinhart, & Reininghaus, 2016; Reininghaus, Depp, & Myin-Germeys, 2016) to identify symptoms or patterns in real time and offer individually tailored feedback to change thoughts or behavior. An example in individuals with anhedonia was conducted by Van Roekel et al. (2017). They used ESM data to provide individually tailored lifestyle advice, which successfully increased positive affect and pleasure at 1-month follow-up. A promising next step would be to implement this feedback in daily life, where it could directly target real-life social interactive capacities. Further, a recent study by Pot-Kolder et al. (2018) used VR for improving daily-life social impairments in patients with psychosis. In their intervention study in 116 patients with a psychotic disorder, they used VR, combined with elements of cognitive behavioral therapy (CBT). Patients in the treatment group were exposed to social situations in VR (e.g., a supermarket), while engaging in typical CBT exercises, such as challenging and exploring suspicious thoughts, and dropping safety behaviors. Using this set-up, a significant drop in momentary paranoid ideation and momentary anxiety was observed posttreatment and at 6-month follow-ups. However, the authors observed no significant change in actual social participation defined as time spent with others (measured with ESM). Although actual interaction with avatars was restricted, this study is promising in combining ESM and VR.

There are some additional considerations with regard to the more ecologically valid methods such as VR and ESM. As mentioned, the generalizability criterion has not been assessed in most VR studies of social cognition or social functioning. Therefore, it is unclear whether disturbances found in VR actually translate to daily life. One way of addressing generalizability would be to determine whether findings found in VR studies converge with expected real-life counterparts. This was done, for example, in the supermarket task of Greenwood et al. (2016). For ESM studies, internal validity remains challenging. While ESM usually relies entirely on self-report in dynamically changing and uncontrolled environments, it is difficult to pinpoint what exactly drives results from ESM studies. One way of addressing these issues, characteristic of both VR and ESM, would be to examine whether studies employing both techniques converge or show similar patterns. For example, ESM could be an excellent candidate for checking the generalizability of VR findings. Conversely, because every aspect of the environment can be controlled in VR, hypotheses obtained through ESM studies could be replicated in a VR environment. This would be a fruitful avenue for future work, as there are no studies using this combination of methods in psychosis or any other field in mental health research.

In conclusion, we argue that VR and ESM would both be promising and necessary additions to laboratory studies assessing social cognition. Incorporating these two methods would cover a broader spectrum of components of social functioning than what has presently been covered. Finally, more research on elucidating correlations between traditional social cognition tasks and social functioning in daily life is warranted, as well as a discussion of their meaning for psychotic symptoms in daily life.

## References

Barch, D. M., Pagliaccio, D., & Luking, K. (2016). Mechanisms underlying motivational deficits in psychopathology: similarities and differences in depression and schizophrenia. *Current Topics in Behavioral Neurosciences*, *27*, 411–449. https://doi.org/10.1007/7854_2015_376.

Bebbington, P. E., Angermeyer, M., Azorin, J. M., Brugha, T., Kilian, R., Johnson, S., … EuroSC Research Group (2005). The European Schizophrenia Cohort (EuroSC)—a naturalistic prognostic and economic study. *Social Psychiatry and Psychiatric Epidemiology*, *40*(9), 707–717. https://doi.org/10.1007/s00127-005-0955-5.

Birchwood, M., Smith, J., Cochrane, R., Wetton, S., & Copestake, S. (1990). The social functioning scale—the development and validation of a new scale of social-adjustment for use in family intervention programs with schizophrenic-patients. *British Journal of Psychiatry*, *157*, 853–859. https://doi.org/10.1192/bjp.157.6.853.

Brissos, S., Molodynski, A., Videira Dias, V., & Luisa Figueira, M. (2011). The importance of measuring psychosocial functioning in schizophrenia. *Annals of General Psychiatry*, *10*(18).

Brown, L. H., Silvia, P. J., Myin-Germeys, I., & Kwapil, T. R. (2007). When the need to belong goes wrong: the expression of social anhedonia and social anxiety in daily life. *Psychological Science*, *18*(9), 778–782. https://doi.org/10.1111/j.1467-9280.2007.01978.x.

Brunet-Gouet, E., Oker, A., Martin, J. -C., Grynszpan, O., & Jackson, P. L. (2016). Editorial: advances in virtual agents and affective computing for the understanding and remediation of social cognitive disorders. *Frontiers in Human Neuroscience*, *9*, 697. https://doi.org/10.3389/fnhum.2015.00697.

Burgess, P. W., Alderman, N., Forbes, C., Costello, A., Coates, L. M. A., Dawson, D. R., … Channon, S. (2006). The case for the development and use of "ecologically valid" measures of executive function in experimental and clinical neuropsychology. *Journal of the International Neuropsychological Society*, *12*(2), 194–209. https://doi.org/10.1017/S1355617706060310.

Burns, T., & Patrick, D. (2007). Social functioning as an outcome measure in schizophrenia studies. *Acta Psychiatrica Scandinavica, 116*(6), 403–418. https://doi.org/10.1111/j.1600-0447.2007.01108.x.

Carruthers, P. (2009). How we know our own minds: the relationship between mindreading and metacognition. *Behavioral and Brain Sciences, 32*(2). https://doi.org/10.1017/S0140525x09000545 121–+.

Couture, S. M., Penn, D. L., & Roberts, D. L. (2006). The functional significance of social cognition in schizophrenia: a review. *Schizophrenia Bulletin, 32*(Suppl. 1), S44–S63. https://doi.org/10.1093/schbul/sbl029.

Derryberry, D., & Reed, M. A. (1997). Motivational and attentional components of personality. In G. Matthews (Ed.), *Cognitive science perspectives on personality and emotion* (pp. 443–473). Amsterdam: Elsevier.

Dyck, M., Winbeck, M., Leiberg, S., Chen, Y., & Mathiak, K. (2010). Virtual faces as a tool to study emotion recognition deficits in schizophrenia. *Psychiatry Research, 179*(3), 247–252. https://doi.org/10.1016/j.psychres.2009.11.004.

Eichenberg, C., & Wolters, C. (2012). Virtual realities in the treatment of mental disorders: a review of the current state of research. In *Virtual reality in psychological, medical and pedagogical applications* (pp. 35–64). London: IntechOpen Limited. https://doi.org/10.5772/50094.

Endicott, J., Spitzer, R. L., Fleiss, J. L., & Cohen, J. (1976). Global assessment scale—procedure for measuring overall severity of psychiatric disturbance. *Archives of General Psychiatry, 33*(6), 766–771.

Fett, A. K. J., Viechtbauer, W., Dominguez, M. d. G., Penn, D. L., van Os, J., & Krabbendam, L. (2011). The relationship between neurocognition and social cognition with functional outcomes in schizophrenia: a meta-analysis. *Neuroscience and Biobehavioral Reviews, 35*, 573–588.

Fiszdon, J. M., & Reddy, L. F. (2012). Review of social cognitive treatments for psychosis. *Clinical Psychology Review, 32*(8), 724–740. https://doi.org/10.1016/j.cpr.2012.09.003.

Franzen, M. D., & Wilhelm, K. L. (1996). Conceptual foundations of ecological validity in neuropsychological assessment. In *Ecological validity of neuropsychological testing* (pp. 91–112). Delray Beach, FL/England: Gr Press/St Lucie Press, Inc.

Freeman, D., Dunn, G., Fowler, D., Bebbington, P., Kuipers, E., Emsley, R., … Garety, P. (2013). Current paranoid thinking in patients with delusions: the presence of cognitive-affective biases. *Schizophrenia Bulletin, 39*(6), 1281–1287. https://doi.org/10.1093/schbul/sbs145.

Freeman, D., Gittins, M., Pugh, K., Antley, A., Slater, M., & Dunn, G. (2008). What makes one person paranoid and another person anxious? The differential prediction of social anxiety and persecutory ideation in an experimental situation. *Psychological Medicine. 38*(08). https://doi.org/10.1017/S0033291708003589.

Freeman, D., Reeve, S., Robinson, A., Ehlers, A., Clark, D., Spanlang, B., & Slater, M. (2017). Virtual reality in the assessment, understanding, and treatment of mental health disorders. *Psychological Medicine, 47*, 2393–2400.

Froese, T., Iizuka, H., & Ikegami, T. (2015). Embodied social interaction constitutes social cognition in pairs of humans: a minimalist virtual reality experiment. *Scientific Reports, 4*, 3672. https://doi.org/10.1038/srep03672.

Green, M. F., Horan, W. P., & Lee, J. (2015). Social cognition in schizophrenia. *Nature Reviews: Neuroscience, 16*(10), 620–631. https://doi.org/10.1038/nrn4005.

Green, M. F., Horan, W. P., Lee, J., McCleery, A., Reddy, L. F., & Wynn, J. K. (2018). Social disconnection in schizophrenia and the general community. *Schizophrenia Bulletin, 44*(2), 242–249. https://doi.org/10.1093/schbul/sbx082.

Green, M. F., Penn, D. L., Bentall, R., Carpenter, W. T., Gaebel, W., Gur, R. C., ... Heinssen, R. (2008). Social cognition in schizophrenia: an NIMH workshop on definitions, assessment, and research opportunities. *Schizophrenia Bulletin, 34*(6), 1211–1220. https://doi.org/10.1093/schbul/sbm145.

Greenwood, K. E., Morris, R., Smith, V., Jones, A. M., Pearman, D., & Wykes, T. (2016). Virtual shopping: a viable alternative to direct assessment of real life function? *Schizophrenia Research, 172*(1–3), 206–210. https://doi.org/10.1016/j.schres.2016.02.029.

Gutiérrez-Maldonado, J., Rus-Calafell, M., Márquezrejón, S., & Ribas-Sabaté, J. (2012). Associations between facial emotion recognition, cognition and alexithymia in patients with schizophrenia: comparison of photographic and virtual reality presentations. *Annual Review of CyberTherapy and Telemedicine, 10*, 88–92. https://doi.org/10.3233/978-1-61499-121-2-88.

Hektner, J. M., Schmidt, J. A., & Csikszentmihalyi, M. (2007). *Experience sampling method. Measuring the quality of everyday life*. Sage Publications Inc.

Henry, J. D., Cowan, D. G., Lee, T., & Sachdev, P. S. (2015). Recent trends in testing social cognition. *Current Opinion in Psychiatry, 28*(2), 133–140. https://doi.org/10.1097/YCO.0000000000000139.

Henry, M., Joyal, C. C., & Nolin, P. (2012). Development and initial assessment of a new paradigm for assessing cognitive and motor inhibition: the bimodal virtual-reality Stroop. *Journal of Neuroscience Methods, 210*(2), 125–131. https://doi.org/10.1016/j.jneumeth.2012.07.025.

Hooker, C., & Park, S. (2002). Emotion processing and its relationship to social functioning in schizophrenia patients. *Psychiatry Research, 112*(1), 41–50. Pii S0165-1781(02)00177-4. https://doi.org/10.1016/S0165-1781(02)00177-4

Ickes, W. J. (1997). Empathic accuracy. *Journal of Personality, 61*, 587–610. https://doi.org/10.1111/j.1467-6494.1993.tb00783.x.

Ickes, W., Stinson, L., Bissonnette, V., & Garcia, S. (1990). Naturalistic social cognition: empathic accuracy in mixed-sex dyads. *Journal of Personality and Social Psychology, 59*(4), 730–742. https://doi.org/10.1037/0022-3514.59.4.730.

Irani, F., Seligman, S., Kamath, V., Kohler, C., & Gur, R. C. (2012). A meta-analysis of emotion perception and functional outcomes in schizophrenia. *Schizophrenia Research, 137*(1–3), 203–211. https://doi.org/10.1016/j.schres.2012.01.023.

Janssens, M., Lataster, T., Simons, C. J., Oorschot, M., Lardinois, M., van Os, J., ... Group (2012). Emotion recognition in psychosis: no evidence for an association with real world social functioning. *Schizophrenia Research, 142*(1–3), 116–121. https://doi.org/10.1016/j.schres.2012.10.003.

Kakela, J., Panula, J., Oinas, E., Hirvonen, N., Jaaskelainen, E., & Miettunen, J. (2014). Family history of psychosis and social, occupational and global outcome in schizophrenia: a meta-analysis. *Acta Psychiatrica Scandinavica, 130*(4), 269–278. https://doi.org/10.1111/acps.12317.

Kasanova, Z., Oorschot, M., & Myin-Germeys, I. (2018). Social anhedonia and asociality in psychosis revisited. An experience sampling study. *Psychiatry Research, 270*, 375–381. https://doi.org/10.1016/j.psychres.2018.09.057.

Keysers, C., & Gazzola, V. (2007). Integrating simulation and theory of mind: from self to social cognition. *Trends in Cognitive Sciences, 11*, 194–196.

Kim, K., Kim, J. -J., Kim, J., Park, D. -E., Jang, H. J., Ku, J., ... Kim, S. I. (2007). Characteristics of social perception assessed in schizophrenia using virtual reality. *Cyber Psychology & Behavior, 10*(2), 215–219. https://doi.org/10.1089/cpb.2006.9966.

Ku, J., Cho, W., Kim, J. -J., Peled, A., Wiederhold, B. K., Wiederhold, M. D., ... Kim, S. I. (2003). A virtual environment for investigating schizophrenic patients' characteristics: assessment of cognitive and navigation ability. *Cyber Psychology & Behavior, 6*(4), 397–404. https://doi.org/10.1089/109493103322278781.

Kwapil, T. R. (1998). Social anhedonia as a predictor of the development of schizophrenia-spectrum disorders. *Journal of Abnormal Psychology, 107*(4), 558–565. https://doi.org/10.1037//0021-843x.107.4.558.

Lahera, G., Herrera, S., Reinares, M., Benito, A., Rullas, M., Gonzalez-Cases, J., & Vieta, E. (2015). Hostile attributions in bipolar disorder and schizophrenia contribute to poor social functioning. *Acta Psychiatrica Scandinavica, 131*(6), 472–482. https://doi.org/10.1111/acps.12399.

Lincoln, T. M., Mehl, S., Kesting, M. L., & Rief, W. (2011). Negative symptoms and social cognition: identifying targets for psychological interventions. *Schizophrenia Bulletin, 37* (Suppl. 2), S23–S32. https://doi.org/10.1093/schbul/sbr066.

MacLeod, C. M. (1991). Half a century of research on the Stroop effect: an integrative review. *Psychological Bulletin, 109*(2), 163–203. https://doi.org/10.1037/0033-2909.109.2.163.

McGurk, S. R., Twamley, E. W., Sitzer, D. I., McHugo, G. J., & Mueser, K. T. (2007). A meta-analysis of cognitive remediation in schizophrenia. *American Journal of Psychiatry, 164*(12), 1791–1802. https://doi.org/10.1176/appi.ajp.2007.07060906.

Myin-Germeys, I., Kasanova, Z., Vaessen, T., Vachon, H., Kirtley, O., Viechtbauer, W., & Reininghaus, U. (2018). Experience sampling methodology in mental health research: new insights and technical developments. *World Psychiatry, 17*(2), 123–132. https://doi.org/10.1002/wps.20513.

Myin-Germeys, I., Klippel, A., Steinhart, H., & Reininghaus, U. (2016). Ecological momentary interventions in psychiatry. *Current Opinion in Psychiatry, 29*(4), 258–263. https://doi.org/10.1097/YCO.0000000000000255.

Myin-Germeys, I., Oorschot, M., Collip, D., Lataster, J., Delespaul, P., & van Os, J. (2009). Experience sampling research in psychopathology: opening the black box of daily life. *Psychological Medicine, 39*(9), 1533–1547. https://doi.org/10.1017/S0033291708004947.

Oorschot, M., Lataster, T., Thewissen, V., Lardinois, M., Wichers, M., van Os, J., … Myin-Germeys, I. (2013). Emotional experience in negative symptoms of schizophrenia—no evidence for a generalized hedonic deficit. *Schizophrenia Bulletin, 39*(1), 217–225. https://doi.org/10.1093/schbul/sbr137.

Palmier-Claus, J. E., Myin-Germeys, I., Barkus, E., Bentley, L., Udachina, A., Delespaul, P. A., … Dunn, G. (2011). Experience sampling research in individuals with mental illness: reflections and guidance. *Acta Psychiatrica Scandinavica, 123*(1), 12–20. https://doi.org/10.1111/j.1600-0447.2010.01596.x.

Pan, X., & Hamilton, A. F. d. C. (2018). Why and how to use virtual reality to study human social interaction: the challenges of exploring a new research landscape. *British Journal of Psychology*, 1–23. https://doi.org/10.1111/bjop.12290.

Park, S. -H., Ku, J., Kim, J. -J., Jang, H. J., Kim, S. Y., Kim, S. H., … Kim, S. I. (2009). Increased personal space of patients with schizophrenia in a virtual social environment. *Psychiatry Research, 169*(3), 197–202. https://doi.org/10.1016/j.psychres.2008.06.039.

Parsons, T. D. (2015). Virtual reality for enhanced ecological validity and experimental control in the clinical, affective and social neurosciences. *Frontiers in Human Neuroscience, 9*, 660. https://doi.org/10.3389/fnhum.2015.00660.

Penn, D. L., Sanna, L. J., & Roberts, D. L. (2008). Social cognition in schizophrenia: an overview. *Schizophrenia Bulletin, 34*(3), 408–411. https://doi.org/10.1093/schbul/sbn014.

Pfeiffer, U. J., Timmermans, B., Vogeley, K., Frith, C. D., & Schilbach, L. (2013). Towards a neuroscience of social interaction. *Frontiers in Human Neuroscience*, 7, 1–2. https://doi.org/10.3389/fnhum.2013.00022.

Pinkham, A. E., Harvey, P. D., & Penn, D. L. (2018). Social cognition psychometric evaluation: results of the final validation study. *Schizophrenia Bulletin, 44*(4), 737–748.

Pinkham, A. E., Penn, D. L., Green, M. F., & Harvey, P. D. (2016). Social cognition psychometric evaluation: results of the initial psychometric study. *Schizophrenia Bulletin, 42* (2), 494–504.

Pot-Kolder, R. M. C. A., Geraets, C. N. W., Veling, W., van Beilen, M., Staring, A. B. P., Gijsman, H. J., ... van der Gaag, M. (2018). Virtual-reality-based cognitive behavioural therapy versus waiting list control for paranoid ideation and social avoidance in patients with psychotic disorders: a single-blind randomised controlled trial. *The Lancet Psychiatry, 5*(3), 217–226. https://doi.org/10.1016/s2215-0366(18)30053-1.

Premkumar, P., Cooke, M. A., Fannon, D., Peters, E., Michel, T. M., Aasen, I., ... Kumari, V. (2008). Misattribution bias of threat-related facial expressions is related to a longer duration of illness and poor executive function in schizophrenia and schizoaffective disorder. *European Psychiatry, 23*(1), 14–19. https://doi.org/10.1016/j.eurpsy.2007.10.004.

Priebe, S. (2007). Social outcomes in schizophrenia. *British Journal of Psychiatry, 191*, S15–S20. https://doi.org/10.1192/bjp.191.50.s15.

Reeder, C., Pile, V., Crawford, P., Cella, M., Rose, D., Wykes, T., ... Callard, F. (2016). The feasibility and acceptability to service users of CIRCuiTS, a computerized cognitive remediation therapy programme for schizophrenia. *Behavioural and Cognitive Psychotherapy, 44*(3), 288–305. https://doi.org/10.1017/S1352465815000168.

Reininghaus, U., Depp, C. A., & Myin-Germeys, I. (2016). Ecological interventionist causal models in psychosis: targeting psychological mechanisms in daily life. *Schizophrenia Bulletin, 42*(2), 264–269. https://doi.org/10.1093/schbul/sbv193.

Reininghaus, U., Kempton, M. J., Valmaggia, L., Craig, T. K., Garety, P., Onyejiaka, A., ... Morgan, C. (2016). Stress sensitivity, aberrant salience, and threat anticipation in early psychosis: an experience sampling study. *Schizophrenia Bulletin, 42*(3), 712–722. https://doi.org/10.1093/schbul/sbv190.

Rus-Calafell, M., Garety, P., Sason, E., Craig, T. J. K., & Valmaggia, L. R. (2017). Virtual reality in the assessment and treatment of psychosis: a systematic review of its utility, acceptability and effectiveness. *Psychological Medicine, 48*(03), 1–30. https://doi.org/10.1017/S0033291717001945.

Savla, G. N., Vella, L., Armstrong, C. C., Penn, D. L., & Twamley, E. W. (2013). Deficits in domains of social cognition in schizophrenia: a meta-analysis of the empirical evidence. *Schizophrenia Bulletin, 39*(5), 979–992. https://doi.org/10.1093/schbul/sbs080.

Schilbach, L., Timmermans, B., Reddy, V., Costall, A., Bente, G., Schlicht, T., & Vogeley, K. (2013). Toward a second-person neuroscience. *Behavioral and Brain Sciences, 36*(04), 393–414. https://doi.org/10.1017/S0140525X12000660.

Schneider, M., Myin, E., & Myin-Germeys, I. n.d. Is social cognition a prerequisite for social interaction? A study in psychotic disorder (under review).

Schneider, M., Reininghaus, U., van Nierop, M., Janssens, M., Myin-Germeys, I., & Investigators, G. (2017). Does the social functioning scale reflect real-life social functioning? An experience sampling study in patients with a non-affective psychotic disorder and healthy control individuals. *Psychological Medicine, 47*(16), 2777–2786. https://doi.org/10.1017/S0033291717001295.

Schonherr, J. (2017). What's so special about interaction in social cognition? *Review of Philosophy and Psychology, 8*(2), 181–198. https://doi.org/10.1007/s13164-016-0299-y.

Shiffman, S., Stone, A. A., & Hufford, M. R. (2008). Ecological momentary assessment. *Annual Review of Clinical Psychology, 4*(1), 1–32. https://doi.org/10.1146/annurev.clinpsy.3.022806.091415.

Simons, C. J., Bartels-Velthuis, A. A., & Pijnenborg, G. H. (2016). Cognitive performance and long-term social functioning in psychotic disorder: a three-year follow-up study. *PLoS One, 11*(4), e0151299. https://doi.org/10.1371/journal.pone.0151299.

Sprong, M., Schothorst, P., Vos, E., Hox, J., & Van Engeland, H. (2007). Theory of mind in schizophrenia—meta-analysis. *British Journal of Psychiatry*, *191*, 5–13. https://doi.org/10.1192/bjp.bp.107.035899.

Stroop, J. R. (1935). Studies of interference in serial verbal reactions. *Journal of Experimental Psychology*, *18*, 643–662. https://doi.org/10.1037/h0054651.

Valmaggia, L. (2017). The use of virtual reality in psychosis research and treatment. *World Psychiatry*, *16*, 246–247.

Valmaggia, L. R., Day, F., & Rus-Calafell, M. (2016). Using virtual reality to investigate psychological processes and mechanisms associated with the onset and maintenance of psychosis: a systematic review. *Social Psychiatry and Psychiatric Epidemiology*, *51*, 921–936.

Van Roekel, E., Vrijen, C., Heininga, V. E., Masselink, M., Bos, E. H., & Oldehinkel, A. J. (2017). An exploratory randomized controlled trial of personalized lifestyle advice and tandem skydives as a means to reduce anhedonia. *Behavior Therapy*, *48*(1), 76–96. https://doi.org/10.1016/j.beth.2016.09.009.

Veling, W., Pot-Kolder, R., Counotte, J., Van Os, J., & Van Der Gaag, M. (2016). Environmental social stress, paranoia and psychosis liability: a virtual reality study. *Schizophrenia Bulletin*, *42*(6), 1363–1371. https://doi.org/10.1093/schbul/sbw031.

Vrijen, C., Hartman, C. A., & Oldehinkel, A. J. (2018). Reward-related attentional bias at age 16 predicts onset of depression during nine years of follow-up. *Journal of the American Academy of Child and Adolescent Psychiatry*. https://doi.org/10.1016/j.jaac.2018.06.009.

Woolverton, C. B., Bell, E. K., Moe, A. M., Harrison-Monroe, P., & Breitborde, N. J. (2017). Social cognition and the course of social functioning in first-episode psychosis. *Early Intervention in Psychiatry*. https://doi.org/10.1111/eip.12432.

Wykes, T., Huddy, V., Cellard, C., McGurk, S. R., & Czobor, P. (2011). A meta-analysis of cognitive remediation for schizophrenia: methodology and effect sizes. *American Journal of Psychiatry*, *168*(5), 472–485. https://doi.org/10.1176/appi.ajp.2010.10060855.

Wykes, T., Joyce, E., Velikonja, T., Watson, A., Aarons, G., Birchwood, M., … Upthegrove, R. (2018). The CIRCuiTS study (Implementation of cognitive remediation in early intervention services): protocol for a randomised controlled trial. *Trials*. *19*, 183. https://doi.org/10.1186/s13063-018-2553-3.

Zaki, J., & Ochsner, K. (2009). The need for a cognitive neuroscience of naturalistic social cognition. *Annals of the New York Academy of Sciences*, *1167*, 16–30. https://doi.org/10.1111/j.1749-6632.2009.04601.x.

CHAPTER ELEVEN

# Social cognitive interventions

**Joanna M. Fiszdon*,†, Charles A. Davidson‡**
*VA Connecticut Healthcare System, Department of Psychology, West Haven, CT, United States
†Department of Psychiatry, Yale University School of Medicine, New Haven, CT, United States
‡Department of Clinical Psychology, Mercer University College of Health Professions, Atlanta, GA, United States

## Origins of social cognitive interventions

The relatively recent interest in interventions targeting social cognitive impairments in individuals with psychosis is founded on the notion that social cognitive impairments are closely linked to functioning, and that by targeting social cognition, we may affect functional improvement. Work conducted over the past 20 years supports this notion by indicating that social cognition: (1) predicts functional outcomes (Brekke, Hoe, Long, & Green, 2007), (2) contributes unique predictive power above and beyond variance accounted for by neurocognitive ability (Couture, Penn, & Roberts, 2006), and (3) may mediate the well-known relationship between neurocognition and functioning (Pinkham & Penn, 2006). Importantly, as demonstrated in the subsequent Proof of Concept section, there is also evidence that social cognitive skills are malleable among people with schizophrenia.

Social cognitive interventions distinguish themselves from other psychosocial treatments in their unique focus on deficits and biases associated with processing social information, as opposed to targeting psychiatric symptoms more broadly. Nevertheless, many social cognitive interventions still share certain characteristics of other treatment approaches. Similar to social skills training, social cognitive interventions focus on training specific skills, though the types of skills targeted by the two approaches vary. Also similar to cognitive behavioral therapy, social cognitive interventions may include explicit analyses of data in support of certain conclusions, and both approaches focus on improving how information is being processed, although the content area differs by approach. Given the significant heterogeneity among social cognitive treatments, the amount of overlap with the older, more established interventions varies considerably.

*Social Cognition in Psychosis*
https://doi.org/10.1016/B978-0-12-815315-4.00011-2

## Proof of concept

Some of the earliest work on social cognitive treatments for individuals with psychosis includes proof-of-concept studies demonstrating that specific techniques could engage social cognitive abilities such as social perception, theory of mind (ToM), attributional bias (AB), and emotion perception.

### Social perception

In one of the earliest proof-of-concept trials, Corrigan, Hirschbeck, and Wolfe (1995) examined the impact of semantic elaboration memory training on social perception. While a relationship had previously been reported between cue perception, vigilance, and recall memory, researchers had originally assumed that impairments in recall memory were due to poor vigilance, and that focused vigilance training would remediate these recall memory impairments, leading to improved social cue perception. However, there was also some evidence to suggest that recall memory impairments persisted following vigilance training. To better understand the unique contributions of these two types of training, Corrigan and colleagues examined the impact of a single 60-minute session of either visual vigilance training only, or a combined vigilance and semantic elaboration memory training on social cue perception. Participants randomized to the vigilance plus semantic elaboration training were not only cued to pay attention to brief video vignettes of social situations, but were also asked to describe the story lines of these vignettes. As hypothesized, the combined vigilance and memory recall training was superior to the vigilance-only training in a number of social cues identified in the video vignettes, as well as on a separate measure of social perception, demonstrating the unique contribution of gistful recall memory to social perception.

### ToM/attributional bias

A few years after the Corrigan and colleagues study, a French group examined the impact of verbal elaboration on ToM performance (Sarfati, Passerieux, & Hardy-Bayle, 2000). Their proof-of-concept study was based on the theory that ToM abilities are impacted by ineffective processing of information: namely trouble differentiating between and extracting the more and less relevant components of social scenes, along with prior research suggesting that asking participants to provide verbal justification for their responses while performing an information processing task improves performance. To test their hypothesis that verbal elaboration forces participants to focus on and process the relevant components of context, leading to improved ToM, they administered two versions of a cartoon character intention strip. In the standard pictorial-only version, participants were presented with a series of three cartoon panels showing a character performing a specific action, and then asked to select from several options the correct pictorial panel that reflected what the character was most likely to do next. In the enhanced version, instead of cartoons, the answer choices consisted of short sentences describing what the character might do next. As had been hypothesized, the introduction of verbal material in the modified procedure was associated with improved ToM performance, not only in patients with schizophrenia, but also in controls.

This same group (Kayser, Sarfati, Besche, & Hardy-Bayle, 2006) later evaluated the impact of a related verbal mediation procedure, in which over two 1-hour sessions, participants were trained in ToM skills by being asked to describe and analyze social interactions they viewed in brief video clips. As needed, study therapists further cued participants to verbalize their interpretations of what was happening and why. Compared with an untrained group, the trained group showed significant improvement on the previously described pictorial cartoon character intention strip, as well as on a communication disorder rating scale, adding further support for the positive impact of verbal elaboration on ToM performance.

Few proof-of-concept studies have targeted AB in schizophrenia. However, preliminary results suggest interpretation biases related to anxiety in schizophrenia may be engaged with cognitive bias modification (CBM), a technique with a large evidence base for nonpsychotic disorders. CBM typically involves rote computerized classical conditioning where the participant learns alternative, more positive interpretations or emotional responses for stimuli that were previously associated with negative responses. Steel et al. (2010) tested individually-administered CBM utilizing visual imagery of computer-administered audio vignettes in a sample of individuals with schizophrenia and comorbid anxiety. No overall effects were found for interpretation biases or state anxiety, but utilizing visual imagery was associated with larger changes in interpretation bias. In a subsequent small study testing a similar technique (Turner et al., 2011), three participants improved in negative interpretation bias, while the remaining three did not change.

## Emotion perception

There have also been a number of brief trials focused on evaluating specific methods to improve emotion perception. Penn and Combs (2000) examined the impact of contingent monetary reinforcement, which would ostensibly improve performance by engaging motivational systems and hence improving performance, versus the impact of facial feedback, where based on research suggesting that expressing an emotion on one's face leads to enhanced experience of that emotion, mimicking an observed expression may help the person better recognize the underlying emotion. Forty inpatients diagnosed with schizophrenia were randomized to a single session of one of four conditions: an active control condition entailing repeated practice in identifying emotions, a condition where participants received monetary reinforcement for each correct emotion identification trial, a facial feedback condition in which participants were asked to mimic the target expression prior to making a decision about identifying the correct emotion type, and a condition combining facial feedback with monetary reinforcement. All three experimental conditions were associated with improvements in affect recognition, though there was no generalization to affect discrimination. There was also some indication that improvements in affect recognition persisted at 1 week follow-up. Results of this trial suggested that both monetary reinforcement and facial feedback approaches could be used to improve affect recognition in individuals with schizophrenia.

There have also been several successful trials that evaluated the impact of shaping attention to the face. These trials were developed based on the literature that emotion perception correlates more highly with attention and vigilance than with other cognitive abilities, and visual scanning data suggesting that people with schizophrenia tend to focus on nonessential facial features. In a 2008 trial, Combs et al. (2008) randomized 60 inpatients with schizophrenia to a single session of one of three conditions: computer-based attention shaping (attention prompts to the face) accompanied by monetary reinforcement for correct emotion identification responses, a monetary reinforcement for a correct responses only condition, and a simple repeated administration of emotion recognition trials condition. Significant improvements in emotion identification were reported only for the attention shaping combined with monetary reinforcement condition, with improvements maintained at 1 week. In contrast to the earlier 2000 Penn and Combs trial, there were no significant improvements in emotion recognition for the monetary contingency only condition.

In a follow-up study (Combs, Chapman, Waguspack, Basso, & Penn, 2011) with the same attention-shaping procedure, this time without the monetary reinforcement, participants were randomized to receive one, three, or five attention-shaping sessions. While there were significant improvements in emotion identification across all three conditions, improvements were greatest for those who received five sessions, suggesting a training dose-response effect. Interestingly, eye tracking data collected during the study indicated that while participants in the single session condition increased their time scanning essential facial features, those in the three and five session conditions decreased their scanning time in these regions, which the authors cautiously interpreted as the intervention either having a primary impact on early, whole face processing (versus essential features), or the training resulting in greater efficiency (and hence less time needed) in processing facial information.

## Targeted single-domain interventions

Based in part on some of the earlier proof-of-concept trials, a number of interventions have emerged that also focus on improving skills in some specific social cognitive domains. These interventions are described in the following sections.

### Emotion perception

From among the numerous social cognitive interventions, emotion perception training has by far been the most studied. It has also been the domain with the most consistently reported improvements following training. One of the earliest trials of emotion perception training comes from Hodel and colleagues who, in the late 1990s, developed Emotional Management Therapy (EMT), an intervention intended to not only improve affect perception, but also coping with stress, with the assumption that this will lead to improvements in social adjustment and psychopathology. EMT was delivered in small groups, ranging from 11 to 24 sessions, and used role-plays and in-vivo exercises to initially train how to perceive emotions in oneself and in other group members, then helping participants identify strategies currently used to cope with distress, and later teaching efficient emotion management strategies. There have been three evaluations of this intervention, and findings have been mixed. In the first multicenter, single-group study with outpatients with schizophrenia, published in 1997 (as described in Hodel,

Kern, & Brenner, 2004), EMT was reportedly associated with improvements in emotion perception, social adjustment, and psychopathology. In a second study with outpatients with first-episode psychosis (Hodel, Brenner, Merlo, & Teuber, 1998), compared with those allocated to a control group, there were no treatment-related improvements in emotion perception or social functioning at the end of EMT treatment or at an 8-month follow-up. In the most recent trial of inpatients with treatment-resistant schizophrenia consecutively assigned to EMT or an active control (Hodel et al., 2004), EMT was associated with significant improvements in emotion perception, social adjustment, and psychiatric symptoms, with gains in social adjustment and symptoms maintained at 4-month follow-up, while emotion perception performance at follow-up returned to baseline levels.

Perhaps the best known and most widely tested affect perception training comes from the work of Wolver and Frommann. Originally introduced in 2003 (Frommann, Streit, & Wolwer, 2003), Training in Affect Recognition (TAR) is a 12-session, small-group, manualized intervention consisting of three phases in which participants learn to identify and discriminate specific features associated with prototypical facial expressions, make and justify holistic judgments of emotion, and identify and discriminate emotional expressions in situational contexts. The training consists of drill and practice exercises, along with training of strategies such as verbalization, self-instruction, and use of associations. To date, there have been seven trials of this intervention, four of which are randomized controlled trials (RCTs). Results indicate that TAR leads to significant improvements in facial, as well as prosodic, affect recognition (Drusch, Stroth, Kamp, Frommann, & Wolwer, 2014; Frommann et al., 2003; Habel et al., 2010; Sachs et al., 2012; Wolwer et al., 2005; Wolwer & Frommann, 2009, 2011), with some evidence that facial affect recognition (FAR) gains are durable over a month or more (Wolwer & Frommann, 2009). There is also some evidence that TAR may be associated with improved social functioning (Sachs et al., 2012; Wolwer & Frommann, 2011), and that TAR is associated with increased activation in several brain areas (Habel et al., 2010). There are conflicting reports of how improvement in affect recognition corresponds to concurrent changes in cognition and symptoms (Sachs et al., 2012; Wolwer et al., 2005), along with data suggesting that while TAR is associated with changes in fixation on salient facial features (eyes, nose, mouth), these fixation changes are not directly related to the amount of improvement in affect recognition.

Recognizing the unique advantages of therapist-led and computer-administered training, namely goal formulation, motivational enhancement and individualized training for the therapist-led interventions, and more controlled, standardized training with opportunity for massed practice and graded increases in task difficulty for computer-administered training, Gaudelus, Virgile, Geliot, Team, and Franck (2016) developed a computer-aided program focused on remediating affect recognition impairments. The GAIA s-face intervention consists of initial social cognitive assessment and development of personalized training goals, followed by a combination of therapist-aided clinic plus at-home "transfer" sessions that focus on exercises training affect recognition and discrimination using photographs and videos, followed by a generalization phase in which additional exercises are assigned based on the trainees' social cognitive profiles and individual requests. In a preliminary evaluation of this relatively new intervention, inpatients and outpatients were randomized to 30 sessions, over 10 weeks, of either GAIA s-face, or a computer-assisted training focused on selective attention. While both groups showed improvements in affect recognition, the group receiving GAIA s-face training had significantly greater gains. GAIA s-face condition was also associated with significant reductions in symptoms and improvements in social functioning. These improvements did not generalize to ToM or cognitive empathy, neither of which was targeted by the intervention.

In the first study of fully computerized affect perception training, Silver, Goodman, Knoll, and Isakov (2004) evaluated the efficacy of a commercially available emotion training program. During the training, participants were shown photographs of faces, scenes, or objects and asked questions about what emotions were displayed and what emotion a character may experience in a particular situation based on that character's likes, needs, and/or thoughts. Correct responses were reinforced, while incorrect responses resulted in re-administration of a trial. Three 15-minute sessions of the training led to significant improvement in pre-post emotion recognition (though not emotion discrimination) along with reductions in errors across the training sessions, suggesting that even without explicit direction, brief practice with these stimuli led to learning.

Several other studies examined the effects of a single session of a commercially available computerized micro-expressions training program, called Micro-Expressions Training Tool (METT). During the training, participants are shown video examples of commonly confused facial expressions, along with instruction on what differences in key facial features can be used to correctly identify them, and are then administered multiple trials in which they themselves have to correctly identify these emotions. In a 2006 trial (Russell, Chu, & Phillips, 2006) in which the training was administered to outpatients with schizophrenia and healthy controls, both groups showed significant improvements in affect recognition, with posttraining performance in the schizophrenia sample comparable to pretraining healthy control performance. In a subsequent RCT of the same program (Russell, Green, Simpson, & Coltheart, 2008), the training again led to significant improvements in emotion recognition accuracy, as well as an increase in eye movements directed toward salient facial features. Gains in emotion recognition were maintained at 1-week follow-up. In yet another report by the same group (Marsh et al., 2010), there were posttraining improvements in affect recognition for static photos of both the trained as well as novel faces, though no posttraining generalization to affect recognition in dynamic, video-clip stimuli of social interactions. The authors also report that posttraining improvement in affect recognition was greater for the outpatient than the inpatient subsamples (though both benefitted), and that amount of posttraining improvement was predicted by lower baseline anhedonia/asociality, higher social functioning, better facial identity recognition, and higher working memory. In supplementary analyses from the same trial (Marsh, Luckett, Russell, Coltheart, & Green, 2012), the authors found that while the training was associated with improvements in visual scanning (i.e., more attention to salient features), the degree of this improvement did not significantly correlate with amount of improvement in affect recognition.

And finally, in the most recently published trial of a targeted affect recognition intervention (Tsotsi, Kosmidis, & Bozikas, 2017), FAR training was compared with facial features training (AFF). FAR separately targeted each of the basic emotions, and progressed through an initial phase in which trainees made guesses (until correct) about an emotion displayed in photographs of faces, to a second phase where they identified differences between photos displaying an emotion or a neutral expression, along with reinforcement and feedback on additional differences, and a final phase in which they practiced identifying differences between emotional and neutral faces on their own, with no assistance from the therapist. During AFF, participants were trained to attend to and recognize differences in facial features by first indicating whether pairs of morphed photographs were identical, or differed in the size or shape of one of the salient features, with corrective feedback and reinforcement as appropriate, then identifying specific differences between pairs of morphed photographs (again with feedback and reinforcement), and finally by matching a target neutral face to the correct neutral photo of the same person at a different age. Compared with AFF and a control condition, only FAR was associated with improvements in affect recognition, with posttraining performance on par with that of healthy controls. However, despite randomization, those in the FAR condition had significantly better baseline affect recognition than those in either AFF or the control condition, making it difficult to ascertain whether improvements were due solely to the type of training.

## Social perception/knowledge

Relative to interventions targeting emotion perception, there have been far fewer targeted treatments of social perception/social knowledge. Two of these studies specifically evaluated the Social Perception module of Integrated Psychological Therapy (IPT). During this 21-session, semi-structured module, therapist-led small groups learn social context processing skills by initially identifying relevant details in still photographs of social situations, then interpreting these situations, and finally debating with other group members about potential inferences that can be made about these situations. In a small trial with 20 outpatients with schizophrenia, those randomized to the intervention improved in social perception, though not in selective attention, functioning, or symptoms (García, Fuentes, Ruíz, Gallach, & Roder, 2003). In a subsequent small RCT with 18 outpatients with schizophrenia, these same findings were replicated, and improvements in social perception were maintained at 6-month follow-up (Fuentes, Garcia, Ruiz, Soler, & Roder, 2007).

Another treatment targeting social perception, known as Social Cognitive Enhancement Training (SCET), was evaluated by Choi and Kwon (2006). This intervention is also administered in small, therapist-led groups. Participants train to evaluate social cues by sequentially ordering cartoons depicting social situations, and then providing explanations for their choices. There are three sequential complexity levels, each administered over 12 sessions, for a total treatment duration of 6 months. In a small RCT, the authors reported improvements on a picture arrangement measure, with amount of improvement increasing over the three complexity levels. The authors also reported some improvement on a social behavior sequencing task at 2 months into the treatment, but group differences were no longer significant at the end of the treatment.

## ToM/attributional bias

We are not aware of any trials that solely targeted AB. As AB and ToM impairments are likely interrelated, and as many trials of ToM also address AB, the results of these trials are grouped together here.

Based on prior work indicating that social skills training leads to enhanced social competence, and that metacognitive interventions can aid in remediating deficient information processing strategies, Roncone et al. (2004) hypothesized that the Instrumental Enrichment Program (IEP) originally developed as a method to promote "the enrichment of structural cognitive modification by helping the subject's capacity to mentally anticipate actions" (Fuerstein, 1980, p. 423), because of its focus on evaluating social situations, may also lead to improved ToM skills. The intervention was administered in small weekly groups, over the course of 22 weeks. At the end of the intervention period, compared with those randomized to usual treatment, participants in IEP had a significant reduction in negative symptoms, improvement in verbal fluency, planning, strategic thinking, first and second level ToM abilities, social function, and the recognition of negatively-valenced emotions. While there are no subsequent evaluations of this intervention's effects on ToM, the authors have noted that it shares many qualities with the well-known, though much more time intensive and broad-based, Cognitive Enhancement Therapy (CET; Hogarty, 2000).

This same group later developed another intervention, known as Cognitive Emotional Rehabilitation (REC), which focused on understanding and applying specific thinking strategies when trying to understand and interpret social situations. This intervention relies on psychoeducation and cognitive-behavioral techniques such as Socratic questioning to teach strategies to recognize emotions and infer mental states based on observed behaviors. This group-based intervention is delivered weekly, over the course of 6 months. In a 2011 trial of this intervention (Veltro et al., 2011), participants were randomized to REC or the well-known Problem Solving Therapy (PST). Consistent with the focus of each intervention, only those in the REC condition improved in first order ToM, and only those in PST improved in problem solving and working memory. Both groups improved in emotion attribution, though these improvements were greater for the REC condition. Finally, both groups improved in overall psychiatric symptoms and social functioning, though those in REC had significantly greater improvements on interpersonal communication and social engagement subscales. While providing some evidence for the specificity of proximal training effects, this study also highlighted that both social cognition focused, as well as problem-solving focused methods may be similarly effective in producing improvements in more distal, though arguably more important symptom and functioning domains.

Hypothesizing that observation and imitation of emotions can lead to improvements in ToM and empathy, Mazza and colleagues developed and evaluated a 12-week, group-based, Emotion and ToM Imitation Training (ETIT; Mazza et al., 2010). The training progressed from initially observing the direction of cartoon characters' gazes, to observing and imitating emotions seen in stills, and then guessing the emotion, to replicating emotions they thought matched those of a character in a vignette, and then to observing emotions portrayed by cartoon characters and inferring their mental states and likely future actions. Compared with those randomized to active control of PST, ETIT was associated with significantly greater improvements in ToM, emotion recognition, empathy, social functioning, and positive symptoms. ETIT was also associated with activation changes in medio-frontal areas specific for facial movement, which the authors interpreted as indicating that ETIT led to engagement of the mirror neuron system. While there were also posttreatment improvements in several aspects of neurocognition, these improvements were similar across the two training conditions.

The Mental State Reasoning Training for Social Cognitive Impairment (SoCog-MSRT) encourages perspective-taking and helping participants consider alternative interpretations and viewpoints associated with social situations. It is administered in small groups, with training stimuli consisting of games and short films centering on social vignettes. The feasibility and preliminary efficacy of SoCog-MSRT was initially evaluated in 2013 (Marsh et al., 2013), in a sample of 17 participants with schizophrenia. Compared with pretraining performance, participants improved on two of the three administered ToM measures, along with a self-report measure of social understanding. Improvements did not generalize to an affect recognition measure. Social cognitive improvements were smaller for individuals with poorer baseline working memory and premorbid IQ. SoCog-MSRT was again evaluated in 2016 (Marsh et al., 2016), this time in a quasirandom study comparing it with targeted emotion recognition training (SocCog-ERT—the previously described METT training with additional exercises). The interventions were administered over the course of 12 group training sessions, conducted over 6 weeks, with assessments at pre (T1), post (T2), and for a subsample, at 3-month follow-up (T3). At posttreatment, neither of the groups improved on a measure of mental state attribution, though an improvement did become apparent, for both groups, at T3. While the MSRT group did not improve in emotion recognition at T2 or T3, there was a trend for improvement in the SoCog-ERT condition at T2, which reached significance at T3. The SoCog-MSRT group did show posttraining improvements in false belief reasoning; however, these were not maintained at T3. Surprisingly, the SoCog-ERT group also improved in false belief reasoning, with improvements maintained at T3. While these findings suggest some efficacy for both interventions, contrary to authors' expectations, the pattern of improvements did not neatly segregate by targeted social cognitive ability.

Several interventions have also been developed that expressly took into consideration cognitive impairments common in individuals with psychosis. For example, Roberts, Kleinlein, and Stevens (2012) speculated that cognitive impairments in this population may make it more challenging to spontaneously generate alternative interpretations about social situations, resulting in the person becoming more entrenched in his or her initial interpretation, and less likely to consider viable alternatives, making CBT-based approaches difficult to execute successfully. The intervention developed by this group, Mary/Eddie/Bill (abbreviated "MEB"), was specifically designed to make it less taxing to consider varied interpretations of social situations by introducing three prototypical characters, each representing a unique attributional style. These characters were My Fault Mary, who consistently blames herself for negative events, Easy Eddie, who always assumes that negative events are due to accidental, environmental factors, and Blaming Bill, who always blames other people for causing negative events. In an uncontrolled trial, outpatients attended group sessions in which they practiced identifying these characters in short stories, video clips, and examples provided from their own social interactions. They also practiced using these three prototypes to generate alternative interpretations of how others may interpret social situations. Among treatment completers (i.e., attended at least five of six group sessions), there were significant improvements in ToM and jumping to conclusions bias (the latter assessed using the authors' Social Cognition Screening Questionnaire overconfidence scale, described as a proxy measure for jumping to conclusions bias), providing preliminary support for this more structured method of identifying disparate mental states.

Similar to Roberts and colleagues, Fiszdon et al. (2016) also speculated that interventions targeting complex, higher-order social cognitive processes such as ToM and AB may require special adaptations to make them effective for individuals with cognitive impairments. To address this, they developed a brief, four-module, individually administered intervention called Understanding Social Situations (USS). In the first three modules targeting ToM, stills, video clips, and short vignettes of social situations were used to train how to separate social facts from guesses, how to make probability judgments, and how to combine these two processes to infer others' mental states. Verbal elaboration was used to encourage sequenced processing of relevant social features, and principles such as errorless learning, scaffolding, performance-based increases in task difficulty, and massed drill and practice were applied to initially train fundamental components of ToM, later moving on to more complex ToM skills. The fourth module focused on training a positive interpretive bias for ambiguous situations, and was based on prior laboratory work on CBM in anxiety (Constans, Penn, Ihen, & Hope, 1999; Yiend, Mackintosh, & Mathews, 2005), which had shown that an automatic positive interpretive bias could be induced by presenting ambiguous social scenes that are then disambiguated as positive events by presenting additional information. The intervention was administered over 7–10 sessions, using a double-baseline, within-subjects design. While designed primarily as a feasibility trial, the training was associated with large improvements on a content knowledge test, medium improvements on one of the primary measures of AB, and small improvements on several exploratory measures of ToM. Importantly, baseline levels of cognitive impairment did not correlate with degree of social cognitive improvement, providing some support for the notion that using principles from neurocognitive remediation may minimize the purported negative impact of cognitive impairments on learning higher-order social cognitive skills.

## Comprehensive, multidomain interventions

While there have been a number of interventions that target multiple social cognitive domains, many of them were either administered as part of larger psychosocial programs, or themselves encompassed additional training in neurocognitive remediation or social skills training, and as such are not included in this review. The two best known and most researched comprehensive social cognitive treatments are Social Cognition Interaction Training (SCIT) and Social Cognitive Skill Training (SCST).

### Social Cognition Interaction Training

SCIT is a manualized CBT-based group psychotherapy that targets emotion perception, attributional style, and ToM to enhance social functioning and recovery (Penn et al., 2005). SCIT utilizes exercises, group process, psychoeducation, and video-based techniques over three progressive phases: (1)

understanding emotions; (2) social cognitive biases; and (3) integration, delivered over 18–24 sessions. For example, AB are targeted in the second phase using the MEB technique described herein. Consistent with CBT, participants practice specific social observation skills, separating facts from guesses, avoiding jumping to conclusions biases, weighing evidence, and identifying ones' own and others' emotions and AB throughout the sessions (Roberts, Penn, & Combs, 2016).

SCIT was initially developed with inpatient participants with schizophrenia-spectrum disorders in the United States (Combs et al., 2007; Penn et al., 2005). The first formal RCT was completed in the United States in 2014 (Roberts et al., 2014), which was followed by an RCT in Israel (Hasson-Ohayon, Kravetz, Levy, & Roe, 2014) and England (Gordon et al., 2018). SCIT has also had positive outcomes in several other nations, and with other languages in addition to English (Bartholomeusz et al., 2013; Tas, Danaci, Cubukcuoglu, & Brune, 2012; Voutilainen, Kouhia, Roberts, & Oksanen, 2016; Wang et al., 2013; Xu, Zhu, Zhangzi, & Zhen, 2011). It has been successfully adapted to outpatient (Parker, Foley, Walker, & Dark, 2013; Roberts & Penn, 2009; Roberts, Penn, Labate, Margolis, & Sterne, 2010) and forensic settings (Taylor et al., 2016), first-episode schizophrenia (Bartholomeusz et al., 2013; Cacciotti-Saija et al., 2015; Xu et al., 2011), and subclinical levels of psychotic-like symptoms (Chan et al., 2010). Investigators in Turkey have also expanded family involvement for the "practice partner" element of outside-session practice in SCIT (Tas et al., 2012).

At least 12 controlled trials of SCIT have been reported, including more than 480 participants. Results generally suggest that SCIT training is associated with improvements in emotion perception and ToM across the psychosis spectrum, and in different settings. Results are promising, but less consistent with quality of life, social engagement, social functioning, or AB, partially due to psychometric limitations (Pinkham, Harvey, & Penn, 2017). One of the main draws of SCIT is that it tends to be enjoyable, engaging, and perceived as helpful for participants and group facilitators (Gordon et al., 2018), although there are challenges and adaptations required to implement SCIT in different settings (Roberts et al., 2010).

## Social Cognitive Skill Training

The other well-known comprehensive intervention, developed by Horan and colleagues, is SCST. It shares several of the components of SCIT, namely training in how to separate facts from guesses and how to use this information to make informed decisions about social situations. This is supplemented by facial mimicry and affect perception training exercises similar to TAR, along with training in social cue perception, social norms, and how to identify sarcasm, white lies, and deception. Skills taught increase in complexity over the course of the training, and repeated practice is emphasized. Each group-based, manualized session includes review of content presented at the last session, a segment introducing new social cognitive skills, plus practice and role plays. In the first evaluation of what was initially a 12-session intervention, outpatients with schizophrenia were randomized to SCST, or an active control, illness management training (Horan et al., 2009). SCST was associated with significant posttraining improvement in facial emotion identification, but not social perception, AB, or ToM. Both conditions were associated with medium-level improvements on a composite measure of neurocognition.

In the second evaluation of SCST, the authors sought to better understand the unique effects of social cognitive versus neurocognitive remediation (NR), as well as whether NR is a necessary precondition for the success of social cognitive interventions (Horan et al., 2011). Outpatients with schizophrenia were randomized to either an expanded 24 session SCST, or time-matched NR, a hybrid SCST and neurocognitive intervention, or a standard illness management active control condition (ST). For affect perception, improvements were greatest in the SCST group, with the hybrid group showing some improvement, though not significantly different from any of the other groups. For emotion management, SCST was associated with greater improvements relative to the NR and hybrid conditions, with ST improving at trend level. All groups improved on social perception, ToM, and neurocognition to a similar degree. There were no significant improvements for symptoms or functional capacity. These results were interpreted to indicate that SCST shows some specificity for affect perception, and that this impact can be achieved solely with social cognitive training, without need for NR. While the authors did not find greater effects for the hybrid condition, it should be noted that the social cognitive component of the hybrid condition was shortened to 12 sessions, and that the neurocognitive component was administered concurrently with social cognitive training (while some recommend that neurocognition be targeted before social cognition).

The efficacy of SCST has also been evaluated in an Egyptian sample (Gohar, Hamdi, El Ray, Horan, & Green, 2013), where outpatients were randomized to an abbreviated and culturally adapted version of SCST, or an active control consisting of illness management and leisure skills training. Compared with control, SCST was associated with significant improvements in emotional intelligence, though not neurocognition. While the assessment battery was small and did not include other social cognitive measures nor more distal measures of functioning, this was nevertheless an important study, providing preliminary evidence that SCST may be efficacious in non-Western samples.

In the most recent trial of SCST (Horan et al., 2017), the investigators sought to enhance the intervention's effects on functional outcomes, and supplemented the 24-session intervention with 6 in-vivo community-based practice sessions. Participants were randomized to SCST with in-vivo training, SCST with additional clinic-based training, or a time-matched illness management control condition. Compared with active control, the SCST conditions evidenced significant improvements in affect perception, which were maintained at 3-month follow-up. While there was some evidence of additional SCST-associated gains in emotional intelligence, AB, empathic accuracy, and ToM, these were either at trend-level, or were not significant once corrected for multiple comparisons, and were absent at 3-month follow-up. There were no effects on neurocognition or symptoms. Contrary to hypotheses, no improvement in functioning was found for the in-vivo condition.

## Computerized comprehensive social cognitive interventions

Given the considerable time commitment associated with comprehensive social cognitive interventions, along with the need for trained staff, difficulty in scheduling lab-based group sessions, and challenges of adapting group-administered content to individual needs, interest has surged in computerized, remotely delivered training methods. Two such approaches have been evaluated to date, e-Motional Training (ET) and SocialVille.

ET is a 12 session, self-administered online intervention. The initial four sessions focus on emotion recognition and begin with didactic content about how salient facial features can be used to distinguish between different emotions, and then move on to micro-expression training. The subsequent eight sessions are devoted to training ToM, social processing, and AB, with trainees queried about thoughts and intentions of characters seen interacting in a video clip, followed by feedback and metacognitive suggestions as needed. In an initial pilot trial (Vasquez-Campo, Marono, Lahera,

Mateos, & Garcia-Caballero, 2016), the intervention was associated with improvements in affect recognition, ToM, and symptoms, with participants in both ET and control conditions improving in AB, and neither condition improving in emotional intelligence. While promising, assessors were not blind to condition, and reported results appear to be within-group comparisons, so more rigorous evaluations will be necessary before any conclusions can be drawn about the efficacy of this program.

The other fully computerized intervention, SocialVille, was developed by the same group that developed the well-known and studied, commercially available, BrainHQ computerized neurocognitive training. Similar to BrainHQ, SocialVille training is based on principles of neuroplasticity-based learning (Nahum et al., 2014). Unlike most social cognitive interventions, SocialVille exercises target "impaired brain systems underlying social cognition rather than the impaired social behaviors per-se that are targeted by molar social skills training approaches" (p. e12). The computerized exercises purportedly train affect perception, social cue perception, ToM and self-referential processing, and include tasks such as identifying and matching gaze direction, matching faces based on displayed emotion, selecting emotions that correspond to specific social situations, and answering questions about social interactions presented during brief video clips. Individual exercises are brief, and adjust in difficulty based on performance, and many include a processing speed and/or a working memory component. In the first evaluation of SocialVille, after completion of 24 hours of training, participants had improved on the training tasks themselves, with large gains on speeded tasks and moderate gains on tasks that had a working memory component. On untrained outcome measures, there was a trend for improvement in interpreting emotions conveyed by changes in tone of voice, and a significant improvement in behavioral inhibition, anticipatory pleasure, and one of the social functioning measures. There were also significant improvements in median reaction times (though not correct responses) on a measure of facial memory. Additional data on the efficacy of SocialVille is anticipated from an ongoing, multisite RCT with a more comprehensive assessment battery (Rose et al., 2015).

## Discussion and conclusions

In this chapter, we have provided an overview of interventions focused solely on improving social cognition for people with psychosis. Overall, there is clear and convincing evidence that treatments targeting social cognition can be efficacious (i.e., engage and improve social cognitive abilities and characteristics) and effective (i.e., implemented in real-world settings with positive effects on social functioning) for people with psychosis.

Research suggests the relationship between social cognition and functioning is only partly accounted for by neurocognition (Barbato et al., 2013). However, it is straightforward to assume neurocognitive deficits can bottleneck social cognitive abilities, and for some people, neurocognitive deficits explain social cognitive impairment (Fanning, Bell, & Fiszdon, 2012). Neurocognitive abilities are necessary for participants to engage productively in and generalize from the tasks of social cognitive treatment, such as learning heuristics of Roberts' MEB or the thinking abstractly for role play in SCST. Some social cognitive treatments have shown positive effects on neurocognition, and positive outcomes have been documented independently of changes in neurocognition and symptoms (Horan et al., 2009, 2011). However, it has been suggested that interventions that only target social cognition to the exclusion of neurocognition may be less likely to lead to sustained, generalizable effects (Roberts & Velligan, 2012). As such, a two-pronged approach is often recommended, and many treatments are designed with added or blended neurocognitive and social cognitive treatments. Further, generalizable and durable change is more likely when interventions such as social cognitive treatment are part of a more comprehensive approach to treatment and recovery, such as psychiatric rehabilitation (Brekke et al., 2007).

The evidence base for social cognitive treatment for schizophrenia has several limitations. Many studies lack sufficient sample size, independent replication, randomization, control conditions, generalization assessment, and/or durability assessment. Additionally, measures used as social cognitive outcomes are typically untrained, but may be quite proximal to the training. Change in proximal measures suggests efficacy of target engagement, but limit interpretation of "change in social cognition" generally. For example, if someone is trained to identify emotional expressions in photos of actors, they may show improvement in identifying emotions in photos of actors, but we cannot assume they have improved in the general target construct: emotion perception. Similarly, many studies may have only one measure per social cognitive domain, or all outcome measures may use the same method (e.g., self-report survey), lacking multimethod assessment.

Beyond the preceding limitations that are inherent to any clinical research with finite resources, social cognitive treatment research also suffers from some psychometric problems. These interventions are intended to change social cognition that is relevant to functioning, but the social cognition measures we have are limited in internal and test-retest reliability, and may not actually be related to functioning, especially after accounting for neurocognition (Pinkham et al., 2017). While some complex psychological constructs, such as clinical levels of depression or general wellbeing, can be measured efficiently with adequate psychometrics, more complex or "higher-level" social cognitive constructs are typically not measured well, yet. The most reliably measured social cognitive constructs are abilities (rather than characteristics) whose tests are performance-based, which may be limited in generalizability and differentiation from neurocognition. Changes in complex constructs such as AB or social perception have to be interpreted with a grain of salt. Performance-based measures of related constructs, such as reasoning or interpretation biases, have shown more adequate psychometrics, although their generalizability is debatable (Hasson-Ohayon et al., 2015).

Similar to measures of social functioning and negative symptoms, measures based on observed behaviors or checklists of related behaviors may also hold promise, compared with attempting to self-report symptoms, or validly engage in vignettes (Davidson, Lesser, Parente, & Fiszdon, 2017; Granholm, Ben-Zeev, Link, Bradshaw, & Holden, 2012). Similarly, dynamic assessment of potential to respond to social cognitive techniques might provide more generalizable or valid information, at least for treatment purposes, as dynamic testing is designed to assess capacity to respond to an intervention (Davidson, Johannesen, & Fiszdon, 2016; Grigorenko, 2009). To our knowledge, only one recent study (Clayson et al., 2018) has been published utilizing this method with social cognitive measures. The results did not show incremental validity for prediction of between-person differences in community functioning or phase of illness, although prediction of response to intervention was not tested. Regardless of psychometrics and generalizability, clinical psychosis research is hampered by a lack of a firm understanding or consensus on the heterogeneous profiles and causal links between different abilities and characteristics, including questionable domain-specificity of social cognitive constructs and overlap with lower-level constructs, such as neurocognition, and higher-level constructs, such as metacognition and psychotic symptoms (Browne et al., 2016; Hasson-Ohayon et al., 2015; Mancuso, Horan, Kern, & Green, 2011). It is unlikely there will ever be a blood test for perceiving others' social behaviors as ill-intentioned, tasks that pick up on such a bias will not be perfect, and the causal dynamics over a lifetime are not likely amenable to comprehensive assessment on an individual basis. These limitations are the challenge of human research, and suggest more work to do the best we can, not discounting the excellent work that forms our foundation. In summary, while demonstrating evidence for functionally-significant change in social behavior and complex psychological constructs is difficult, the preponderance of evidence that social cognitive training can be useful for people with psychosis suggests these treatment approaches are valuable and warrant further research and development.

New programs addressing social cognition are constantly being developed and evaluated, taking advantage of the extensive literature foundation described herein. New programs are taking advantage of modern technology, such as online programs (Vasquez-Campo et al., 2016), virtual reality (Peyroux & Franck, 2014), and mobile assessment (Granholm et al., 2012). Theoretical and developmental work will help clarify the interplay between different levels of psychopathology and social cognition (Howes & Murray, 2014; Riehle & Lincoln, 2018). Some research has also suggested that certain social cognitive targets, such as ToM, may be more strongly associated with functioning (Fett et al., 2011), and further research will help to demonstrate the most efficient training to improve functional outcomes. Social cognitive treatment is a fast-moving field. We need to improve measurement to better evaluate treatments, generalization of social cognitive skills to functioning, and penetration of evidence-based treatments into accessible healthcare. A plethora of new research will be completed to further this cause, and we look forward to benefitting from this work and applying it to enhance rehabilitation and recovery for people with psychosis.

## References

Barbato, M., Liu, L., Penn, D. L., Keefe, R. S. E., Perkins, D. O. W., Woods, S. W., & Addington, J. (2013). Social cognition as a mediator between neurocognition and functional outcome in individuals at clinical high risk for psychosis. *Schizophrenia Research, 150*(2–3), 542–546.

Bartholomeusz, C. F., Allott, K., Killackey, E., Liu, P., Wood, S. J., & Thompson, A. (2013). Social cognition training as an intervention for improving functional outcome in first-episode psychosis: a feasibility study. *Early Intervention in Psychiatry*, 7(4), 421–426.

Brekke, J. S., Hoe, M., Long, J., & Green, M. F. (2007). How neurocognition and social cognition influence functional change during community-based psychosocial rehabilitation for individuals with schizophrenia. *Schizophrenia Bulletin, 33*(5), 1247–1256.

Browne, J., Penn, D. L., Raykov, T., Pinkham, A. E., Kelsven, S., Buck, B., & Harvey, P. D. (2016). Social cognition in schizophrenia: factor structure of emotion processing and theory of mind. *Psychiatry Research, 242*, 150–156.

Cacciotti-Saija, C., Langdon, R., Ward, P. B., Hickie, I. B., Scott, E. M., Naismith, S. L., … Guastella, A. J. (2015). A double-blind randomized controlled trial of oxytocin nasal spray and social cognition training for young people with early psychosis. *Schizophrenia Bulletin, 41*(2), 483–493.

Chan, R. C. K., Gao, X. -J., Li, X. -Y., Li, H. -H., Cui, J. -F., Deng, Y. -Y., & Wang, Y. (2010). The Social Cognition and Interaction Training (SCIT): an extension to individuals with schizotypal personality features. *Psychiatry Research, 178*, 208–210.

Choi, K. H., & Kwon, J. H. (2006). Social cognition enhancement training for schizophrenia: a preliminary randomized controlled trial. *Community Mental Health Journal, 42*(2), 177–187.

Clayson, P. E., Kern, R. S., Nuechterlein, K. H., Knowlton, B. J., Bearden, C. E., Cannon, T. D., & Green, M. F. (2018). Social vs. non-social measures of learning potential for predicting community functioning across phase of illness in schizophrenia. *Schizophrenia Research*,https://doi.org/10.1016/j.schres.2018.07.046.

Combs, D. R., Adams, S. D., Penn, D. L., Roberts, D., Tiegreen, J., & Stem, P. (2007). Social Cognition and Interaction Training (SCIT) for inpatients with schizophrenia spectrum disorders: preliminary findings. *Schizophrenia Research, 91*(1–3), 112–116.

Combs, D. R., Chapman, D., Waguspack, J., Basso, M. R., & Penn, D. L. (2011). Attention shaping as a means to improve emotion perception deficits in outpatients with schizophrenia and impaired controls. *Schizophrenia Research, 127*, 151–156.

Combs, D. R., Tosheva, A., Penn, D. L., Basso, M. R., Wanner, J. L., & Laib, K. (2008). Attentional-shaping as a means to improve emotion perception deficits in schizophrenia. *Schizophrenia Research, 105*(1–3), 68–77.

Constans, J. I., Penn, D. L., Ihen, G. H., & Hope, D. A. (1999). Interpretive biases for ambiguous stimuli in social anxiety. *Behaviour Research and Therapy, 37*(7), 643–651.

Corrigan, P. W., Hirschbeck, J. N., & Wolfe, M. (1995). Memory and vigilance training to improve social perception in schizophrenia. *Schizophrenia Research, 17*(3), 257–265.

Couture, S. M., Penn, D. L., & Roberts, D. L. (2006). The functional significance of social cognition in schizophrenia: a review. *Schizophrenia Bulletin, 32*(Suppl. 1), S44–S63.

Davidson, C. A., Johannesen, J. K., & Fiszdon, J. M. (2016). Role of learning potential in cognitive remediation: construct and predictive validity. *Schizophrenia Research, 171*, 117–124.

Davidson, C. A., Lesser, R., Parente, L. T., & Fiszdon, J. M. (2017). Psychometrics of social cognitive measures for psychosis treatment research. *Schizophrenia Research, 193*, 51–57.

Drusch, K., Stroth, S., Kamp, D., Frommann, N., & Wolwer, W. (2014). Effects of Training of Affect Recognition on the recognition and visual exploration of emotional faces in schizophrenia. *Schizophrenia Research, 159*(2–3), 485–490.

Fanning, J. R., Bell, M. D., & Fiszdon, J. M. (2012). Is it possible to have impaired neurocognition but good social cognition in schizophrenia? *Schizophrenia Research, 135*, 68–71.

Fett, A. -K. J., Viechtbauer, W., Dominguez, M. -d. -G., Penn, D. L., van Os, J., & Krabbendam, L. (2011). The relationship between neurocognition and social cognition with functional outcomes in schizophrenia: a meta-analysis. *Neuroscience and Biobehavioral Reviews, 35*(3), 573–588.

Fiszdon, J. M., Roberts, D. L., Penn, D. L., Choi, K. H., Tek, C., Choi, J., & Bell, M. D. (2016). Understanding Social Situations (USS): a proof-of-concept social-cognitive intervention targeting theory of mind and attributional bias in individuals with psychosis. *Psychiatric Rehabilitation Journal, 40*(1), 12–20.

Frommann, N., Streit, M., & Wolwer, W. (2003). Remediation of facial affect recognition impairments in patients with schizophrenia: a new training program. *Psychiatry Research, 117*(3), 281–284.

Fuentes, I., Garcia, S., Ruiz, J., Soler, M., & Roder, V. (2007). Social perception training in schizophrenia: a pilot study. *International Journal of Psychology and Psychological Therapy, 7*(1), 1–12.

Fuerstein, R. (1980). *Instrumental enrichment*. Baltimore: University Park Press.

García, S., Fuentes, I., Ruíz, J., Gallach, E., & Roder, V. (2003). Application of the IPT in a Spanish sample: evaluation of the 'social perception subprogramme'. *International Journal of Psychology and Psychological Therapy*, *3*, 299–310.

Gaudelus, B., Virgile, J., Geliot, S., Team, T. G. R. S., & Franck, N. (2016). Improving facial emotion recognition in schizophrenia: a controlled study comparing specific and attentional focused cognitive remediation. *Frontiers in Psychiatry*. 7(105), https://doi.org/10.3389/psyt.2016.00105.

Gohar, S. M., Hamdi, E., El Ray, L. A., Horan, W. P., & Green, M. F. (2013). Adapting and evaluating a social cognitive remediation program for schizophrenia in Arabic. *Schizophrenia Research*, *148*(1–3), 12–17.

Gordon, A., Davis, P. J., Patterson, S., Pepping, C. A., Scott, J. G., Salter, K., & Connell, M. (2018). A randomized waitlist control community study of Social Cognition and Interaction Training for people with schizophrenia. *British Journal of Clinical Psychology*, *57*(1), 116–130.

Granholm, E., Ben-Zeev, D., Link, P. C., Bradshaw, K. R., & Holden, J. L. (2012). Mobile Assessment and Treatment for Schizophrenia (MATS): a pilot trial of an interactive text-messaging intervention for medication adherence, socialization, and auditory hallucinations. *Schizophrenia Bulletin*, *38*, 414–425.

Grigorenko, E. L. (2009). Dynamic assessment and response to intervention: two sides of one coin. *Journal of Learning Disabilities*, *42*, 111–132.

Habel, U., Koch, K., Kellermann, T., Reske, M., Frommann, N., Wolwer, W., … Schneider, F. (2010). Training of affect recognition in schizophrenia: neurobiological correlates. *Social Neuroscience*, *5*(1), 92–104.

Hasson-Ohayon, I., Avidan-Msika, M., Mashiach-Eizenberg, M., Kravetz, S., Rozencwaig, S., Shalev, H., & Lysaker, P. H. (2015). Metacognitive and social cognition approaches to understanding the impact of schizophrenia on social quality of life. *Schizophrenia Research*, *161*, 389–391.

Hasson-Ohayon, I., Kravetz, S., Levy, I., & Roe, D. (2014). Metacognitive and interpersonal interventions for persons with severe mental illness: theory and practice. *Israel Journal of Psychiatry & Related Sciences*, *46*(2), 141–148.

Hodel, B., Brenner, H. D., Merlo, M. C. G., & Teuber, J. F. (1998). Emotional management therapy in early psychosis. *British Journal of Psychiatry*, *172*(Suppl. 33), 128–133.

Hodel, B., Kern, R. S., & Brenner, H. D. (2004). Emotion Management Training (EMT) in persons with treatment-resistant schizophrenia: first results. *Schizophrenia Research*, *68*(1), 107–108.

Hogarty, G. E. (2000). Cognitive rehabilitation of schizophrenia. *Harvard Mental Health Letter*, *17*(2), 4–6.

Horan, W. P., Dolinksy, M., Lee, J., Kern, R. S., Hellemann, G., Sugar, C. A., … Green, M. F. (2017). Social cognitive skills training for psychosis with community-based training exercises: a randomized controlled trial. *Schizophrenia Bulletin*. https://doi.org/10.1093/schbul/sbx167.

Horan, W. P., Kern, R. S., Shokat-Fadai, K., Sergi, M. J., Wynn, J. K., & Green, M. F. (2009). Social cognitive skills training in schizophrenia: an initial efficacy study of stabilized outpatients. *Schizophrenia Research*, *107*(1), 47–54.

Horan, W. P., Kern, R. S., Tripp, C., Hellemann, G., Wynn, J. K., Bell, M., … Green, M. F. (2011). Efficacy and specificity of Social Cognitive Skills Training for outpatients with psychotic disorders. *Journal of Psychiatric Research*, *45*(8), 1113–1122.

Howes, O. D., & Murray, R. M. (2014). Schizophrenia: an integrated sociodevelopmental-cognitive model. *The Lancet*, *383*(9929), 1677–1687.

Kayser, N., Sarfati, Y., Besche, C., & Hardy-Bayle, M. C. (2006). Elaboration of a rehabilitation method based on a pathogenetic hypothesis of "theory of mind" impairment in schizophrenia. *Neuropsychological Rehabilitation*, *16*(1), 83–95.

Mancuso, F., Horan, W. P., Kern, R. S., & Green, M. F. (2011). Social cognition in psychosis: multidimensional structure, clinical correlates, and relationship with functional outcome. *Schizophrenia Research, 125*(2–3), 143–151.
Marsh, P. J., Green, M. J., Russell, T. A., McGuire, J., Harris, A., & Coltheart, M. (2010). Remediation of facial emotion recognition in schizophrenia: functional predictors, generalizability, and durability. *American Journal of Psychiatric Rehabilitation, 13*(2), 143–170.
Marsh, P. J., Langdon, R., McGuire, J., Harris, A., Polito, V., & Coltheart, M. (2013). An open clinical trial assessing a novel training program for social cognitive impairment in schizophrenia. *Australasian Psychiatry, 21*(2), 122–126.
Marsh, P. J., Luckett, G., Russell, T. A., Coltheart, M., & Green, M. J. (2012). Effects of facial emotion recognition remediation on visual scanning of novel face stimuli. *Schizophrenia Research, 141*(2–3), 234–240.
Marsh, P. J., Polito, V., Singh, S., Coltheart, M., Langdon, R., & Harris, A. W. (2016). A quasi-randomized feasibility pilot study of specific treatments to improve emotion recognition and mental-state reasoning impairments in schizophrenia. *BMC Psychiatry, 16*(1), 360. https://doi.org/10.1186/s12888-016-1064-6.
Mazza, M., Lucci, G., Pacitti, F., Pino, M. C., Mariano, M., Casacchia, M., & Roncone, R. (2010). Could schizophrenic subjects improve their social cognition abilities only with observation and imitation of social situations? *Neuropsychological Rehabilitation, 20*(5), 675–703.
Nahum, M., Fisher, M., Loewy, R., Poelke, G., Ventura, J., Nuechterlein, K. H., … Vinogradov, S. (2014). A novel, online social cognitive training program for young adults with schizophrenia: a pilot study. *Schizophrenia Research: Cognition, 1*(1), e11–e19.
Parker, S., Foley, S., Walker, P., & Dark, F. (2013). Improving the social cognitive deficits of schizophrenia: a community trial of Social Cognition and Interaction Training (SCIT). *Australasian Psychiatry, 21*(4), 346–351.
Penn, D. L., & Combs, D. (2000). Modification of affect perception deficits in schizophrenia. *Schizophrenia Research, 46*(2–3), 217.
Penn, D. L., Roberts, D. L., Munt, E. D., Silverstein, E., Jones, N., & Sheitman, B. (2005). A pilot study of social cognition and interaction training (SCIT) for schizophrenia. *Schizophrenia Research, 80*(2–3), 357–359.
Peyroux, E., & Franck, N. (2014). RC2S: a cognitive remediation program to improve social cognition in schizophrenia and related disorders. *Frontiers in Human Neuroscience, 8*(400) https://doi.org/10.3389/fnhum.2014.00400.
Pinkham, A. E., Harvey, P. D., & Penn, D. L. (2017). Social cognition psychometric evaluation: results of the final validation study. *Schizophrenia Bulletin*, https://doi.org/10.1093/schbul/sbx117 [ePub ahead of print].
Pinkham, A. E., & Penn, D. L. (2006). Neurocognitive and social cognitive predictors of interpersonal skill in schizophrenia. *Psychiatry Research, 143*(2–3), 167–178.
Riehle, M., & Lincoln, T. M. (2018). Investigating the social costs of schizophrenia: facial expressions in dyadic interactions of people with and without schizophrenia. *Journal of Abnormal Psychology, 127*(2), 202–215.
Roberts, D. L., Combs, D. R., Willoughby, M., Mintz, J., Gibson, C., Rupp, B., & Penn, D. L. (2014). A randomized, controlled trial of Social Cognition and Interaction Training (SCIT) for outpatients with schizophrenia spectrum disorders. *British Journal of Clinical Psychology, 53*(3), 281–298.
Roberts, D. L., Kleinlein, P., & Stevens, B. (2012). An alternative to generating alternative interpretations in social cognitive therapy for psychosis. *Behavioural and Cognitive Psychotherapy, 40*(4), 491–495.
Roberts, D. L., & Penn, D. L. (2009). Social cognition and interaction training (SCIT) for outpatients with schizophrenia: a preliminary study. *Psychiatry Research, 166*(2–3), 141–147.

Roberts, D. L., Penn, D. L., & Combs, D. R. (2016). *Social Cognition and Interaction Training (SCIT): Group psychotherapy for schizophrenia and other psychotic disorders, clinician guide*. New York, NY: Oxford University Press.

Roberts, D. L., Penn, D. L., Labate, D., Margolis, S. A., & Sterne, A. (2010). Transportability and feasibility of Social Cognition And Interaction Training (SCIT) in community settings. *Behavioural and Cognitive Psychotherapy, 38*(1), 35–47.

Roberts, D. L., & Velligan, D. I. (2012). Can social functioning in schizophrenia be improved through targeted social cognitive intervention? *Rehabilitation Research and Practice, 8*, https://doi.org/10.1155/2012/742106.

Roncone, R., Mazza, M., Frangou, I., De Risio, A., Ussorio, D., Tozzini, C., & Casacchia, M. (2004). Rehabilitation of theory of mind deficit in schizophrenia: a pilot study of metacognitive strategies in group treatment. *Neuropsychological Rehabilitation, 14*(4), 421–435.

Rose, A., Vinogradov, S., Fisher, M., Green, M. F., Ventura, J., Hooker, C., … Nahum, M. (2015). Randomized controlled trial of computer-based treatment of social cognition in schizophrenia: the TRuSST trial protocol. *BMC Psychiatry, 15*, 142.

Russell, T. A., Chu, E., & Phillips, M. L. (2006). A pilot study to investigate the effectiveness of emotion recognition remediation in schizophrenia using the micro-expression training tool. *British Journal of Clinical Psychology, 45*(Pt 4), 579–583.

Russell, T. A., Green, M. J., Simpson, I., & Coltheart, M. (2008). Remediation of facial emotion perception in schizophrenia: concomitant changes in visual attention. *Schizophrenia Research, 103*(1–3), 248–256.

Sachs, G., Winklbaur, B., Jagsch, R., Lasser, I., Kryspin-Exner, I., Frommann, N., & Wolwer, W. (2012). Training of affect recognition (TAR) in schizophrenia—impact on functional outcome. *Schizophrenia Research, 138*(2–3), 262–267.

Sarfati, Y., Passerieux, C., & Hardy-Bayle, M. (2000). Can verbalization remedy the theory of mind deficit in schizophrenia? *Psychopathology, 33*(5), 246–251.

Silver, H., Goodman, C., Knoll, G., & Isakov, V. (2004). Brief emotion training improves recognition of facial emotions in chronic schizophrenia. A pilot study [erratum appears in Psychiatry Res. 2004 Nov 30;129(1):113]. *Psychiatry Research, 128*(2), 147–154.

Steel, C., Wykes, T., Ruddle, A., Smith, G., Shah, D. M., & Holmes, E. A. (2010). Can we harness computerised cognitive bias modification to treat anxiety in schizophrenia? A first step highlighting the role of mental imagery. *Psychiatry Research, 178*(3), 451–455.

Tas, C., Danaci, A. E., Cubukcuoglu, Z., & Brune, M. (2012). Impact of family involvement on social cognition training in clinically stable outpatients with schizophrenia—a randomized pilot study. *Psychiatry Research, 195*(1), 32–38.

Taylor, R., Cella, M., Csipke, E., Heriot-Maitland, C., Gibbs, C., & Wykes, T. (2016). Tackling social cognition in schizophrenia: a randomized feasibility trial. *Behavioural and Cognitive Psychotherapy, 44*(3), 306–317.

Tsotsi, S., Kosmidis, M. H., & Bozikas, V. P. (2017). Improved facial affect recognition in schizophrenia following an emotion intervention, but not training attention-to-facial-features or treatment-as-usual. *Psychiatry Research, 254*, 135–142.

Turner, R., Hoppitt, L., Hodgekins, J., Wilkinson, J., Mackintosh, B., & Fowler, D. (2011). Cognitive bias modification in the treatment of social anxiety in early psychosis: a single case series. *Behavioural and Cognitive Psychotherapy, 39*(3), 341–347.

Vasquez-Campo, M., Marono, Y., Lahera, G., Mateos, R., & Garcia-Caballero, A. (2016). e-Motional Training: pilot study on a novel online training program on social cognition for patients with schizophrenia. *Schizophrenia Research: Cognition, 4*, 10–17.

Veltro, F., Mazza, M., Vendittelli, N., Alberti, M., Casacchia, M., & Roncone, R. (2011). A comparison of the effectiveness of problem solving training and of cognitive-emotional rehabilitation on neurocognition, social cognition and social functioning in people with schizophrenia. *Clinical Practice and Epidemiology in Mental Health*, 7, 123–132.

Voutilainen, G., Kouhia, T., Roberts, D. L., & Oksanen, J. (2016). Social Cognition and Interaction Training (SCIT) for adults with psychotic disorders: a feasibility study in Finland. *Behavioural and Cognitive Psychotherapy*, *44*(6), 711–716.
Wang, Y., Roberts, D. L., Xu, B., Cao, R., Yan, M., & Jiang, Q. (2013). Social cognition and interaction training for patients with stable schizophrenia in Chinese community settings. *Psychiatry Research*, *210*(3), 751–755.
Wolwer, W., & Frommann, N. (2009). The training of affect recognition (TAR): efficacy, functional specificity and generalization of effects. *Schizophrenia Bulletin*, *35*(Suppl. 1), 351.
Wolwer, W., & Frommann, N. (2011). Social-cognitive remediation in schizophrenia: generalization of effects of the training of affect recognition (TAR). *Schizophrenia Bulletin*, *37*(Suppl. 2), S63–S70.
Wolwer, W., Frommann, N., Halfmann, S., Piaszek, A., Streit, M., & Gaebel, W. (2005). Remediation of impairments in facial affect recognition in schizophrenia: efficacy and specificity of a new training program. *Schizophrenia Research*, *80*(2–3), 295–303.
Xu, W., Zhu, J., Zhangzi, J. Z., & Zhen, L. D. (2011). Social Cognition and Interaction Training: early rehabilitation effects in schizophrenia. *Chinese Journal of Rehabilitation Medicine*, *26*(11), 1062–1064.
Yiend, J., Mackintosh, B., & Mathews, A. (2005). Enduring consequences of experimentally induced biases in interpretation. *Behaviour Research and Therapy*, *43*(6), 779–797.

CHAPTER TWELVE

# Psychosocial interventions for social dysfunction in psychosis

**Kathryn E. Lewandowski*[,†], Raquelle I. Mesholam-Gately[†,‡], Matcheri S. Keshavan[†,‡]**
*McLean Hospital, Psychotic Disorders Division, Belmont, MA, United States
[†]Harvard Medical School, Department of Psychiatry, Boston, MA, United States
[‡]Massachusetts Mental Health Center Public Psychiatry Division, Beth Israel Deaconess Medical Center, Boston, MA, United States

In this chapter, we provide an overview of psychosocial interventions that incorporate social engagement or social skills targets for patients with psychosis. We briefly review early formulations of psychosis with regard to sociality, and how they have informed psychosocial intervention approaches. We then describe social functioning as a target in broad-based psychosocial treatments, and interventions targeting social functioning directly. We will then discuss evidence regarding efficacy of these interventions on social cognition and social function. Psychosocial interventions that target social cognition as the sole therapeutic aim are covered in a separate chapter (Fiszdon & Davidon, Chapter 11).

## Early interventions for social deficits in psychosis

Impairment in interpersonal functioning has long been considered a hallmark feature of schizophrenia, and is present in both prodromal and residual phases of illness, perhaps long before symptom onset (e.g., Schiffman et al., 2004; Strauss, Kokes, Klorman, & Sacksteder, 1977; Tarbox et al., 2014). Poorer premorbid social adjustment is among the strongest predictors of prognosis (Mueser, Bellack, Douglas, & Morrison, 1991; Strauss et al., 1977), and deficits in social cognition are highly predictive of overall community functioning, quality of life, and even morbidity/mortality (Brekke, Kay, Lee, & Green, 2005; Cacioppo, Cacioppo, Capitanio, & Cole, 2015; Cacioppo, Hughes, Waite, Hawkley, & Thisted, 2006; Fett et al., 2011; Holt-Lunstad, Smith, Baker, Harris, & Stephenson, 2015). Interventions aimed at improving social functioning have been available for decades, developed based on evolving theories of the nature and origin of social dysfunction in psychosis.

*Social Cognition in Psychosis*
https://doi.org/10.1016/B978-0-12-815315-4.00012-4

## Social skills deficits and training

The behavioral model of social skills (Argyle, Bryant, & Trower, 1974; Goldstein, Gershaw, & Sprafkin, 1975; Hersen & Bellack, 1976; Liberman, 1982; Liberman, Wallace, Falloon, & Vaughn, 1981) posits that deficits in social functioning in patients with schizophrenia result from poor integration and maintenance of social skills, due either to failure to integrate them during development, or to loss of skills resulting from loss of social opportunity. Based on this model, a number of social skills training (SST) programs were developed aimed at the teaching and practice of skills related to successful interpersonal interactions (e.g., Bellack & Hersen, 1979; Bellack, Mueser, Gingerich, & Agresta, 2004; Eisler & Frederiksen, 1980; Morrison & Bellack, 1987; Wallace et al., 1980), and became one of the most widely used behavioral interventions in schizophrenia (Morrison & Bellack, 1987). Most social skills paradigms focus on explicit behavioral efforts to build specific, appropriate social behaviors (and reduce inappropriate behaviors) using techniques such as psychoeducation, modeling, role play, practice, and feedback (Morrison & Bellack, 1987). Areas of training vary considerably by program, but may include process-related aspects of social skills such as eye contact, vocal volume and prosody, and gesturing, or content-related topics such as making "I" statements, assertiveness (Field & Test, 1975; Hersen, Turner, Edelstein, & Pinkston, 1975b), or conversation skills (Hansen, St Lawrence, & Christoff, 1985; Minkin et al., 1976). SST programs report improvements in acquisition of specific skills and patient self-report of efficacy; however, the external validity of assessments of the efficacy of these programs has been largely lacking (e.g., Bellack, 1983; Liberman, 1982), and improvements on specifically-trained skills often fail to transfer to untrained skills, community outcomes, or quality of life (Hogarty, Goldberg, & Schooler, 1974; Patterson & Leeuwenkamp, 2008; Trower, 1980; Wallace et al., 1980).

## Social motivation and reward

Another hypothesized contributor to social functioning deficits in psychosis involves the role of anhedonia and amotivation. Since early descriptions of schizophrenia (e.g., Bleuler, 1911; Krapelin, 1919), lack of involvement in and motivation for pleasurable activities has been identified as a hallmark of psychosis, and asociality in the context of treatment and relapse prevention has been identified as an early indicator of symptom exacerbation, and suggested as an assessment target in the rehabilitative milieu (Lukoff,

Liberman, & Nuechterlein, 1986). More recent findings suggest that patients with schizophrenia do, in fact, report a desire for social affiliation (Blanchard, Park, Catalano, & Bennett, 2015; Gard, Fisher, Garrett, Genevsky, & Vinogradov, 2009; McCarthy et al., 2018). However, evidence also suggests that patients with schizophrenia and related disorders may experience reduced anticipatory pleasure regarding social experiences (Campellone & Kring, 2018; Campellone, Truong, Gard, & Schlosser, 2018; Engel, Fritzsche, & Lincoln, 2016; Mueser et al., 1991), reduced positive and increased negative affect in the context of social situations (McCarthy et al., 2018), and both reduced social approach behaviors and increased social avoidance behaviors (de la Asuncion, Docx, Sabbe, Morrens, & de Bruijn, 2015; Radke, Pfersmann, & Derntl, 2015). Fulford, Campellone, and Gard (2018) proposed a model in which diminished social approach and reward sensitivity, and heightened social avoidance and punishment sensitivity, may interact in social situations to produce decreased social approach and anticipatory pleasure prior to social engagement, disruptions in reward learning and prediction errors in the social setting, and negative consolidation after the social encounter is over. Because abnormalities in social reward processing and motivation are thought to be distinct from skills deficits (Fulford et al., 2018), some intervention efforts, including therapeutic communities, milieu therapy, and group therapy have sought to increase social exposure and reframe aspects of social reward and learning in a therapeutic setting.

### *Therapeutic communities and milieu therapy*

Based largely on the early observations of Bleuler (1911), the role of therapeutic community settings for patients with schizophrenia was to harness social and group processes as part of the treatment itself (Becker & Kösters, 2012). These models emphasize community-style settings, or milieus, as active elements of treatment, as opposed to "asylum-style" treatment settings in which patients are treated largely in isolation, and social engagement is not a focus. A systematic review (Lees, Manning, & Rawlings, 1999) found support for the efficacy and effectiveness of therapeutic community treatment in terms of symptoms and social outcomes. Similarly, Soteria, a therapeutic milieu environment started in the 1970s in the United States and continued in Europe, emphasized social networks and communal responsibilities in open treatment settings. These facilities are staffed largely by nonmedical staff, and emphasize no or low-dose medication interventions. Soterias have been found to reduce psychotic symptoms similar to antipsychotic effects, and improve social adjustment more than traditional medical intervention models (Mosher, 1999).

### *Group therapy*

Group therapy interventions for patients with schizophrenia have historically been oriented toward providing a supportive, safe environment in which patients could gain exposure to social settings, and practice and develop experience with social engagement (Mosher & Keith, 1979). While early findings were mixed, some evidence suggests that group therapy as part of larger integrated treatment could be effective in both inpatient and outpatient settings, and may offer particular benefit for socialization and interpersonal skills outcomes (Claghorn, Johnstone, Cook, & Itschner, 1974; Mosher & Keith, 1979; O'Brien et al., 1972). More recent evidence supports the utility of group therapy for social deficits through dynamic skills and reinforcement learning (Elis, Caponigro, & Kring, 2013; Kopelowicz, Liberman, & Zarate, 2006; Orfanos, Banks, & Priebe, 2015). A large metaanalysis found that group-based interventions were superior to treatment as usual in reducing negative symptoms and improving social functioning regardless of group orientation (Orfanos et al., 2015), suggesting that there are some nonspecific benefits to group-format treatment for patients beyond explicit skills training. Group interventions may target both skills deficits and aberrant reward and motivation processes by providing modeling, practice, and feedback experiences, as well as in vivo opportunities to evaluate and modify expectancies, prediction errors, and reinforcement learning in a relatively controlled and safe environment.

## Psychosocial interventions that target social cognition and/or social functioning directly: Targets and outcomes

In recent years, a number of group-based interventions have taken a more targeted approach to improving social skills, social cognition, and social functioning. In this section, our predominant focus is on broad-based multimodal interventions for psychosis, with a strong evidence base. We do not review pure SST or brief proof of concept social cognition approaches but integrated, multielement psychosocial interventions that target and report on social cognition and/or social functioning outcomes as part(s) of the interventions. The vast majority of these approaches includes cognitive remediation. The combination of cognitive remediation with social cognitive or social rehabilitative approaches was driven largely by findings that both neurocognitive deficits and social cognitive impairments are predictive of poor community functioning (e.g., Ospina et al., 2018). Based on models hypothesizing that some impairments in social functioning may be related to cognitive impairment, the combined approach of strengthening neurocognition to augment patients' ability to maximally benefit from social cognition and social rehabilitation interventions has been studied in several paradigms.

In addition to significantly improving cognition, including social cognition, metaanalyses (McGurk, Twamley, Sitzer, McHugo, & Mueser, 2007; Wykes, Huddy, Cellard, McGurk, & Czobor, 2011) have shown that, as a heterogeneous group, cognitive remediation programs significantly enhance psychosocial functioning, particularly when provided with adjunctive psychosocial rehabilitation. Moreover, combining social cognition training with traditional neurocognitive approaches to cognitive remediation, especially in group settings, is especially beneficial for functional outcome (Mueller, Schmidt, & Roder, 2015). As such, we begin our review with integrated cognitive remediation plus rehabilitation approaches, followed by social cognitive interventions that incorporate other models, such as cognitive-behavioral therapy, psychoeducation, and metacognitive training.

## Integrated Psychological Therapy

Integrated Psychological Therapy (IPT), originally developed in Germany (Brenner et al., 1994; Brenner, Stramke, Mewes, Liese, & Seeger, 1980), is one of the first and most extensively used and studied group-based cognitive remediation interventions for schizophrenia. IPT is a comprehensive approach that combines group-based neurocognitive and social cognitive remediation therapies. Five integrated treatment components are delivered in a group setting over the course of 12–18 weeks in a bottom-up fashion, starting with basic neurocognitive abilities, followed sequentially and hierarchically by social perception, verbal communication, social skills, and interpersonal problem-solving. Groups are small, typically consisting of five to eight participants who meet for 30–90 minutes per session.

Unlike many other cognitive remediation programs, the neurocognitive training in IPT is not computerized, and focuses on group-based strategic planning and problem-solving rather than task repetition. Groups work collaboratively on basic cognitive exercises similar to those introduced in computerized programs. The second IPT component, social cognition training in social perception, emphasizes the accurate identification of relevant social information and reducing environmental distractions not relevant for social understanding that may lead to overinterpreting or drawing conclusions not consistent with reality. The third component, verbal communication, builds on the prior neurocognitive and social-cognitive training to focus on improving attention to and using reciprocity in conversations. The final two treatment components, social skills and interpersonal problem-solving, rely on more traditional behavioral skills interventions with established SST and problem-solving methods to improve independent living, vocational and interpersonal skills, and the analysis and application of various solutions to social situations.

In a metaanalysis of 30 studies of IPT in schizophrenia (Roder, Mueller, Mueser, & Brenner, 2006), IPT was associated with significant and medium-sized improvements in neurocognition (Cohen's $d=0.54$), positive symptoms ($d=0.46$), negative symptoms ($d=0.41$), and psychosocial function ($d=0.41$), with beneficial effects lasting for as long as 8 months.

## Cognitive Enhancement Therapy

Cognitive Enhancement Therapy (CET) (Hogarty et al., 2004; Hogarty & Greenwald, 2006) is another widely-used, well-studied integrated intervention combining neurocognitive and social-cognitive training. CET is a developmental approach, recognizing the importance of brain developmental processes around adolescence for the emergence and maturation of cognition and secondary socialization processes, and, thus aiming to "jump-start" cognitive development through enriched environmental and socialization experiences (Hogarty & Flesher, 1999; Keshavan, 1999). Like IPT, which CET was inspired by, the treatment values the power of the group modality and the importance of addressing both the social and non-social deficits that limit functional recovery in schizophrenia.

CET is an 18-month intervention comprised of 60 hours of computer-based neurocognitive training, along with 45 social-cognitive group sessions. The intervention begins with the neurocognitive training, which includes three components in the following bottom-up, hierarchical order: attention training (months 0–4), memory training (months 5–11), and problem-solving training (months 12–18). Unique to CET, the neurocognitive training is administered in participant pairs with the assistance of a therapist/coach. During the 1-hour sessions, the participant pair takes turns working on the computerized cognitive exercise or keeping score for the other partner while providing support and encouragement.

A 5-minute break is taken mid-session for relaxation and further promotion of socialization. The therapist/coach sets up and introduces the cognitive exercises, helps participants with strategic problem-solving for the neurocognitive exercises, applies concepts learned in the social-cognitive group to neurocognitive training and everyday life, facilitates socialization, and provides guidance and further support. This partnering and collaboration in this triad provides a meaningful social component, enhancing neurocognitive training with frequent opportunities for practicing social cognition skills, strategizing, empowerment, improving treatment engagement and motivation, and building familiarity and teamwork among the small three-person group who ultimately becomes part of the larger social-cognitive group.

The 1.5 hour social-cognitive group sessions, which are comprised of three to four participant pairs and their coaches, start after approximately 3 months of neurocognitive training in attention. These 45 sessions are highly structured, including psychoeducation, in-group exercises, and homework assignments. The group curriculum spans three modules: Basic Concepts (months 4–8), Social Cognition (months 9–14), and CET Applications (months 15–18). Basic Concepts includes education about schizophrenia and its management, medication, stress and its regulation, aspects of neurocognition and strategies for improvement, verbal communication such as gistful thinking, and motivation. The Social Cognition module addresses important aspects of social-cognitive abilities, such as discerning emotional and other nonverbal cues, providing support, perspective-taking, and social context appraisal. The last group module, CET Applications, is directed toward applying and generalizing learned social-cognitive skills to everyday situations, including social interactions, vocational settings, and other meaningful activities.

In addition to the neurocognitive and group social-cognitive training, CET tailors the intervention to the participant's specific goals and challenges through weekly one-on-one meetings between the participant and the therapist/coach. Individual recovery plans are then developed from these goals and needs. The rehabilitation approach is further personalized through assessment of the participant's cognitive style(s), a concept introduced by Hogarty et al. (2004) to better characterize the greatest areas of cognitive impairment the person is experiencing. Based on literature reviews of the heterogeneous presentation of cognitive deficits in schizophrenia and their own clinical experience, they identified three primary cognitive styles: unmotivated, disorganized, and inflexible. Although these styles are viewed as dimensional and potentially overlapping, one cognitive style is typically most prominent and helps to guide and adapt the CET treatment toward the participant's primary cognitive challenges.

CET has been listed by the US Substance Abuse and Mental Health Services Administration (SAMHSA) as an evidence-based practice for schizophrenia. Its efficacy has been shown across randomized clinical trials (RCTs) of both chronic (Hogarty et al., 2004) and early course schizophrenia (Eack et al., 2009), with significant and medium-to-large improvements demonstrated in neurocognition and social cognition, as well as functioning (see Fig. 1). Notably, of these outcomes assessed, the greatest gains from CET were observed in social cognition at 24-month follow-ups, particularly for individuals with early course schizophrenia. CET also significantly improves negative symptoms with medium-sized effects ($d=0.61$) (Eack, Mesholam-Gately, Greenwald, Hogarty, & Keshavan, 2013), and it has been shown to be an effective treatment in people with schizophrenia who misuse alcohol and/or cannabis, with significant and large improvements found in social cognition ($d=1.13$), social adjustment ($d=0.92$), and neurocognition ($d=0.86$) (Eack et al., 2015).

## Neuropsychological Educational Approach to Remediation

The Neuropsychological Educational Approach to Remediation (NEAR) is an entirely group-based, clinician-led cognitive remediation treatment for a range of psychiatric conditions including schizophrenia (Medalia, Revheim, & Casey, 2002; Medalia, Revheim, & Herlands, 2009). Computerized training is integrated with "bridging groups" of three to nine participants to apply learned skills to daily functioning. NEAR, unlike the majority of other cognitive remediation interventions, is not wedded to nor does it require use of specific software programs for the computerized training, but allows for utilization of any relevant and engaging, commercially available software for targeting neurocognitive deficits.

While NEAR does not specifically target social cognition, the bridging groups incorporate discussions and in-group exercises for generalizing trained neurocognitive abilities to everyday life, including social relationships. The bridging groups format, as with other group approaches, may also enhance social functioning. Moreover, NEAR uniquely uses peer leaders during both the computerized training and bridging groups. Because NEAR has rolling admissions, peer leaders are those who have more experience with the computerized and bridging group activities than newer

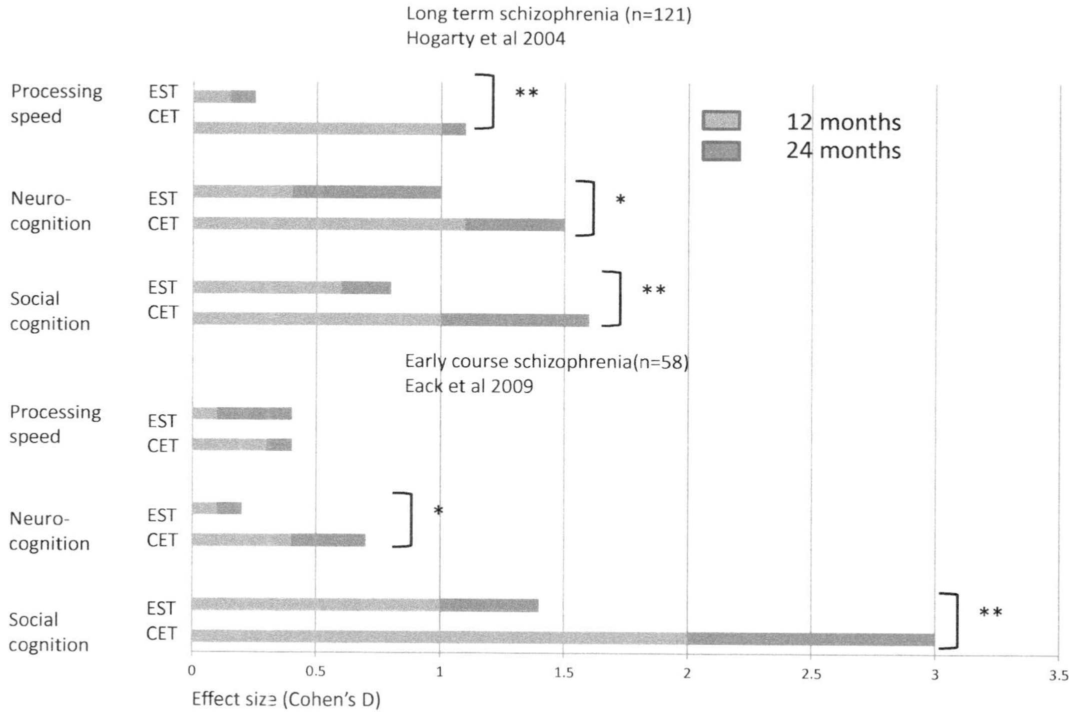


**Fig. 1** Effect sizes of Cognitive Enhancement Therapy in chronic and early course schizophrenia. *From Hogarty, G. E., Flesher, S., Ulrich, R., Carter, M., Greenwald, D., Pogue-Geile, M., …, & Zoretich, R. (2004). Cognitive enhancement therapy for schizophrenia: effects of a 2-year randomized trial on cognition and behavior.* Archives of General Psychiatry 61*(9), 866–876 and Eack, S. M., Greenwald, D. P., Hogarty, S. S., Cooley, S. J., DiBarry, A. L., Montrose, D. M., & Keshavan M. S. (2009). Cognitive enhancement therapy for early-course schizophrenia: effects of a two-year randomized controlled trial.* Psychiatric Services 60*(11), 1468–1476.*

participants. These peer leaders can thus provide additional training and guidance, serve as mentors, and illustrate the benefits of the intervention for the newer NEAR participants, as well as facilitate motivation and engagement, and build self-confidence and social skills for both the newer participants and themselves.

Given NEAR's flexible approach, it has been broadly implemented in various community agencies, and its efficacy has been examined in a number of studies; though, at times, the rigor of the evidence base has been limited by small sample sizes, or not being published in peer-reviewed literature (Keefe et al., 2012; Medalia & Freilich, 2008). While most studies have evaluated cognition and functional outcomes, one randomized controlled trial and a few community-based investigations have examined the impact of NEAR on measures of psychosocial functioning. In a multisite randomized waitlist control trial conducted in Australia (Rogers & Redoblado-Hodge, 2006), individuals with schizophrenia exposed to 20–30 sessions of NEAR over the course of 15 weeks showed significant improvements in social and occupational functioning relative to a control group. NEAR was also found to improve treatment engagement and goals in a sample of mixed-diagnosis outpatients at an inner city Intensive Psychiatric Rehabilitation Treatment program (Revheim, Kamnitzer, Casey, & Medalia, 2001).

## Functional Remediation

Functional Remediation (FR) is an intervention developed for people with bipolar disorder, including presentations with psychotic symptoms. As the name suggests, FR's primary aim is improving functional outcomes, including psychosocial functioning, rather than just cognition (Vieta, Torrent, & Martinez-Aran, 2014). Originally developed in Spain (Martínez-Arán et al., 2011), this manualized, evidence-based intervention focuses on psychoeducation about neurocognitive deficits and their impact on everyday life, as well as the application of learned cognitive skills and other strategies to improve daily functioning, particularly in social and occupational domains. FR includes 21 weekly sessions, each lasting 90 minutes, and is designed for a group format, but can also be adapted by clinicians for individual work with patients.

Sessions include practical daily life examples, individual and group tasks, and homework. They cover a broad range of topics such as cognitive skills, interpersonal relationships, autonomy, communication and stress management, as well as varied approaches such as modeling techniques, role-playing, self-instructions, verbal instructions, positive reinforcement, and meta-cognitive cues. Strategies and tasks are designed for performance and repetition in the clinic and at home, and can involve family or other caregivers to help facilitate and reinforce at-home practice.

The first RCT of FR was a multisite investigation in Spain conducted with 239 euthymic outpatients with bipolar I or II disorder, including 61% in the FR group who had a lifetime history of psychotic symptoms (Torrent et al., 2013). After finishing the 21-week intervention, with 183 completers, the FR group showed a large and significantly improved functional outcome ($d=0.93$) compared with a treatment-as-usual (TAU) group. The functional gains were specifically observed in interpersonal and occupational domains. These positive effects of FR lasted for 6 months after the final session of the intervention (1 year after baseline) (Bonnin et al., 2016); however, when functional outcomes were analyzed in detail, the improvements were only maintained for global functioning and the autonomy domain indices.

## Neurocognitive Enhancement Therapy

Neurocognitive Enhancement Therapy (NET) is a 26-week intervention comprised of three components: up to 5 hours per week of computer-based neurocognitive training, feedback on cognitive performance in the workplace, and a weekly social processing group. While NET has been studied in multiple investigations, only one (Bell, Bryson, Greig, Corcoran, & Wexler, 2001) examined social functioning or social cognition, specifically affect recognition, as one of the outcomes. Results showed that NET plus work therapy, compared with work therapy alone, significantly improved affect recognition in a sample of individuals with schizophrenia-spectrum disorders.

## Combined theory of mind and cognitive remediation therapy

This integrated approach was designed to address theory of mind (ToM) difficulties in schizophrenia through a combination of computerized cognitive remediation therapy (CRT) (via CogPack Software, http://www.markersoftware.com), and one of two recently developed, clinician-led social cognitive group interventions: Social Cognitive Training (SCT) or Theory of Mind Intervention (ToMI) (Bechi et al., 2012, 2013, 2015). SCT uses short videos showing social interactions to address emotion recognition and ToM abilities, and is conducted in 1-hour sessions/week over the course of 12 weeks. The ToMI uses comic strips, also showing social interactions, to target cognitive and affective aspects of ToM, and is implemented in 1-hour sessions/twice a week for a total of 18 sessions. Both SCT and ToMI include groups of about five members, and sessions end with guided discussions of hypothesized interpretations of the social interactions. The CRT is comprised of two 1-hour sessions/week over a 12-week period.

The efficacy of this approach for improving ToM in schizophrenia has been demonstrated in two prospective controlled studies compared with usual rehabilitative interventions (Bechi et al., 2012, 2013) and an RCT using an active control group (newspaper discussion) comparison (Bechi et al., 2015). The RCT further showed that the ToM improvements were not related to baseline neurocognitive or social cognitive performance, nor to the positive changes in these domains after the interventions.

## Cognitive Adaptation Training

Considered an adaptation or compensatory approach, Cognitive Adaptation Training (CAT) (Velligan et al., 2000) uses a combination of individualized cognitive-behavioral assessment, skills training, and personally-tailored environmental supports or modifications to circumvent some of the cognitive impairments associated with schizophrenia. Examples of environmental supports include signs, checklists, calendars, reminder notes, hygiene supplies, and pill containers. CAT can be particularly beneficial or well-suited for those with lower intellect who may experience more challenges with many social cognition training programs that often place more demand on cognitive functioning or rely on more advanced vocabulary or conceptual understanding. CAT has been shown to produce significant improvements in global and social functioning, motivation, medication adherence, and positive symptoms, as well as reductions in psychiatric relapse in schizophrenia (Velligan et al., 2000, 2006).

## Cognitive Behavioral Social Skills Training

Cognitive Behavioral Social Skills Training (CBSST) is the first integrated psychosocial approach systematically combining elements of cognitive behavioral therapy (CBT) and SST (Granholm et al., 2005; McQuaid et al., 2000). Originally developed to improve social functioning and quality of life in older individuals with chronic schizophrenia, symptoms are considered secondary targets. CBSST is flexible and personalized, and guided by a treatment manual with descriptions of skills and homework assignments (Granholm, McQuaid, & Holden, 2016).

The intervention is divided into three modules: CBSST Thought Challenging or Healthy Thinking Module, CBSST Social Skills Training Module, and the CBSST Problem Solving Module. These modules consist of age- and illness-sensitive psychoeducation in cognitive and behavioral coping skills (e.g., for health and financial challenges), social functioning, problem-solving, and compensatory supports for neurocognitive deficits (written materials, mnemonic strategies, repetition, etc.), as well as techniques such as role-playing of communication skills, thought challenging (e.g., simple steps such as the 3 Cs—Catch It, Check It, Change It), and problem-solving training. There are typically 18 weekly sessions in either a group or individual format, though the exact number of sessions can vary, and, because the modules are self-contained, the intervention can be designed with rolling admissions for participants to enter at the beginning of any module.

Like CET, CBSST has been recognized by SAMHSA as an evidence-based practice for schizophrenia. It is a well-studied intervention, and its efficacy has been demonstrated across multiple investigations, including several RCTs with comparisons to TAU, as well as active goal-focused supportive contact groups. The first RCT pilot study was conducted in 2002 by Granholm, McQuaid, McClure, Pedrelli, and Jeste (2002), but only included 15 people, and just focused on CBSST's beneficial effects on positive and depressive symptoms. Outcomes of subsequent larger RCTs have shown that middle-aged to older individuals with schizophrenia who received CBSST performed significantly better on measures of social functioning and independent living skills (Granholm et al., 2005, 2007; Granholm, Holden, Link, & McQuaid, 2014; Granholm, Holden, Link, McQuaid, & Jeste, 2013), as well as cognitive insight (Granholm et al., 2005), experiential negative symptoms (amotivation/asociality), and defeatist performance attitudes (Granholm et al., 2014) relative to TAU and supportive contact control groups. Medium-sized effects were typically observed across the different assessment measures, and it is notable that one of these studies (Granholm et al., 2014) examined a broader age range, 18–65. Furthermore, the beneficial impact of CBSST on social functioning and independent living skills has been shown to persist for up to 1 year after the completion of CBSST (Granholm et al., 2007, 2014).

While CBSST was developed for and mostly studied in samples of older, chronically ill people with schizophrenia, a preliminary feasibility study of

CBSST for first-episode psychosis (Herman, Shireen, Bromley, Yiu, & Granholm, 2018) showed that CBSST group participation was associated with significant and medium effects on measures of adaptive functioning (including social competence) and independent living skills, as well as large effects on a global index of social and occupational functioning. These positive changes were also maintained at 3-month follow-up. The effects of CBSST for young individuals at risk for psychosis are currently being studied.

## Functional Adaptation Skills Training

Like CBSST, Functional Adaptation Skills Training (FAST) integrates components of CBT and SST for older populations of English-speaking (FAST) and Spanish-speaking (PEDAL) individuals with schizophrenia and other psychoses (Patterson et al., 2003, 2005, 2006). FAST consists of 24 clinician-led, semiweekly, 120-minute group sessions, which rely on a highly structured group curriculum comprised of psychoeducation, homework, and in-session practice and behavioral modeling using role play.

FAST has been found to improve social function and negative symptoms, with effects lasting up to 6 months (Patterson et al., 2005). When FAST was combined with cognitive remediation in an RCT (Bowie, McGurk, Mausbach, Patterson, & Harvey, 2012), significantly greater functional improvements were observed than those demonstrated from either FAST or cognitive remediation alone. Such findings again point to the unique and powerful benefits of broad, multielement psychosocial interventions.

## Metacognitive training

Metacognitive training (MCT) for psychosis (Moritz et al., 2014) aims to improve awareness, insight and understanding of cognitive biases or distortions (e.g., jumping to conclusions) that are the basis for delusions, and uses a hybrid of CBT and psychoeducation methods, including progressive practice of metacognitive exercises, to correct these biases. MCT also concentrates on enhancing awareness of cognitive strengths and challenges, as well as understanding the impacts of cognitive difficulties on everyday life and methods for mitigating these impacts. While MCT is similar to CBT, MCT is typically more focused on social cognition and neurocognitive processes, especially the "how" of thinking.

The MCT manualized training program is available at no cost in a group format (http://www.uke.de/mct) (Moritz, Veckenstedt, Bohn, Kother, & Woodward, 2013; Moritz, Vitzthum, Randjbar, Veckenstedt, & Woodward, 2010), and in an individualized format (MCT+; http://www.uke.de/mct_plus) (Moritz, Veckenstedt, Randjbar, Vitzthum, & Woodward, 2011). MCT is comprised of eight modules: Attribution, Jumping to Conclusions I, Changing Beliefs, To Empathize I, Memory, To Empathize II (ToM second order), Jumping to Conclusions II, and Mood and Self-Esteem. While, as noted, the primary target of MCT is positive psychotic symptoms, MCT+ also addresses negative symptoms, as well as individual symptoms and recovery plans. Moreover, the latest versions of MCT, as well as MCT+, incorporate exercises for managing social difficulties.

Multiple studies, including recent metaanalyses (Eichner & Berna, 2016; Liu, Tang, Hung, Tsai, & Lin, 2018; van Oosterhout et al., 2016), have demonstrated the efficacy of MCT; however, the primary outcomes have been positive symptoms, particularly delusions, jumping to conclusions, and acceptance/subjective effectiveness. Fewer studies have examined social cognition or social functioning outcomes, though some promising findings have emerged with MCT showing positive impacts on satisfaction with interpersonal relations (Moritz et al., 2011), ToM (Ochoa et al., 2017; Rocha & Queirós, 2013), and social perception, emotion recognition, and social functioning (Rocha & Queirós, 2013). In the latter MCT study (Rocha & Queirós, 2013), MCT was combined with Social Cognition and Interaction Training.

## Summary, limitations, and future directions

### Summary

Interventions targeting social functioning have largely been guided by prevailing frameworks for understanding social deficits in psychosis, and the recognition of social engagement as a key treatment target in this population. Early formulations focused on skills training based on the behavioral model of social skills deficits in patients with psychosis. More recently-developed social cognitive interventions have incorporated research findings showing complex interactions between social cognition, neurocognition, metacognition, and reward and motivation related to social engagement, together with social skills rehabilitation and social context models, to create multidimensional interventions aimed at targeting multiple aspects of sociality. Overall there is good evidence for the efficacy of a number of interventions, and newer programs have found more convincing evidence of transfer of treatment effects to community-based measures and recovery goals. However, some limitations remain (Horan & Green, 2017).

## Limitations

First, how outcomes should be measured in terms of efficacy has been a longstanding challenge that remains unresolved. Early studies of SST found that skills were acquired successfully; however, assessments often lacked psychometric validation (Morrison & Bellack, 1987). For example, early assessments of social skills often relied on judges or raters to assess whether a patient was "skilled" or "unskilled" based on a social interaction (e.g., Argyle et al., 1974; Eisler & Frederiksen, 1980; Hersen, Turner, Edelstein, & Pinkston, 1975a), lacking specificity of which component behaviors were contributing to the assessment. Measurement of social treatment effects continues to be a concern due to limited psychometric support, despite development of a number of tests of various aspects of social cognition. The Social Cognition Psychometric Evaluation (SCOPE) study found that limited psychometric data are available for most measures of social cognition (Pinkham et al., 2014). As described in Chapter 9, only two measures evaluated by the SCOPE study—the Bell Lysaker Emotion Recognition Task and the Hinting task—had strong psychometric properties. Three others—the Emotion Recognition Task, the Reading the Mind in the Eyes Task, and The Awareness of Social Inferences Test—had moderate psychometric properties, and further studies were recommended. Other tasks were found to show poor psychometrics, and their use was cautioned (Pinkham, Penn, Green, & Harvey, 2016).

In addition to limitations in psychometrics, many measures lack clear evidence of generalizability to real-world settings (Figueira & Brissos, 2011; Priebe, 2007). Early assessments often failed to account for environmental effects on social functioning, and lacked external validity (Bellack, 1983; Liberman, 1982; Morrison & Bellack, 1987). Additionally, inadequate attention was paid to the generalization of skills acquisition to daily functioning (Morrison & Bellack, 1987). Consistent with these early reports, recent studies have shown that social functional capacity, as measured in a laboratory setting under optimal conditions, may not correlate with real world social functional behavior (Bowie et al., 2012; Gupta, Holshausen, Mausbach, Patterson, & Bowie, 2012; Leifker, Bowie, & Harvey, 2009; Menendez-Miranda et al., 2015). Thus, improvements in social processing or functioning on a lab measure may not translate to improved social functioning in daily life.

## Future directions

Careful development and validation of assessments of social cognition, social capacity, and social functioning are still needed. Novel methods for assessment of social cognitive ability and real-world social functioning, such as experience sampling or ecological momentary assessment (EMA) techniques, have been used to capture not only what patients are able to do in a laboratory setting, but what they are actually doing in the real world, accounting for contextual aspects of sociality without the confounds of retrospective reporting or controlled laboratory and testing settings (e.g., Granholm, Loh, & Swendsen, 2008; Schneider et al., 2017; Chapter 10). EMA also permits evaluation of the of the social environment as a key variable. EMA is feasible in patients with psychosis (Granholm et al., 2008), and recent findings have shown complex interactions among patient characteristics, mood, behavior, and context (e.g., Depp et al., 2016; Granholm, Ben-Zeev, Fulford, & Swendsen, 2013). Last, precisely defining meaningful functional outcomes is key. As noted herein, social functional capacity may not be strongly associated with social functional behavior in real-world settings. Studies must decide which outcome is being targeted, select assessments that are valid measures of that target, and interpret findings in the context of the assessments that are being used.

Novel intervention techniques targeting social cognition and social functioning are continuing to be developed based on an evolving literature regarding frameworks for understanding social cognitive deficits in psychosis. For example, the importance of ecological validity is not just an issue in assessment, but may also limit generalizability of interventions to real world settings. The use of mobile devices to deliver interventions in contextually relevant settings may help patients to learn and apply skills and concepts in ways that bridge the gap between the clinic and the real world (Depp et al., 2010; Myin-Germeys, Klippel, Steinhart, & Reininghaus, 2016). For example, a recent randomized controlled trial of a virtual-reality based CBT intervention targeting paranoid ideation to increase social engagement was feasible, and showed significant effects on paranoia and anxiety, although the authors did not find transfer to increased socialization (Pot-Kolder et al., 2018).

As can be seen in the preceding sections, with the exception of FR (Martínez-Arán et al., 2011), which was developed explicitly for patients with bipolar disorder, the vast majority of these programs were developed based on models of social deficits in schizophrenia, and subsequently applied to this population. Interventions developed based on models of social cognitive and functional impairments in related populations, such as patients with affective psychosis, are needed. Indeed, some evidence suggests that social cognitive deficits may differ between patients with affective and non-affective psychosis, highlighting this need.

Interventions targeting social cognition and social functioning for patients with psychosis have been developed, and are evolving with advancing models of social cognitive dysfunction in this population. The evidence supports the use of these treatments in patients, and continued development of programming for patients with related illnesses and with enhanced ecological validity.

## References

Argyle, M., Bryant, B., & Trower, P. (1974). Social skills training and psychotherapy: a comparative study. *Psychological Medicine*, *4*, 435–443.

Bechi, M., Bosia, M., Spangaro, M., Buonocore, M., Cocchi, F., Pigoni, A., ... Cavallaro, R. (2015). Combined social cognitive and neurocognitive rehabilitation strategies in schizophrenia: neuropsychological and psychopathological influences on Theory of Mind improvement. *Psychological Medicine*, *45*(15), 3147–3157.

Bechi, M., Riccaboni, R., Ali, S., Fresi, F., Buonocore, M., Bosia, M., ... Cavallaro, R. (2012). Theory of mind and emotion processing training for patients with schizophrenia: preliminary findings. *Psychiatry Research*, *198*(3), 371–377.

Bechi, M., Spangaro, M., Bosia, M., Zanoletti, A., Fresi, F., Buonocore, M., ... Cavallaro, R. (2013). Theory of Mind intervention for outpatients with schizophrenia. *Neuropsychological Rehabilitation*, *23*(3), 383–400.

Becker, T., & Kösters, M. (2012). Psychosocial interventions in people with severe mental illness: a Bleulerian perspective. *Neuropsychobiology*, *66*(1), 70–75.

Bell, M., Bryson, G., Greig, T., Corcoran, C., & Wexler, B. E. (2001). Neurocognitive enhancement therapy with work therapy: effects on neuropsychological test performance. *Archives of General Psychiatry*, *58*(8), 763–768.

Bellack, A. S. (1983). Recurrent problems in the behavioral assessment of social skills. *Behaviour Research and Therapy*, *21*, 29–42.

Bellack, A. S., & Hersen, M. (1979). *Research and practice in social skills training*. New York: Plenum Press.

Bellack, A. S., Mueser, K. T., Gingerich, S., & Agresta, J. (2004). *Social skills training for schizophrenia: A step-by-step guide* (2nd ed.). New York, NY: Guilford Press.

Blanchard, J. J., Park, S. G., Catalano, L. T., & Bennett, M. E. (2015). Social affiliation and negative symptoms in schizophrenia: examining the role of behavioral skills and subjective responding. *Schizophrenia Research*, *168*(1–2), 491–497.

Bleuler, E. (1911). *Dementia praecox or the group of schizophrenias*. Oxford: International Universities Press.

Bonnin, C. M., Torrent, C., Arango, C., Amann, B. L., Solé, B., González-Pinto, A., … CIBERSAM, F. R. G. (2016). Functional remediation in bipolar disorder: 1-year follow-up of neurocognitive and functional outcome. *The British Journal of Psychiatry*, *208*(1), 87–93.

Bowie, C. R., McGurk, S. R., Mausbach, B., Patterson, T. L., & Harvey, P. D. (2012). Combined cognitive remediation and functional skills training for schizophrenia: effects on cognition, functional competence, and real-world behavior. *The American Journal of Psychiatry*, *169*(7), 710–718.

Brekke, J., Kay, D. D., Lee, K. S., & Green, M. F. (2005). Biosocial pathways to functional outcome in schizophrenia. *Schizophrenia Research*, *80*(2–3), 213–225.

Brenner, H. D., Roder, V., Hodel, B., Kienzle, N., Reed, D., & Liberman, R. P. (1994). *Integrated psychological therapy for schizophrenic patients (IPT)*. Seattle, WA: Hogrefe & Huber Publishers.

Brenner, H. D., Stramke, W. G., Mewes, J., Liese, F., & Seeger, G. (1980). A treatment program, based on training of cognitive and communicative functions, in the rehabilitation of chronic schizophrenic patients (author's transl). *Der Nervenarzt*, *51*(2), 106–112.

Cacioppo, J. T., Cacioppo, S., Capitanio, J. P., & Cole, S. W. (2015). The neuroendocrinology of social isolation. *Annual Review of Psychology*, *66*, 733–767.

Cacioppo, J. T., Hughes, M. E., Waite, L. J., Hawkley, L. C., & Thisted, R. A. (2006). Loneliness as a specific risk factor for depressive symptoms: cross-sectional and longitudinal analyses. *Psychology and Aging*, *21*(1), 140–151.

Campellone, T. R., & Kring, A. M. (2018). Anticipated pleasure for positive and negative social interaction outcomes in schizophrenia. *Psychiatry Research*, *259*, 203–209.

Campellone, T. R., Truong, B., Gard, D., & Schlosser, D. A. (2018). Social motivation in people with recent-onset schizophrenia spectrum disorders. *Journal of Psychiatric Research*, *99*, 96–103.

Claghorn, J. L., Johnstone, E. E., Cook, T. H., & Itschner, L. (1974). Group therapy and maintenance treatment of schizophrenics. *Archives of General Psychiatry*, *31*(3), 361–365.

de la Asuncion, J., Docx, L., Sabbe, B., Morrens, M., & de Bruijn, E. R. (2015). Converging evidence of social avoidant behavior in schizophrenia from two approach-avoidance tasks. *Journal of Psychiatric Research*, *69*, 135–141.

Depp, C. A., Mausbach, B., Granholm, E., Cardenas, V., Ben-Zeev, D., Patterson, T. L., … Jeste, D. V. (2010). Mobile interventions for severe mental illness: design and preliminary data from three approaches. *The Journal of Nervous and Mental Disease*, *198*(10), 715–721.

Depp, C. A., Moore, R. C., Perivoliotis, D., Holden, J. L., Swendsen, J., & Granholm, E. L. (2016). Social behavior, interaction appraisals, and suicidal ideation in schizophrenia: the dangers of being alone. *Schizophrenia Research*, *172*(1–3), 195–200.

Eack, S. M., Greenwald, D. P., Hogarty, S. S., Cooley, S. J., DiBarry, A. L., Montrose, D. M., & Keshavan, M. S. (2009). Cognitive enhancement therapy for early-course schizophrenia: effects of a two-year randomized controlled trial. *Psychiatric Services*, *60*(11), 1468–1476.

Eack, S. M., Hogarty, S. S., Greenwald, D. P., Litschge, M. Y., McKnight, S. A., Bangalore, S. S., … Cornelius, J. R. (2015). Cognitive Enhancement Therapy in substance misusing schizophrenia: results of an 18-month feasibility trial. *Schizophrenia Research*, *161*(2–3), 478–483. https://doi.org/10.1016/j.schres.2014.11.017.

Eack, S. M., Mesholam-Gately, R. I., Greenwald, D. P., Hogarty, S. S., & Keshavan, M. S. (2013). Negative symptom improvement during cognitive rehabilitation: results from a 2-year trial of Cognitive Enhancement Therapy. *Psychiatry Research*, *209*(1), 21–26.

Eichner, C., & Berna, F. (2016). Acceptance and efficacy of metacognitive training (MCT) on positive symptoms and delusions in patients with schizophrenia: a meta-analysis taking into account important moderators. *Schizophrenia Bulletin*, *42*(4), 952–962.

Eisler, R. M., & Frederiksen, L. W. (1980). *Perfecting social skills: A guide to interpersonal behavior development*. New York: Plenum Press.

Elis, O., Caponigro, J. M., & Kring, A. M. (2013). Psychosocial treatments for negative symptoms in schizophrenia: current practices and future directions. *Clinical Psychology Review*, *33*(8), 914–928.

Engel, M., Fritzsche, A., & Lincoln, T. M. (2016). Anticipation and experience of emotions in patients with schizophrenia and negative symptoms. An experimental study in a social context. *Schizophrenia Research*, *170*(1), 191–197.

Fett, A. K., Viechtbauer, W., Dominguez, M. D., Penn, D. L., van Os, J., & Krabbendam, L. (2011). The relationship between neurocognition and social cognition with functional outcomes in schizophrenia: a meta-analysis. *Neuroscience and Biobehavioral Reviews*, *35*(3), 573–588.

Field, C. D., & Test, M. A. (1975). Group assertiveness for severely disturbed patients. *Journal of Behavior Therapy and Experimental Psychiatry*, *6*, 129–134.

Figueira, M. L., & Brissos, S. (2011). Measuring psychosocial outcomes in schizophrenia patients. *Current Opinion in Psychiatry*, *24*(2), 91–99.

Fulford, D., Campellone, T., & Gard, D. E. (2018). Social motivation in schizophrenia: how research on basic reward processes informs and limits our understanding. *Clinical Psychology Review*, *63*, 12–24.

Gard, D. E., Fisher, M., Garrett, C., Genevsky, A., & Vinogradov, S. (2009). Motivation and its relationship to neurocognition, social cognition, and functional outcome in schizophrenia. *Schizophrenia Research*, *115*(1), 74–81.

Goldstein, A. P., Gershaw, N. J., & Sprafkin, R. P. (1975). Structured learning therapy—skill training for schizophrenics. *Schizophrenia Bulletin*, (14), 83–86.

Granholm, E., Ben-Zeev, D., Fulford, D., & Swendsen, J. (2013). Ecological Momentary Assessment of social functioning in schizophrenia: impact of performance appraisals and affect on social interactions. *Schizophrenia Research*, *145*(1–3), 120–124.

Granholm, E., Holden, J., Link, P. C., & McQuaid, J. R. (2014). Randomized clinical trial of cognitive behavioral social skills training for schizophrenia: improvement in functioning and experiential negative symptoms. *Journal of Consulting and Clinical Psychology*, *82*(6), 1173–1185.

Granholm, E., Holden, J., Link, P. C., McQuaid, J. R., & Jeste, D. V. (2013). Randomized controlled trial of cognitive behavioral social skills training for older consumers with schizophrenia: defeatist performance attitudes and functional outcome. *The American Journal of Geriatric Psychiatry*, *21*(3), 251–262.

Granholm, E., Loh, C., & Swendsen, J. (2008). Feasibility and validity of computerized ecological momentary assessment in schizophrenia. *Schizophrenia Bulletin*, *34*(3), 507–514.

Granholm, E., McQuaid, J. R., & Holden, J. (2016). *Cognitive behavioral social skills training for schizophrenia: A practical treatment guide*. New York: Guildford Press.

Granholm, E., McQuaid, J. R., McClure, F. S., Auslander, L. A., Perivoliotis, D., Pedrelli, P., … Jeste, D. V. (2005). A randomized, controlled trial of cognitive behavioral social skills training for middle-aged and older outpatients with chronic schizophrenia. *The American Journal of Psychiatry*, *162*(3), 520–529.

Granholm, E., McQuaid, J. R., McClure, F. S., Link, P. C., Perivoliotis, D., Gottlieb, J. D., … Jeste, D. V. (2007). Randomized controlled trial of cognitive behavioral social skills training for older people with schizophrenia: 12-month follow-up. *The Journal of Clinical Psychiatry*, *68*(5), 730–737.

Granholm, E., McQuaid, J. R., McClure, F. S., Pedrelli, P., & Jeste, D. V. (2002). A randomized controlled pilot study of cognitive behavioral social skills training for older patients with schizophrenia. *Schizophrenia Research*, *53*(1–2), 167–169.

Gupta, M., Holshausen, K., Mausbach, B., Patterson, T. L., & Bowie, C. R. (2012). Predictors of change in functional competence and functional behavior after functional adaptation skills training for schizophrenia. *The Journal of Nervous and Mental Disease, 200*(8), 705–711.

Hansen, D. J., St Lawrence, J. S., & Christoff, K. A. (1985). Effects of interpersonal problem-solving training with chronic aftercare patients on problem-solving component skills and effectiveness of solutions. *Journal of Consulting and Clinical Psychology, 53*(2), 167–174.

Herman, Y., Shireen, H., Bromley, S., Yiu, N., & Granholm, E. (2018). Cognitive-behavioural social skills training for first-episode psychosis: a feasibility study. *Early Intervention in Psychiatry, 12*(5), 863–868.

Hersen, M., & Bellack, A. S. (1976). Social skills training for chronic psychiatric patients: rationale, research findings, and future directions. *Comprehensive Psychiatry, 17*(4), 559–580.

Hersen, M., Turner, S. M., Edelstein, B. A., & Pinkston, S. G. (1975a). Effects of phenothiazines and social skills training in a withdrawn schizophrenic. *Journal of Clinical Psychology, 31*(4), 588–594.

Hersen, M., Turner, S. M., Edelstein, B. A., & Pinkston, S. G. (1975b). Effect of practice, instructions, and modeling on components of assertive behavior. *Behaviour Research and Therapy, 11*, 443–451.

Hogarty, G. E., & Flesher, S. (1999). Practice principles of cognitive enhancement therapy for schizophrenia. *Schizophrenia Bulletin, 25*(4), 693–708.

Hogarty, G. E., Flesher, S., Ulrich, R., Carter, M., Greenwald, D., Pogue-Geile, M., … Zoretich, R. (2004). Cognitive enhancement therapy for schizophrenia: effects of a 2-year randomized trial on cognition and behavior. *Archives of General Psychiatry, 61*(9), 866–876.

Hogarty, G. E., Goldberg, S. C., & Schooler, N. R. (1974). Drug and sociotherapy in the aftercare of schizophrenic patients. III. Adjustment of nonrelapsed patients. *Archives of General Psychiatry, 31*(5), 609–618.

Hogarty, G. E., & Greenwald, D. P. (2006). *Cognitive enhancement therapy: The training manual.* Pittsburgh, PA: CET Training, LLC.

Holt-Lunstad, J., Smith, T. B., Baker, M., Harris, T., & Stephenson, D. (2015). Loneliness and social isolation as risk factors for mortality: a meta-analytic review. *Perspectives on Psychological Science, 10*(2), 227–237.

Horan, W. P., & Green, M. F. (2017). Treatment of social cognition in schizophrenia: current status and future directions. *Schizophrenia Research.*

Keefe, R. S., Vinogradov, S., Medalia, A., Buckley, P. F., Caroff, S. N., D'Souza, D. C., … Stroup, T. S. (2012). Feasibility and pilot efficacy results from the multisite Cognitive Remediation in the Schizophrenia Trials Network (CRSTN) randomized controlled trial. *The Journal of Clinical Psychiatry, 73*(7), 1016–1022.

Keshavan, M. S. (1999). Development, disease and degeneration in schizophrenia: a unitary pathophysiological model. *Journal of Psychiatric Research, 33*(6), 513–521.

Kopelowicz, A., Liberman, R. P., & Zarate, R. (2006). Recent advances in social skills training for schizophrenia. *Schizophrenia Bulletin, 32*(Suppl. 1), S12–S23.

Krapelin, E. (1919). *Dementia praecox and paraphrenia.* Edinburgh: E & S Livingstone.

Lees, J., Manning, N., & Rawlings, B. (1999). *Therapeutic community effectiveness: A systematic international review of therapeutic community treatment for people with personality disorders and mentally disordered offenders (CRD Report No. 17).* York: NHS Centre for Reviews and Dissemination, University of York.

Leifker, F. R., Bowie, C. R., & Harvey, P. D. (2009). Determinants of everyday outcomes in schizophrenia: the influences of cognitive impairment, functional capacity, and symptoms. *Schizophrenia Research, 115*(1), 82–87.

Liberman, R. P. (1982). Assessment of social skills. *Schizophrenia Bulletin*, *8*(1), 62–83.
Liberman, R. P., Wallace, C. J., Falloon, I. R., & Vaughn, C. E. (1981). Interpersonal problem-solving therapy for schizophrenics and their families. *Comprehensive Psychiatry*, *22*(6), 627–630.
Liu, Y. C., Tang, C. C., Hung, T. T., Tsai, P. C., & Lin, M. F. (2018). The efficacy of metacognitive training for delusions in patients with schizophrenia: a meta-analysis of randomized controlled trials informs evidence-based practice. *Worldviews on Evidence-Based Nursing*, *15*(2), 130–139.
Lukoff, D., Liberman, R. P., & Nuechterlein, K. H. (1986). Symptom monitoring in the rehabilitation of schizophrenic patients. *Schizophrenia Bulletin*, *12*(4), 578–602.
Martínez-Arán, A., Torrent, C., Solé, B., Bonnín, C. M., Rosa, A. R., Sánchez-Moreno, J., & Vieta, E. (2011). Functional remediation for bipolar disorder. *Clinical Practice and Epidemiology in Mental Health*, 7, 112–116.
McCarthy, J. M., Bradshaw, K. R., Catalano, L. T., Garcia, C. P., Malik, A., Bennett, M. E., & Blanchard, J. J. (2018). Negative symptoms and the formation of social affiliative bonds in schizophrenia. *Schizophrenia Research*, *193*, 225–231.
McGurk, S. R., Twamley, E. W., Sitzer, D. I., McHugo, G. J., & Mueser, K. T. (2007). A meta-analysis of cognitive remediation in schizophrenia. *The American Journal of Psychiatry*, *164*(12), 1791–1802.
McQuaid, J. R., Granholm, E., McClure, F. S., Roepke, S., Pedrelli, P., Patterson, T. L., & Jeste, D. V. (2000). Development of an integrated cognitive-behavioral and social skills training intervention for older patients with schizophrenia. *The Journal of Psychotherapy Practice and Research*, *9*(3), 149–156.
Medalia, A., & Freilich, B. (2008). The Neuropsychological Educational Approach to Cognitive Remediation (NEAR) model: practice principles and outcomes studies. *American Journal of Psychiatric Rehabilitation*, *11*(2), 122–143.
Medalia, A., Revheim, N., & Casey, M. (2002). Remediation of problem-solving skills in schizophrenia: evidence of a persistent effect. *Schizophrenia Research*, *57*(2–3), 165–171.
Medalia, A., Revheim, N., & Herlands, T. (2009). *Cognitive remediation for psychological disorders*. Oxford University Press.
Menendez-Miranda, I., Garcia-Portilla, M. P., Garcia-Alvarez, L., Arrojo, M., Sanchez, P., Sarramea, F., … Bobes, J. (2015). Predictive factors of functional capacity and real-world functioning in patients with schizophrenia. *European Psychiatry*, *30*(5), 622–627.
Minkin, N., Braukmann, C. J., Minkin, B. L., Timbers, G. D., Timbers, B. J., Fixsen, D. L., … Wolf, M. M. (1976). The social validation and training of conversational skills. *Journal of Applied Behavior Analysis*, *9*(2), 127–139.
Moritz, S., Veckenstedt, R., Andreou, C., Bohn, F., Hottenrott, B., Leighton, L., … Roesch-Ely, D. (2014). Sustained and "sleeper" effects of group metacognitive training for schizophrenia: a randomized clinical trial. *JAMA Psychiatry*, *71*(10), 1103–1111. https://doi.org/10.1001/jamapsychiatry.2014.1038.
Moritz, S., Veckenstedt, R., Bohn, F., Kother, U., & Woodward, T. S. (2013). Metacognitive training in schizophrenia: theoretical rationale and administration. In D. L. Roberts & D. L. Penn (Eds.), *Social cognition in schizophrenia. From evidence to treatment* (pp. 358–383). New York, NY: Oxford University Press.
Moritz, S., Veckenstedt, R., Randjbar, S., Vitzthum, F., & Woodward, T. S. (2011). Antipsychotic treatment beyond antipsychotics: metacognitive intervention for schizophrenia patients improves delusional symptoms. *Psychological Medicine*, *41*(9), 1823–1832.
Moritz, S., Vitzthum, F., Randjbar, S., Veckenstedt, R., & Woodward, T. S. (2010). Detecting and defusing cognitive traps: metacognitive intervention in schizophrenia. *Current Opinion in Psychiatry*, *23*(6), 561–569.

Morrison, R. L., & Bellack, A. S. (1987). Social functioning of schizophrenic patients: clinical and research issues. *Schizophrenia Bulletin, 13*(4), 715–725.

Mosher, L. R. (1999). Soteria and other alternatives to acute psychiatric hospitalization: a personal and professional review. *The Journal of Nervous and Mental Disease, 187*(3), 142–149.

Mosher, L. R., & Keith, S. J. (1979). Research on the psychosocial treatment of schizophrenia: a summary report. *The American Journal of Psychiatry, 136*(5), 623–631.

Mueller, D. R., Schmidt, S. J., & Roder, V. (2015). One-year randomized controlled trial and follow-up of integrated neurocognitive therapy for schizophrenia outpatients. *Schizophrenia Bulletin, 41*(3), 604–616.

Mueser, K. T., Bellack, A. S., Douglas, M. S., & Morrison, R. L. (1991). Prevalence and stability of social skill deficits in schizophrenia. *Schizophrenia Research, 5*(2), 167–176.

Myin-Germeys, I., Klippel, A., Steinhart, H., & Reininghaus, U. (2016). Ecological momentary interventions in psychiatry. *Current Opinion in Psychiatry, 29*(4), 258–263.

O'Brien, C. P., Hamm, K. B., Ray, B. A., Pierce, J. F., Luborsky, L., & Mintz, J. (1972). Group vs individual psychotherapy with schizophrenics. A controlled outcome study. *Archives of General Psychiatry, 27*(4), 474–478.

Ochoa, S., López-Carrilero, R., Barrigón, M. L., Pousa, E., Barajas, A., Lorente-Rovira, E., … Spanish, M. S. G. (2017). Randomized control trial to assess the efficacy of metacognitive training compared with a psycho-educational group in people with a recent-onset psychosis. *Psychological Medicine, 47*(9), 1573–1584.

Orfanos, S., Banks, C., & Priebe, S. (2015). Are group psychotherapeutic treatments effective for patients with schizophrenia? A systematic review and meta-analysis. *Psychotherapy and Psychosomatics, 84*(4), 241–249.

Ospina, L. H., Nitzburg, G. C., Shanahan, M., Perez-Rodriguez, M. M., Larsen, E., Latifoglu, A., & Burdick, K. E. (2018). Social cognition moderates the relationship between neurocognition and community functioning in bipolar disorder. *Journal of Affective Disorders, 235*, 7–14.

Patterson, T. L., Bucardo, J., McKibbin, C. L., Mausbach, B. T., Moore, D., Barrio, C., … Jeste, D. V. (2005). Development and pilot testing of a new psychosocial intervention for older Latinos with chronic psychosis. *Schizophrenia Bulletin, 31*(4), 922–930.

Patterson, T. L., & Leeuwenkamp, O. R. (2008). Adjunctive psychosocial therapies for the treatment of schizophrenia. *Schizophrenia Research, 100*(1–3), 108–119.

Patterson, T. L., Mausbach, B. T., McKibbin, C., Goldman, S., Bucardo, J., & Jeste, D. V. (2006). Functional adaptation skills training (FAST): a randomized trial of a psychosocial intervention for middle aged and older patients with chronic psychotic disorders. *Schizophrenia Research, 86*(1–3), 291–299.

Patterson, T. L., McKibbin, C., Taylor, M., Goldman, S., Davila-Fraga, W., Bucardo, J., & Jeste, D. V. (2003). Functional adaptation skills training (FAST): a pilot psychosocial intervention study in middle-aged and older patients with chronic psychotic disorders. *The American Journal of Geriatric Psychiatry, 11*(1), 17–23.

Pinkham, A. E., Penn, D. L., Green, M. F., Buck, B., Healey, K., & Harvey, P. D. (2014). The social cognition psychometric evaluation study: results of the expert survey and RAND panel. *Schizophrenia Bulletin, 40*(4), 813–823.

Pinkham, A. E., Penn, D. L., Green, M. F., & Harvey, P. D. (2016). Social cognition psychometric evaluation: results of the initial psychometric study. *Schizophrenia Bulletin, 42*(2), 494–504.

Pot-Kolder, R. M. C. A., Geraets, C. N. W., Veling, W., van Beilen, M., Staring, A. B. P., Gijsman, H. J., … van der Gaag, M. (2018). Virtual-reality-based cognitive behavioural therapy versus waiting list control for paranoid ideation and social avoidance in patients with psychotic disorders: a single-blind randomised controlled trial. *Lancet Psychiatry*, *5*(3), 217–226.

Priebe, S. (2007). Social outcomes in schizophrenia. *The British Journal of Psychiatry. Supplement*, *50*, s15–s20.

Radke, S., Pfersmann, V., & Derntl, B. (2015). The impact of emotional faces on social motivation in schizophrenia. *European Archives of Psychiatry and Clinical Neuroscience*, *265*(7), 613–622.

Revheim, N., Kamnitzer, D., Casey, M., & Medalia, A. (2001). Implementation of a cognitive rehabilitation program in an inpatient setting. *Psychiatric Rehabilitation Skills*, *5*(3), 403–425.

Rocha, N. B., & Queirós, C. (2013). Metacognitive and social cognition training (MSCT) in schizophrenia: a preliminary efficacy study. *Schizophrenia Research*, *150*(1), 64–68.

Roder, V., Mueller, D. R., Mueser, K. T., & Brenner, H. D. (2006). Integrated psychological therapy (IPT) for schizophrenia: is it effective. *Schizophrenia Bulletin*, *32*(Suppl. 1), S81–S93.

Rogers, P., & Redoblado-Hodge, A. (2006). A multi-site trial of cognitive remediation in schizophrenia: an Australian sample. In: *Proceedings from cognitive remediation in psychiatry, New York, NY.*

Schiffman, J., Walker, E., Ekstrom, M., Schulsinger, F., Sorensen, H., & Mednick, S. (2004). Childhood videotaped social and neuromotor precursors of schizophrenia: a prospective investigation. *The American Journal of Psychiatry*, *161*(11), 2021–2027.

Schneider, M., Reininghaus, U., van Nierop, M., Janssens, M., Myin-Germeys, I., & GROUP Investigators. (2017). Does the Social Functioning Scale reflect real-life social functioning? An experience sampling study in patients with a non-affective psychotic disorder and healthy control individuals. *Psychological Medicine*, *47*(16), 2777–2786.

Strauss, J. S., Kokes, R. F., Klorman, R., & Sacksteder, J. L. (1977). Premorbid adjustment in schizophrenia: concepts, measures, and implications. Part I. The concept of premorbid adjustment. *Schizophrenia Bulletin*, *3*(2), 182–185.

Tarbox, S. I., Addington, J., Cadenhead, K. S., Cannon, T. D., Cornblatt, B. A., Perkins, D. O., … Woods, S. W. (2014). Functional development in clinical high risk youth: prediction of schizophrenia versus other psychotic disorders. *Psychiatry Research*, *215*(1), 52–60.

Torrent, C., Bonnin, C. M., Martínez-Arán, A., Valle, J., Amann, B. L., González-Pinto, A., … Vieta, E. (2013). Efficacy of functional remediation in bipolar disorder: a multicenter randomized controlled study. *The American Journal of Psychiatry*, *170*(8), 852–859.

Trower, P. (1980). Situational analysis of the components and processes of behavior of socially skilled and unskilled patients. *Journal of Consulting and Clinical Psychology*, *48*, 327–339.

van Oosterhout, B., Smit, F., Krabbendam, L., Castelein, S., Staring, A. B., & van der Gaag, M. (2016). Metacognitive training for schizophrenia spectrum patients: a meta-analysis on outcome studies. *Psychological Medicine*, *46*(1), 47–57.

Velligan, D. I., Bow-Thomas, C. C., Huntzinger, C., Ritch, J., Ledbetter, N., Prihoda, T. J., & Miller, A. L. (2000). Randomized controlled trial of the use of compensatory strategies to enhance adaptive functioning in outpatients with schizophrenia. *The American Journal of Psychiatry*, *157*(8), 1317–1323.

Velligan, D. I., Mueller, J., Wang, M., Dicocco, M., Diamond, P. M., Maples, N. J., & Davis, B. (2006). Use of environmental supports among patients with schizophrenia. *Psychiatric Services*, *57*(2), 219–224.

Vieta, E., Torrent, C., & Martinez-Aran, A. (2014). *Functional remediation for bipolar disorder.* Cambridge: Cambridge University Press.

Wallace, C. J., Nelson, C. J., Liberman, R. P., Aitchison, R. A., Lukoff, D., Elder, J. P., & Ferris, C. (1980). A review and critique of social skills training with schizophrenic patients. *Schizophrenia Bulletin*, *6*(1), 42–63.

Wykes, T., Huddy, V., Cellard, C., McGurk, S. R., & Czobor, P. (2011). A meta-analysis of cognitive remediation for schizophrenia: methodology and effect sizes. *The American Journal of Psychiatry*, *168*(5), 472–485.

# Index

Note: Page numbers followed by *f* indicate figures and *t* indicate tables.

Kolibri 360 GmbH
Ferdinand-Jühlke-Straße 7
99095 Erfurt, Deutschland
produktsicherheit@kolibri360.de